Natural Language Processing and Large Language Models

Chengqing Zong · Yang Zhao · Yanjun Ma

Natural Language Processing and Large Language Models

Theory, Hand-on Codes, and Case Studies

清華大学出版社
Tsinghua University Press

Springer

Chengqing Zong
Institute of Automation, Chinese Academy of Sciences
Beijing, China

ShanghaiTech University
Shanghai, China

Yang Zhao
Institute of Automation, Chinese Academy of Sciences
Beijing, China

Yanjun Ma
Baidu Inc.
Beijing, China

ISBN 978-981-92-0681-0 ISBN 978-981-92-0682-7 (eBook)
https://doi.org/10.1007/978-981-92-0682-7

Jointly published with Tsinghua University Press
ISBN of the Co-Publisher's edition: 978-730-26-5155-0

This Springer imprint is published by the registered company Springer Nature Singapore Pte Ltd.
The registered company address is: 152 Beach Road, #21-01/04 Gateway East, Singapore 189721, Singapore

Preface

In recent years, Natural Language Processing (NLP) has attracted significant attention. On one hand, NLP has made great progress. It can now solve many real-world problems with good performance, such as machine translation, dialogue systems, question answering, automatic text summarization, and article writing. On the other hand, people's requirements and expectations for NLP have become higher. This is especially true since 2022, when Large Language Models (LLMs) came into being. This important breakthrough not only changed people's traditional views on NLP but also made people hope that LLMs can help them complete various complex tasks. To be honest, NLP was just an ordinary research field in the past. But now, it has quickly developed into a popular discipline. It has even become a required core course in AI programs and departments at most universities. Obviously, NLP has entered its best development period.

The sustainable development of NLP cannot be achieved without a large number of beginners devoting themselves to this field. However, the current state of NLP has both positive and negative impacts on beginners. On the positive side, the boundaries between different data modalities have become increasingly blurred, and the techniques used across various NLP applications have grown more similar. This convergence has made the learning process easier for beginners. Besides, open source has become a key trend in the NLP community. The community now provides a wealth of open resources, including datasets, algorithms, tools, and pre-trained models, which greatly benefits those who are new to the field. On the negative side, the large number of algorithms and models brings unique challenges to beginners. They have to master both traditional NLP methods and the basic theories of LLMs. What's more, NLP and LLMs are highly practical fields; beginners need to learn essential hands-on skills, such as data processing, model deployment, parameter configuration and optimization, and how to use related platforms. It was exactly these challenges that inspired us to write a book to meet these specific needs.

Fortunately, when the Baidu PaddlePaddle team contacted me about co-writing a book on NLP and LLMs that integrates theory with practice, I agreed right away. We first completed a Chinese version of the book. Shortly after finishing it, editors from Springer reached out to us to discuss translating the book into English, with the goal of making its content available to readers around the world. We started the translation and rewriting immediately and expanded some chapters. This book is written under the assumption that readers already have basic knowledge of NLP theories and methods. Therefore, theoretical content is just briefly introduced, and our main focus is on step-by-step practical implementation processes, such as data processing, structure construction, and model optimization. In terms of structure, it begins by introducing fundamental NLP theories and methods. Next, it explores practical implementation approaches for different NLP applications. Finally, it focuses on LLM-based generation systems. Every piece of code provided in the book has undergone strict testing and is hosted on the Baidu AI Studio Community as Jupyter Notebooks. By following the instructions in the book, readers can directly access the free GPU computing resources provided by the AI Studio Xinghe Community to compile and run the code online—this greatly simplifies the process of practicing and learning.

This book was co-authored by Dr. Zhao Yang and the PaddlePaddle team, whose expertise and support were instrumental in its completion. The book project also received funding from the Publishing Center of the University of Chinese Academy of Sciences. We would like to express our gratitude to Ms. Tian Chenchen, Prof. Xiao Jun, and Ms. Qu Xiaochun from the University of Chinese Academy of Sciences for their invaluable assistance; to Prof. Zhao Tiejun, Prof. Wang Houfeng, Prof. Wang Xiaojie, Prof. Huang Minlie, and Prof. Zhang Jiajun for their constructive suggestions on the book's contents; to Dr. Xiang Lu from the Institute of Automation, Chinese Academy of Sciences for invaluable suggestions; and to Editor Chang Lanlan from Springer for her consistent support throughout the publishing process.

Given the tight schedule and the authors' limited expertise, the book may contain inaccuracies or inappropriate expressions. We sincerely welcome feedback and critical suggestions from readers to facilitate future revisions. Thank you.

Beijing, China

Chengqing Zong
Yang Zhao
Yanjun Ma

Competing Interests The authors have no competing interests to declare that are relevant to the content of this manuscript.

Contents

Chapter 1
Introduction

Abstract This chapter serves as an introduction to Natural Language Processing (NLP), defining the field and its core research areas. It provides a historical overview of NLP methods, starting with early template and rule-based approaches, moving to corpus-based statistical methods that include generative and discriminative models, and culminating in the current neural network-based methods, such as Transformers and large language models (LLMs). The chapter outlines the book's structure, which will focus on neural network techniques, covering basic models, key technologies, and application systems. Finally, it introduces the practical implementation platform used for the book's exercises, Baidu's PaddlePaddle, and provides details on its setup, along with the specific APIs and datasets that will be used.

Natural Language Processing (NLP) is a discipline that studies how to use computer technology to process language texts (sentences, articles, discourses, etc.). The research content includes various processing methods and implementation techniques for identifying, classifying, extracting, converting, and generating information such as morphology, syntax, semantics, and pragmatics. This term appeared in the late 1970s or early 1980s. In general, natural language processing includes two major directions: text analysis, understanding, transformation, and generation. In recent years, with the popularization and application of computer networks and mobile communication technologies, natural language processing has played an increasingly important role in various fields, including human society, economy, and national security. Its data processing has also expanded from the single text modality to the integration of multi-modal data such as voice, image, and video.[1]

When learning natural language processing, we not only need to understand the theoretical methods and models, but also need to master the implementation skills and usage of each method. To this end, this book selects a number of classic models and algorithms from recent years. We will analyze them from the perspective of implementing and provide corresponding programs for readers to practice and exercise.

---

[1] See page 1222 of *The Encyclopedia of Computer Science and Technology* (Tsinghua University Press, 2018).

C. Zong et al., *Natural Language Processing and Large Language Models*,
https://doi.org/10.1007/978-981-92-0682-7_1

Firstly, this chapter gives a brief review and overview of natural language processing methods and then introduces the platform and usage of programs in the book. Finally, this chapter gives a brief explanation of the content and organizational structure of the entire book.

1.1 Overview of Natural Language Processing Methods

Before the mid-1980s, natural language processing mainly adopted template-based and rule-based methods, based on the theoretical foundation of syntactic structure theory proposed by linguists represented by N. Chomsky. From the year 1947, when the concept of machine translation was proposed by W. Weaver and A. Booth, to the mid-1980s, the nearly 40-year development stage of natural language processing technology is usually considered to be a rationalist period based on symbolic logical reasoning.

In the mid-1980s, corpus-based statistical natural language processing methods gradually emerged. The n-gram model and the hidden Markov model, or HMM, were introduced into natural language processing and were applied to many fields. Japanese scholar M. Nagao proposed the example-based machine translation method. In 1989, the first corpus-based statistical machine translation system was born. Then the promotion and application of a series of open-source data, tools, platforms, and systems elevated the data-driven empirical method to a new development stage, which lasted for more than 20 years.

Statistical natural language processing methods can be roughly divided into two categories: generative models and discriminative models. The basic idea of the generative model is: we assume that O is the observed value and Q is the model. Then we should firstly establish the probability density model P(O|Q) of samples, and then use this model to make inference and prediction. This method is based on statistics and Bayes' theorem, requiring infinite samples or as many as possible.

The basic idea of the discriminative model is to model the conditional probability (or posterior probability) P(Q|O) and establish a discriminant function with limited samples in order to find the optimal classification between different categories. Based on this idea, many NLP tasks are transformed into classification tasks or sequence labeling tasks.

Certainly, generative models and discriminative models can be integrated for some tasks. Commonly used models in statistical methods are summarized in Table 1.1.

In the early days of statistical methods, almost all NLP tasks were solved by n-gram models. From key technologies such as Pinyin-to-Chinese word conversion, automatic Chinese word segmentation, named entity recognition, to application systems such as machine translation, n-gram models can fulfill nearly almost all the tasks. Later, discriminative methods do not require complex model derivation, and adding or removing features is easy to implement. Therefore, discriminative models (usually represented as classification or sequence labeling models) became popular and were

Table 1.1 Overview of statistical natural language processing methods

Category	Representative model	Typical applications
Generative model	n-gram model	Chinese word segmentation, machine translation, automatic writing
	Hidden Markov Model (HMM)	Chinese word segmentation and part-of-speech tagging, named entity recognition, speech recognition
Discriminative model	Naive Bayes Classifier (NB)	Word sense disambiguation, part-of-speech tagging, various classification tasks
	Decision Tree	Part-of-speech tagging, syntax analysis, and various classification tasks
	K-nearest Neighbor, K-NN	Text classification, clustering, and sentiment analysis
	Maximum entropy, ME	Word sense disambiguation, syntax analysis, and various classifications
	Perceptron	Chinese word segmentation and part-of-speech tagging, various classification tasks
	Support vector machine (SVM)	Various classification tasks, such as text classification and sentiment classification
	Conditional random field, CRF	Chinese word segmentation, named entity recognition, etc., various classification or sequence labeling tasks
Mixed model	N-gram + Conditional random field, CRF, etc	various sequence labeling tasks such as Chinese word segmentation and part-of-speech tagging

used in many areas, just like the n-gram models before or the Transformer model now.

Anyway, the period in which statistical methods were used is a very important historical stage in the development of natural language processing. Although most models have been surpassed by neural network methods in terms of performance on almost all the tasks and are no longer widely used, many modeling ideas, especially n-gram models, have important enlightening significance for the later establishment of neural language models.

After entering the twenty-first century, a series of new models were developed based on the existing feedforward neural network (FNN), convolutional neural network (CNN), and recurrent neural network (RNN), and they were widely applied to various NLP tasks, gradually forming a neural network-based NLP method. From the early neural network language model (NNLM), the word embedding method, sequence-to-sequence (seq2seq) method, attention mechanism, to a series of methods such as Transformer and pre-trained language model, the performance of various NLP tasks has been continuously improved. Various tools and platforms were open to the public for free use, which greatly lowered the entry threshold of the natural language processing field and laid the foundation for breaking down the barriers between methods for processing different modal data. Thus, multi-modal information fusion became possible. The application of natural language processing technology has also evolved from the early focus on analysis and understanding, with generation as a supplement to a new stage where analysis, understanding, and generation are advancing side by side.

An overview of natural language processing methods based on neural networks, representative models, and tools can be summarized in Table 1.2.

With the improvement of neural network models, various models represented by pre-trained language models have shown powerful capabilities. The scale of training data keeps increasing, and the number of model parameters is rapidly expanding. Models' performances are also continuously improving. At the same time, the model interpretability and improvement of reasoning ability have received increasing attention. In particular, the release of OpenAI's ChatGPT in 2022 attracted widespread attention from both academia and industry for research on large-scale language models. Experiments show that when the size of model parameters exceeds a certain amount, pre-trained language models can not only significantly improve the performance of multiple NLP tasks, but also show multiple special abilities, such as in-context learning (Brown et al., 2020), emergent abilities (Wei et al., 2022a), and chain of thought (Wei et al., 2022b). To highlight the importance of increased parameter size, researchers have proposed the concept of "large language models (LLMs) or pre-trained large models" to represent language models with large parameter sizes.

Although we can believe that neural network methods and pre-trained models based on neural networks will not be the best methods for natural language processing, we have to admit that this method has become the mainstream in this field and will play a leading role for a long time. It can even become a stepping stone to solve a certain type of problem.

The main content of this book will focus on neural network methods, from basic models and tools of neural networks to key technologies and application systems. Through different levels, the current mainstream natural language processing techniques will be comprehensively introduced to readers. Each line of code provided in this book has been carefully debugged by the PaddlePaddle team. Through hands-on practice, readers can better understand the implementation methods of natural language processing techniques and flexibly apply them to various tasks.

1.2 Organization of the Book

This book contains 15 chapters. In addition to this chapter, the remaining 14 chapters will introduce the implementation and application systems of n-gram models, conditional random fields, and some classic models listed in Tables 1.1 and 1.2, respectively, in the order from basic theory to key technologies and finally to application systems. The organization of the book is shown in Fig. 1.1.

Table 1.2 Representative models based on neural networks[2]

Model	Time	Representative papers or websites	Usage Description
Long short-term memory (LSTM)	1997	(Hochreiter and Schmidhuber, 1997)	A form of recurrent neural network (RNN) that can alleviate vanishing gradient and exploding problems in long sequence training
Neural network language model (NNLM)	2003	(Bengio et al., 2003)	Learn distributed representation of words and calculate the probability of word sequence
Graph neural network (GNN)	2009	(Scarselli et al., 2009)	The use of structural information to represent text is widely applied in tasks such as knowledge graph, syntactic analysis, semantic role labeling, and relation extraction
Word embedding models: CBOW, Skip-gram	2013	(Mikolov et al., 2013a,b)	Generate distributed vector representation of words
Gated recurrent unit (GRU)	2014	(Cho et al., 2014)	A variant of LSTM, which simplifies the model structure of LSTM, but can maintain excellent performance
Seq2seq	2014	(Sutskever et al., 2014)	Sequence to sequence modeling, where the lengths of input and output sequences are variable
Generative adversarial networks (GAN)	2014	(Goodfellow et al., 2014)	Used for tasks such as dialog systems and robust machine translation system training
Attention mechanism	2015	(Bahdanau et al., 2015)	Select the information most relevant to the current goal for processing
Residual network (ResNet)	2016	(He et al., 2016)	The use of skip connection strategy to achieve cross-layer signal transmission, which alleviates the difficulty of deep network training
Transformer	2017	(Vaswani et al., 2017)	The encoder-decoder conversion module relies entirely on self-attention mechanisms to calculate the representations of inputs and outputs, which has better performances in long sequence modeling and can achieve parallel computing

(continued)

[2] This table is arranged in the order of time when the models were proposed.

Table 1.2 (continued)

Model	Time	Representative papers or websites	Usage Description
ELMo	2018	(Peters et al., 2018), See the webpage[3]	A Pre-trained language model, using the unidirectional traditional language model based on LSTM
GPT	2018 OpenAI	(Radford et al., 2018)	A Pre-trained language model, using the unidirectional traditional language model based on Transformer
BERT	2018 Google	(Devlin et al., 2019)	A Pre-trained language model, using the masked language model based on the bidirectional self-attention mechanism
ERNIE	2019 Baidu	(Sun et al., 2019)	A pre-trained language model with integration of entity knowledge
MASS	2019 Microsoft	(Song et al., 2019)	A pre-trained language model based on the encoder-decoder framework
Pegasus	2020 Google	(Zhang et al., 2020)	A pre-trained language model based on the encoder-decoder framework
GPT-3	2020 OpenAI	(Brown et al., 2020)	A large-scale pre-trained language model, using the Transformer decoder framework, with parameters up to 175 billion
RocketQA	2021 Baidu	(Qu et al., 2021)	An open-domain question-answering language model
ERNIE-UIE	2022 Baidu	(Lu et al., 2022)	A pre-trained language model, using a knowledge-enhanced universal information extraction framework
InstructGPT	2022 OpenAI	(Ouyang et al., 2022)	A large-scale pre-trained language model, using reinforcement learning methods based on instruction learning and human feedback learning
LLaMA	2023 Meta	(Touvron et al., 2023)	A large-scale pre-trained language model, using the Transformer decoder framework, provides models with different parameter sizes ranging from 670 million to 6.5 billion
GPT-4	2023 OpenAI	(OpenAI, 2023)	A large-scale and multi-modal pre-trained language model, capable of accepting text and image inputs and generating texts

Chapter 2 introduces the implementation of basic models of neural networks, including feedforward neural networks, convolution neural networks, traditional recurrent neural networks, and their variants, such as long-short-term memory

[3] https://allenai.org/allennlp/software/elmo.

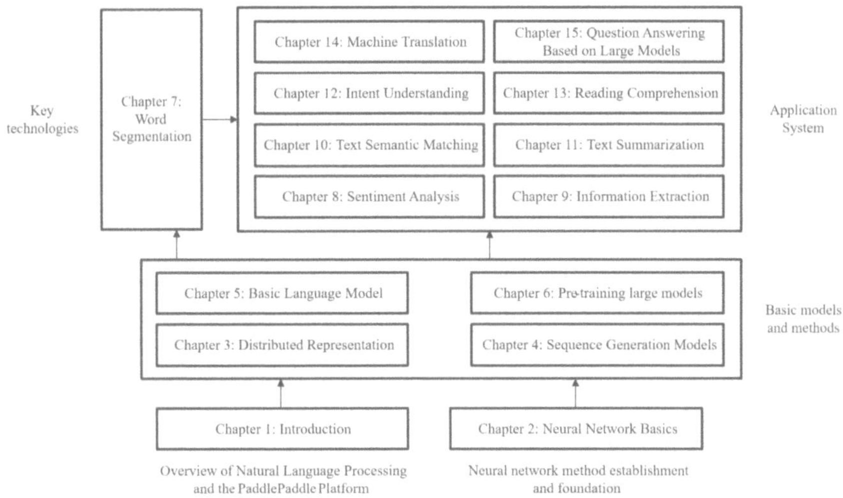

Fig. 1.1 Organization of the book

(LSTM), gated recurrent unit (GRU), and bidirectional long-short-term memory (BiLSTM).

Chapter 3 mainly introduces the implementation of the two most basic word vector (word2vec) generation models: the continuous bag-of-words (CBOW) model and the skip-gram model. This chapter also introduces dynamic word vector methods with ELMo as an example. Based on this, the implementation process of distributed representations of phrases and sentences will also be introduced.

Chapter 4 introduces the implementation of two important sequence generation models, including the sequence-to-sequence (seq2seq) model and the encoder-decoder model (Transformer).

Chapter 5 introduces the implementation of n-gram models and language models based on a feedforward neural network (FNN) and LSTM.

Chapter 6 introduces the implementation of three pre-trained language models (GPT, BERT, and ERNIE) and related technologies for pre-trained large models.

The content introduced in Chaps. 2–6 can be considered as the foundation of statistical methods and neural network methods, and all subsequent technologies are built on these models and methods.

Chapter 7 introduces natural language processing, especially the key technology of Chinese information processing—automatic segmentation of Chinese words.

Chapters 8–15 introduce the implementation of NLP application systems, including sentiment analysis, information extraction, text semantic matching, text summarization, intent understanding in dialogue systems, machine reading comprehension, machine translation, and Q&A based on large models. It should be noted that the implementation methods introduced in these chapters are only based on a certain method or model. The aim is to introduce the implementation process of the task through a specific example, rather than getting a practical system with the best performance.

This chapter only provides a general introduction to statistical and neural network-based NLP methods. Many concepts, models, and methods are not explained in detail. Readers interested in statistical NLP methods can refer to (Zong Chengqing, 2013), and readers who want to understand neural networks and deep learning methods can refer to (Qiu Xipeng, 2020). The references (Zong Chengqing, 2022) and (Zong et al., 2021) provide a more detailed introduction to deep learning-based NLP methods. The reference (Zhang and Teng, 2021) covers both statistical and neural network-based methods. Subsequent chapters will provide corresponding references or websites for the content introduced, and readers can learn the content they are interested in during practice.

1.3 The Platform Used for Practice

In recent years, artificial intelligence technology represented by deep learning has developed rapidly. The neural network has become increasingly complex. The network layers have become deeper, and the size of the parameters has also become larger. Especially with the rise of large models, it is almost impossible to build a neural network from the bottom. At the same time, the challenges of big data and large models to model training have become increasingly severe, requiring a deep learning framework with ultra-large-scale distributed training technology.

The practical cases in this book are developed based on PaddlePaddle from Baidu. PaddlePaddle is China's first open-source deep learning platform with multiple functions and an industrial scale, independently developed by Baidu. It integrates core frameworks, basic model zoos, end-to-end development kits, rich toolkits, and AI Studio. PaddlePaddle achieves building a static and dynamic framework design. It also takes both flexibility and high efficiency into consideration. Besides, PaddlePaddle also supports ultra-large-scale distributed training capabilities and was the first to implement multi-dimensional hybrid parallel strategies and end-to-end self-adaptive distributed training techniques. All these help PaddlePaddle gain an advantage in performance. Besides, PaddlePaddle also supports the efficient training and inference of ERNIE, and the efficiency and effectiveness were improved significantly. Examples include the lightweight reasoning model ERNIE-4.5-21B-A3B-Thinking, the multimodal reasoning model ERNIE-4.5-VL-28B-A3B-Thinking, and the recently released ERNIE 5.0. Readers may access these ERNIE models through open-source repositories on GitHub[4] or the Hugging Face website.[5]

In recent years, the development of PaddlePaddle's open-source environment has made significant progress, and it has been widely used in research institutions, enterprises, public institutions, and individual developers. According to the Deep Learning

[4] https://github.com/PaddlePaddle/ERNIE.

[5] https://huggingface.co/baidu.

Platform Development Report (2022), PaddlePaddle, as a deep learning framework and empowering platform, has ranked first in China's market in terms of its usage in almost all scenarios and fields.[6]

When it comes to practice, this book provides two kinds of operations: operations on personal computers and AI Studio operations. Below, we will introduce the operation methods, as well as the API and datasets mentioned in the book.

1.3.1 Operation on Local Systems

1. **Environment Preparation**

For operations on local systems, readers need to install the PaddlePaddle framework and PaddleNLP. PaddleNLP is a simple, easy-to-use, and powerful natural language processing development library that is developed based on the PaddlePaddle framework. Before installing, it is necessary to confirm the local environment. For example, whether the versions of the operating system, Python, and pip meet the installation requirements of PaddlePaddle. PaddlePaddle supports installations on Windows, macOS, Linux, and other operating systems. This chapter takes the Windows operating system as an example to introduce the quick installation using pip.

Use the following command to check the Python version. Currently, PaddlePaddle supports Python versions 3.9/3.10/3.11/3.12/3.13. If you need to install the latest version of PaddleNLP, it is recommended that the Python version on your local operating system is 3.9 and above.

```
python --version
```

Use the following command to check the pip version. Currently, PaddlePaddle supports pip version 20.2.2 or above.

```
python -m ensurepip

python -m pip --version
```

Use the following command to confirm that Python and pip are 64-bit versions, and the processor architecture is x86_64 (also known as × 64, Intel 64, AMD64).

```
python -c "import platform;print(platform.architecture()[0]);print(platform.machine())"
```

[6] See the Deep Learning Platform Development Report (2022) by China Academy of Information and Communications Technology.

If the first line outputs "64bit" and the second line outputs "x86_64," " × 64," or "AMD64," then the installation requirements for PaddlePaddle are met.

2. **Quick Installation**

(a) **Install PaddlePaddle Framework**

The codes for practice in this book can run on a CPU or GPU. But except for models in Sects. 7.2 and 8.1, the rest of the models are relatively complex and require more computing resources. Running on a GPU can greatly shorten the models' training time. If readers don't have GPUs on their operating systems, it is recommended to choose to run the code on AI Studio. For usage, please refer to Sect. 1.3.2.

The current stable version recommended by PaddlePaddle is 3.3, and it is recommended that readers install the latest version of PaddlePaddle. Use the following command to install the CPU version:

```
python -m pip install paddlepaddle==3.3.0 -i https://pypi.tuna.tsinghua.edu.cn/simple
```

Use the following command to install the GPU version:

```
python -m pip install paddlepaddle-gpu==3.3.0 -i https://pypi.tuna.tsinghua.edu.cn/simple
```

Use the following command to verify the correctness of the installation:

Enter the Python interpreter using the python command. First, enter "import paddle," and then enter "paddle.utils.run_check()." If "PaddlePaddle is installed successfully!" appears, it indicates that PaddlePaddle is successfully installed.

It should be noted that current PaddlePaddle's installation on the Windows operating system supports CUDA versions 11.8/12.6/12.9/13.0. If you need to use other CUDA versions, you need to compile them yourself from the source code. The latest PaddlePaddle version information and installation methods in other environments can be obtained by logging in to the PaddlePaddle official website (https://www.paddlepaddle.org.cn).

(b) **Install PaddleNLP**

Use the following command to install PaddleNLP using pip.

```
pip install --upgrade paddlenlp==3.0.0b4
```

1.3.2 Operation on AI Studio

To facilitate readers' learning and practice, the cases in this book are all hosted in the form of BML Codelabs on AI Studio.[7] The AI Studio, based on PaddlePaddle and ERNIE, integrates open data, open-source algorithms, cloud GPU, and large

[7] https://aistudio.baidu.com/overview.

model development tools, and provides developers with an efficient development environment for models and applications. Readers can directly use the free computing resources provided by AI Studio to compile and run the code in the book online according to the instructions.

The programs and video courses for this book can be found at: https://aistudio.baidu.com/course/introduce/28728.

AI Studio has already installed the latest version of PaddlePaddle by default, and you can install PaddleNLP online, so there is no need to perform the installation operations in Sect. 1.3.1. Readers can choose to run projects in two modes: Basic (CPU) or Advanced (GPU-V100-16 GB/GPU-V100-32 GB/GPU-BI-V150S 32 GB) and earn free GPU computing power by running projects daily, creating or forking projects, and sharing excellent projects.

After entering the project, the page layout of BML CodeLab is shown in Fig. 1.3, which consists of a menu bar, a sidebar, a quick access toolbar, and a code editing area. This section briefly introduces the code editing area and the sidebar. For more detailed usage, please refer to the BML Codelab environment instructions (https://ai.baidu.com/ai-doc/AISTUDIO/Gktuwqf1x).

The code editing area on the BMLCodeLab project page shown in Fig. 1.3 consists of a code writing unit (CodeCell) and a text editing unit (MarkCell). Enter Python code or Linux commands in the code writing unit and click the "Run" button. Then the codes or commands will run on the cloud, and the results will be shown directly on the project page. Enter Markdown-formatted text (such as text, images, mathematical formulas, etc.) in the text editing area to write the project introduction.

Sidebar: Click "Packages" in the sidebar to download the latest version of PaddleNLP. The downloaded file will be automatically saved to /home/aistudio.

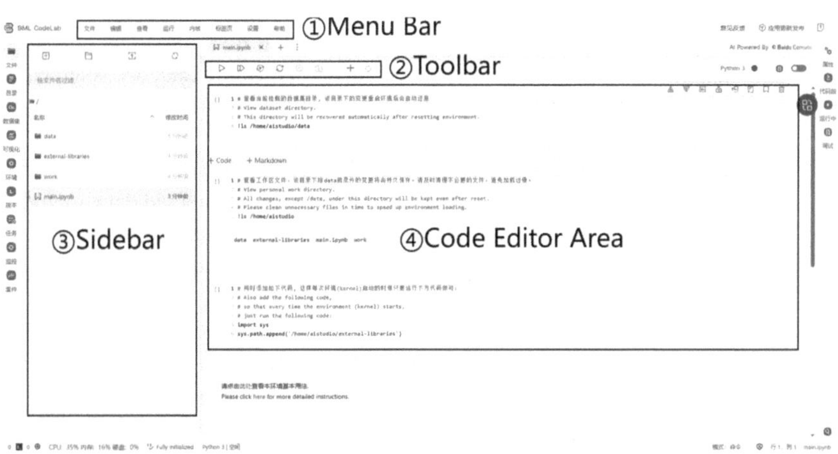

Fig. 1.3 The page layout of the BML CodeLab project

1.3.3 APIs Used in This Book

The APIs and functions used in this book are introduced in Table 1.3.

1.3.4 Datasets Used in This Book

The datasets used in this book are as follows:

(1) icwb2 dataset: It is used in Chap. 7 for the Chinese word segmentation task. icwb2 dataset is jointly built by Academia Sinica, City University of Hong Kong, Peking University, and Microsoft Research Asia. This book uses the dataset provided by Peking University, which contains 18,056 data points in the training set, 1,000 data points in the validation set, and 1,945 data points in the test set.
(2) IMDB dataset: It is used in Sects. 8.1 and 8.2 for the sentiment analysis task. IMDB is a binary classification dataset for sentiment analysis, containing 25,000 data points in the training set and validation set, each with each data point containing users' true rating and sentiment towards a movie.
(3) Baidu's self-built attribute-based sentiment analysis dataset: It is used in Sect. 8.3 for the attribute-based sentiment analysis task. This dataset contains 800 data points in the training set, 100 data points in the validation set, and 100 data points in the test set. Each data includes 3 items: original text, attribute text, and labeled data, in which the attribute text is composed of attributes and opinion words.
(4) CLUENER2020 dataset: It is used in Sect. 9.1 for the named entity recognition task. The CLUENER2020 dataset is based on the text classification dataset THUCTC, open-sourced by Tsinghua University. Some data are selected for fine-grained named entity recognition, including 10,748 data points in the dataset and 1,343 data points in the validation set.
(5) Baidu's self-built information extraction dataset: It is used in Sect. 9.2 for the information extraction task. The dataset has a total of 52 labeled data, including 20 data in the training set, 16 data in the validation set, and 16 data in the test set. Each data contains three entity types: weapon equipment, country, and organization, as well as two relationship types: country of origin and R&D institution.
(6) LCQMC dataset: It is used in Chap. 10 for the text semantic matching task. LCQMC dataset is a large-scale Chinese question matching corpus, which is built by extracting user questions from different fields in Baidu Zhidao (website: https://zhidao.baidu.com/). It contains 238,766 data points in the training set, 4,401 data points in the validation set, and 4,401 data points in the test set.
(7) LCSTS dataset: It is used in Chap. 11 for the text summarization task. The corpus of the LCSTS dataset comes from the large-scale Chinese short text

Table 1.3 Introduction to APIs and their functions used in this book

Tasks	API and examples		Function introduction
Data processing	Dataset		Load data
	AutoTokenizer		Load the interface of different model Tokenizers by specifying the model name
	Paddlenlp.data	DataCollatorWithPadding	Process the same batch of data into the same length for the convenience of model processing
		DataCollatorForSeq2Seq DataCollatorForTokenClassification	NLP task-related data processing API
		paddlenlp.data.Pad paddlenlp.data.Tuple paddlenlp.data.Stack	Data processing-related API
Model building	AutoModel	AutoModel	Load different pre-training models through AutoModel
		ErnieForQuestionAnswering AutoModelForCausalLM AutoModelForTokenClassification AutoModelForSequenceClassification	NLP task-related pre-training model API
	paddle.nn	paddle.nn.Transformer paddle.nn.lstm paddle.nn.GRU paddle.nn.Embedding	Model-related API, including pre-training models such as Transformer and recurrent neural network models
		paddle.nn.functional.dropout paddle.nn.functional.relu paddle.nn.functional.sigmoid paddle.nn.functional.label_smooth	Activation function- related API

(continued)

Table 1.3 (continued)

Tasks	API and examples		Function introduction
Model training	paddle.optimizer	paddle.optimizer.Adam paddle.optimizer.AdamW paddle.optimizer.SGD	Optimization algorithm- related API
	paddle.optimizer.lr	paddle.optimizer.lr.NoamDecay	Learning rate decay- related API
Model configuration	paddle.nn paddle.nn	paddle.nn.CrossEntropyLoss() paddle.nn.BCELoss() paddle.nn.MSELoss() paddle.nn.functional.cross_entropy paddle.nn.functional.mse_loss	Loss function related-API
Model evaluation	paddle.metric	paddle.metric.Accuracy() paddle.metric.Precision paddle.metric.Recall	Evaluation indicators-related API

summarization dataset built by Sina Weibo. This practice has selected part of the corpus, including 8,000 data in the training set, 800 data in the validation set, and 100 data in the test set.

(8) CrossWOZ: It is used in Chap. 12 for the intent understanding task. The CrossWOZ dataset is a large-scale Chinese cross-domain task-oriented dialogue dataset and is jointly built by the Department of Computer Science and Technology and the CoAI group of the Institute for Artificial Intelligence at Tsinghua University. This practice has selected dialogue data about booking hotels and questioning scenic spots to the system, which is a multi-label intent classification dataset with 158 categories, including 44,409 data in the training set, 4,935 data in the validation set, and 4,909 data in the test set.

(9) $DuReader_{robust}$ dataset: It is used in Chap. 13 for the Chinese reading comprehension task. It is the first Chinese dataset that focuses on the robustness of reading comprehension models, with a total of 65,937 data points, including 14,520 data points in the training set, 1,417 data points in the test set, and 50,000 data points in the validation set.

(10) IWSLT 2015 dataset: It is used in Chap. 14 for the machine translation task. The corpus of the IWSLT 2015 dataset comes from TED talks. This practice uses data for English to Chinese and Chinese to English translation, including 2,000 dialogues, 200,000 sentences, and 4,000,000 tokens in the training set. Both the validation set and the test set contain 10–15 dialogues, 1,000–1,500 sentences, and 20,000–30,000 tokens.

Chapter 2
Basics of Neural Network

Abstract This chapter systematically introduces the foundational models of neural networks, beginning with the basic building blocks of neurons and perceptrons. It then explains the role of activation functions before detailing three progressively complex network architectures tailored for specific data types. The text first covers the standard feedforward neural network (FNN), then introduces the convolutional neural network (CNN) as a specialized architecture that uses convolution and pooling to effectively process spatial data like images. Finally, it addresses the challenge of sequential information by presenting the recurrent neural network (RNN) and its advanced, memory-equipped variants, Long Short-Term Memory (LSTM) and the Gated Recurrent Unit (GRU), which are designed to capture temporal relationships in data.

2.1 Overview

Deep learning is a learning method based on neural networks. Multi-level learning models are built by stacking neural networks in the temporal or spatial dimension to complete corresponding application tasks. Practices have proved that the deep learning models built in this way have strong fitting ability. They can automatically learn knowledge from massive data, reduce the dependence on manually extracted features, and achieve better performance in the completed tasks.

This chapter introduces the basic concepts and basic models of neural networks, including feedforward neural networks (FNN), convolutional neural networks (CNN), and recurrent neural networks (RNN), which are widely used in deep learning task modeling. Before introducing the feedforward neural network, we first introduce three important concepts: neuron, perceptron, and activation function, and then we give the most basic structure of the feedforward neural network. Next, we derive the classic convolutional neural network structure from basic operations such as convolution and pooling. Recurrent neural networks are introduced from the simplest recurrent network structure to long short-term memory (LSTM) and gated recurrent unit (GRU).

C. Zong et al., *Natural Language Processing and Large Language Models*,
https://doi.org/10.1007/978-981-92-0682-7_2

2.2 Neuron and Perceptron

2.2.1 Neuron

Neurons are the basic components of the biological nervous system, responsible for receiving and transmitting acquired information. Figure 2.1 shows the basic structure of biological neurons, mainly including dendrites, axons, and cell bodies. Each neuron contains multiple dendrites and one axon, with dendrites used to receive external information and axons used to send information. When the external information received by the neurons accumulates beyond a certain threshold, they will be activated. They will be in an excited state and produce an electrical impulse signal, which will be transmitted backward to the connected neurons through the axon. That is to say, we can consider neurons to have only two states: "yes" (which can be recorded as (1) when activated, and "no" (recorded as 0) when not activated.

In 1943, neuroscientist Warren McCulloch and mathematician Walter Pitts first proposed the artificial neuron, called the MP model, or MP neuron, based on the working characteristics of biological neurons. This opened the prelude to the study of artificial neural networks. Figure 2.2 shows the network structure of artificial neurons. The N inputs $x_1, x_2, x_3, ..., x_N$ received by neurons is similar to the information received by multiple dendrites of biological neurons. The addition operation is equivalent to the accumulation of external information. After reaching a certain threshold, the neuron will be activated.

Specifically, for input $\mathrm{x} = [x_1, x_2, x_3, ..., x_N]$, the calculation formula of a neuron is as follows:

$$z = \sum_{i=1}^{N} w_i x_i + b \tag{2.1}$$

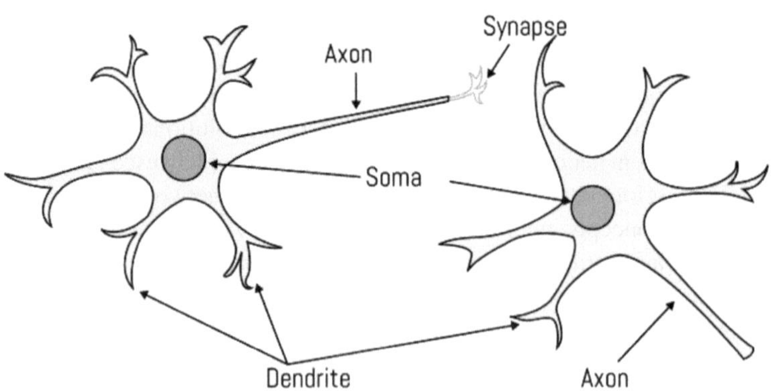

Fig. 2.1 Basic structure of biological neuron

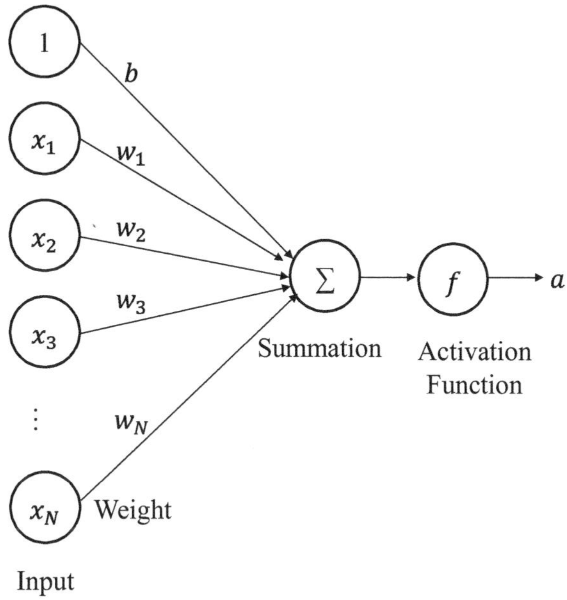

Fig. 2.2 Neuron structure

$$a = f(z) = \begin{cases} z, & z \geq \theta \\ 0, & z < \theta \end{cases} \tag{2.2}$$

In which, w_i is the preset connection weight, denoted as $w = [\mathrm{w}_1, \mathrm{w}_2, \mathrm{w}_3, ..., \mathrm{w}_N]$, used to indicate the influence of its corresponding input data on the output result; b is the bias term. The net activity value z is obtained by summing the weighted input x with the bias term b. The calculation formula of z can be simplified as:

$$z = \boldsymbol{w}^{\mathrm{T}}\boldsymbol{x} + b \tag{2.3}$$

The net activity value is passed through a nonlinear function $f(\cdot)$ to obtain the activity value a of the neuron. The nonlinear function $f(\cdot)$ is called an activation function, which is used here to determine whether the threshold z exceeds the threshold θ.. In the original MP model, the weight takes the value of 0 or 1, and the activation function is a step function represented as 0 or 1. However, in modern neurons (deep learning), the activation function is required to be a continuously differentiable function.

2.2.2 Perceptron

Artificial neurons can simulate the working mode of biological neurons well, but their weight values are preset, which limits the development of neuron technology to some extent, especially when there are many weight parameters. In 1957, Frank Rosenblatt proposed the concept of the perceptron. The structure of the perceptron is the same as that of the MP neuron, and its calculation formula is the same as (2). However, the difference is that the weight of the perceptron is not preset, but is automatically learned by fitting data during the model training process. Due to this ability, perceptrons can be used in more areas, and it also helps lay the foundation for the development of deep learning technology later.

2.2.3 Common Activation Functions

As mentioned above, in MP neurons and perceptrons, the activation function is a step function. With the development of deep learning technology, more and more activation functions are proposed and applied in the learning process of modern neural networks. The activation function is generally a nonlinear curve function, which introduces nonlinear computation into neural networks, thereby enhancing the fitting ability or expression quality of the network. Here are several commonly used activation functions in deep learning.

1. **Sigmoid function**

The meaning of the English word Sigmoid is "S-shaped, S-curved." As the name suggests, the curve of the Sigmoid function is S-shaped. There are two Sigmoid functions commonly used in neural networks: the logistic function and the hyperbolic tangent function (or tanh).

(1) Logistic function

The definition of a logistic function is:

$$\sigma(x) = \frac{1}{1 + \exp(-x)} \tag{2.4}$$

The function curve is shown in Fig. 2.3:

As shown in Fig. 2.3, the function has a value range of (0,1). The smaller the input value x, the closer the value of the logistic function is to 0. As the input value x increases, the function value gradually approaches 1. Compared to the step activation function in the perceptron, the logistic function is continuously differentiable over the entire curve.

The logistic function is a kind of two-sided saturated function. That is to say, the slope of the curve at both ends gradually approaches 0, and these areas are also called saturation zones. Once the value falls into the saturation zone, the gradient

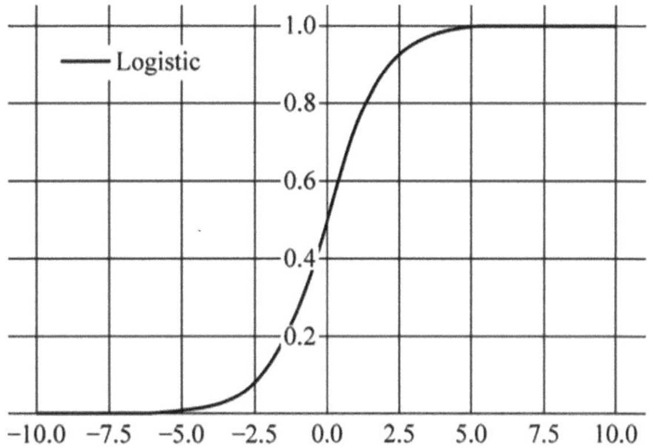

Fig. 2.3 Logistic function curve

will be close to 0. In the process of network training, the vanishing gradient problem may appear. In addition, the value of the logistic function is always greater than 0. In multi-layer neural networks, the non-zero-centered output will cause a bias shift in the input of the neurons in the subsequent layer, and further slow down the convergence speed of gradient descent.

Although the logistic function has problems mentioned above, it is still a commonly used activation function in neural networks due to its excellent properties. The main advantages are as follows:

(1) The logistic function maps the input value to the range (0,1), which can be used for simple binary classification tasks, i.e., classification tasks with two labels of 0 and 1. At the same time, the output value of the function can be regarded as a probability;
(2) The logistic function is used as a gating mechanism to control the output of input information. A commonly used method is to control the forgetting and updating of information in the LSTM model.
(2) tanh function

The definition of the tanh function is:

$$\tanh(x) = \frac{\exp(x) - \exp(-x)}{\exp(x) + \exp(-x)} = 2\sigma(2x) - 1 \tag{2.5}$$

The tanh function is shown in Fig. 2.4:

Unlike the logistic function, the tanh function has a zero-centered curve with a value range of (-1,1), which makes it converge faster than the logistic function during model training.

2. **ReLU function**

ReLU function is defined as

$$\text{ReLU}(x) = \begin{cases} x, \ x \geq 0 \\ 0, \ x < 0 \end{cases} \tag{2.6}$$

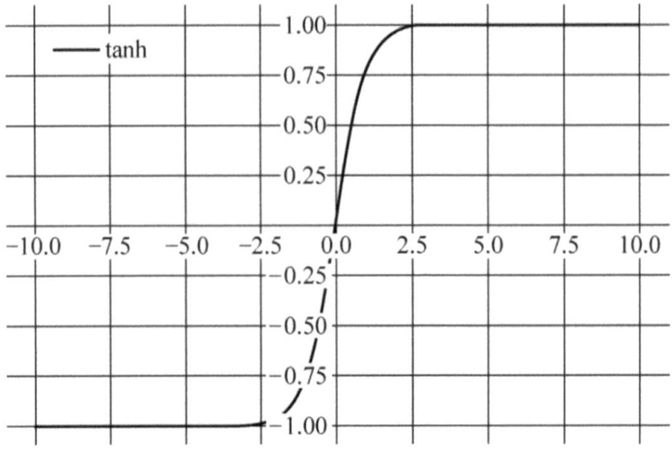

Fig. 2.4 Tanh function

ReLU function is shown in Fig. 2.5:

ReLU is also a commonly used activation function in neural networks, which belongs to the left saturation function. When x is smaller than 0: the derivative is 0, which means that the values on the left side of the origin are suppressed; when x is larger or at 0: the derivative is 1. Compared with the saturation of both ends of the Sigmoid function and tanh function, ReLU can alleviate the vanishing gradient problem to a certain extent. In addition, the ReLU function calculation is simple and does not involve exponentiation, thus speeding up the network calculation efficiency.

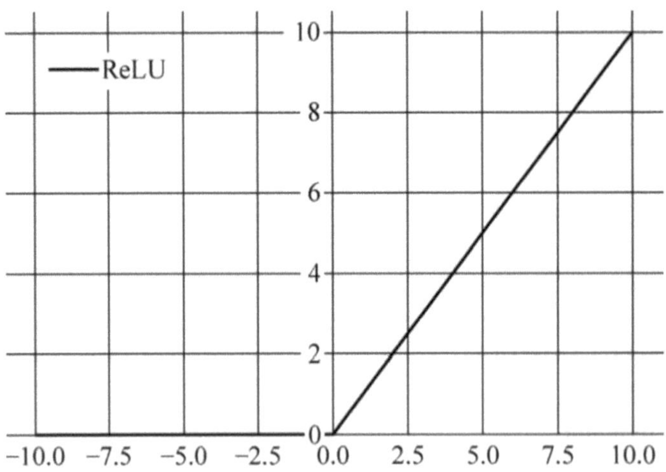

Fig. 2.5 ReLU function

The ReLU activation function also has some disadvantages. First, the ReLU function is non-zero-centered, which will affect the convergence efficiency of the model to some extent during the training process. Second, the ReLU function forces values less than 0 to be set to 0. So if the output value of a ReLU neuron parameter is 0 on all training data after updating during the training process, the gradient of the neuron's parameters in the subsequent training process will be 0 and will never be updated. This phenomenon is called the dying ReLU problem.

2.3 Feedforward Neural Network

The feedforward neural network (FNN), also known as a fully connected neural network, is composed of multiple layers of neurons, as shown in Fig. 2.6. Each layer has multiple neurons. The first layer is the input layer of the network, the last layer is the output layer of the entire network, and the middle layers are hidden layers. Figure 2.6 shows two hidden layers. In the feedforward neural network, information will be unidirectionally transmitted from the input layer to the output layer. The output of the previous layer will be used as the input of the next layer, and the calculation will be completed at the output layer.

Assuming that given input data $x = [x_1, x_2, x_3, \ldots, x_N]$, this data is inputted into the feedforward neural network, and will be gradually transmitted to each hidden layer for calculation. Finally we can get the final result through the output layer. Let's take the l $(1 < l \leq L)$ hidden layer as an example for explanation. We assume that the output of the l-1 layer is $\boldsymbol{a}^{(l-1)}$, which will be inputted into the l layer for the following calculation:

$$z^{(l)} = \boldsymbol{w}^{(l)}\boldsymbol{a}^{(l-1)} + \boldsymbol{b}^{(l)} \tag{2.7}$$

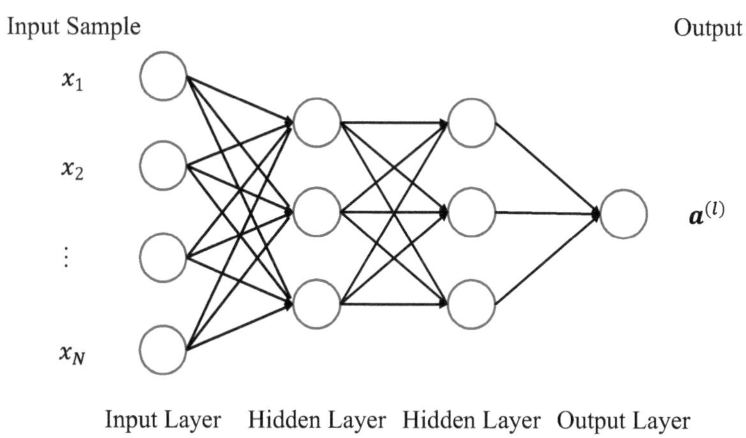

Fig. 2.6 Feedforward neural network structure

$$\boldsymbol{a}^{(l)} = f_l(\boldsymbol{z}^{(l)}) \tag{2.8}$$

In which, $\boldsymbol{z}^{(l)}$ is the net activity of neuron in layer l; $\boldsymbol{a}^{(l)}$ is the output value of neuron in layer l, and in the input layer: $\boldsymbol{a}^{(0)} = \text{x}$; $\boldsymbol{w}^{(l)}$ and $\boldsymbol{b}^{(l)}$ are the weight and bias of layer l respectively; f_l is the activation function used by neuron in layer l. The feedforward neural network finally obtains the output value $\boldsymbol{a}^{(l)}$ by calculating forward layer by layer.

2.4 Convolution Neural Network

In the feedforward neural network, neurons of two adjacent layers are connected in pairs, that is, each neuron of the previous layer is connected to all neurons of the next layer. Such network structure will face many problems in practical applications. Taking image processing as an example, when an image is inputted, feedforward neural network will face the following two problems:

- The model has too many parameters, making it difficult to train. Assuming that an input color image is composed of $256 \times 256 \times 3$ pixels and the number of neurons in the hidden layer is 128, the network in this hidden layer will require $256 \times 256 \times 3 \times 128 = 3 \times 2^{23}$ weight parameters. Such a large number of network parameters will make it difficult to train the network.
- The spatial information contained in the image will be lost. In an image, spatially adjacent pixels often have strong correlation, such as close corresponding RGB values, and the data between each pixel in various RGB channels are also closely related. If the feedforward network is directly used for processing, the image needs to be converted into a one-dimensional vector, which will result in the loss of spatial information (pixels' coordinate positions).

In order to solve the problems mentioned above, people usually use convolution neural network to process images, and use convolution to extract features from input images. Convolution calculation is carried out in the spatial neighborhood of pixel points, and the spatial information of input images can be used. Moreover, convolution has local connection, weight sharing and other characteristics, so the number of parameters in convolution neural network is generally much smaller than that in feedforward neural network.

2.4.1 Convolution

Convolution is an important computing method in neural networks, especially widely used in computer vision, image, and video processing fields. Convolution has many types, such as one-dimensional convolution, two-dimensional convolution, and three-dimensional convolution.

1. **1D convolution and filter**

One-dimensional convolution is the simplest type of convolution, commonly used in signal processing to calculate the cumulative delay of a signal. In the process of signal transmission, signal attenuation usually occurs. Assuming the attenuation rate of the signal is w_k, then the signal x_t generated by a signal generator at time t after attenuation becomes $w_k x_t$. In this way, the signal $y_{t+\Delta}$ received after k-1 times of launches at time $t + \Delta$ should be the accumulation of the signal generated at the initial time t and the signal after multiple attenuations:

$$y_{t+\Delta} = \sum_{k=1}^{K} w_k x_{t-k+1} \tag{2.9}$$

In which, K is the number of launches till the time $t + \Delta$.

We usually name the attenuation rate w_K as convolution kernel, or filter, K as the length of the filter, or the size of the convolution kernel.

2. **Two-dimensional Convolution**

In the field of image processing, since images are two-dimensional structures, convolution calculations need to be extended to two dimensions. Common convolution kernel sizes include 3×3, 5×5, etc.

Assuming an image $\boldsymbol{X} \in \mathbb{M} \times \mathbb{N}$ and a convolution kernel $\boldsymbol{W} \in \mathbb{U} \times \mathbb{V}$, the convolution calculation method is:

$$y_{ij} = \sum_{u=1}^{U} \sum_{v=1}^{V} w_{uv} x_{i+u-1,j+v-1} \tag{2.10}$$

Figure 2.7 shows the convolution calculation process of an image with the size of3×3 and a convolution kernel with size of 2×2.

Assuming that images with the size of 3×3 are numbered from left to right as 0, 1, 2, and from top to bottom as 0, 1, 2, then the pixel coordinate in the upper left corner is (0, 0), the coordinate of the square in the lower right corner is (2, 2), and the coordinates of the remaining pixels are deduced accordingly. The same numbering is used for the convolution kernel with size 2×2. In Fig. 7a, the convolution kernel is aligned with the image in the blue region from (0, 0) to (1, 1). The pixel values in this region are weighted and summed with the weights of the convolution kernel to obtain the value 25, which is placed at the (0, 0) position of the result matrix. In Fig. 7b, the convolution kernel slides to the right and aligns with the image region from (0, 1) to (1, 2), and the convolution calculation yields the value 31, which is placed at the (0, 1) position of the result matrix. In a similar manner, the convolution calculation results for Fig. 7c, d are 43 and 49, respectively, which are placed at the (1, 0) and (1, 1) positions.

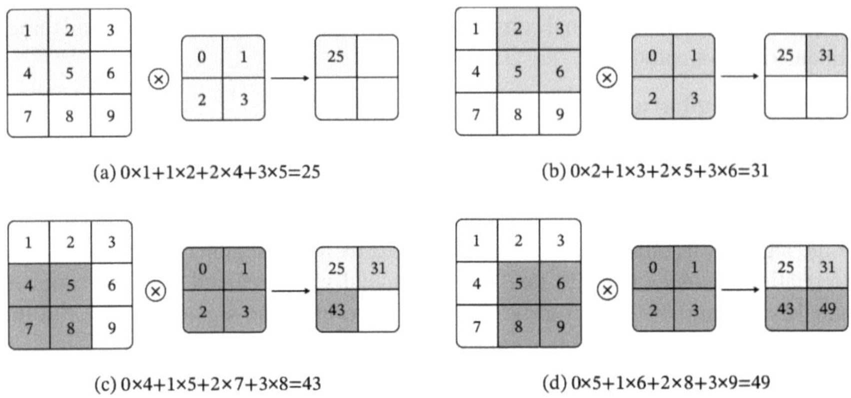

Fig. 2.7 Convolution calculation

From the above example, it can be seen that through the operation of convolution, an image with its original size of 3×3 was transformed into an image with the size of 2×2, and the transformed image is usually called a feature map.

It should be noted that to facilitate understanding, no bias is introduced in the convolution calculation in Fig. 2.7. In practical applications, sometimes, in order to better fit the data, bias needs to be set during the convolution operation. Assuming that the bias is set to 1, the convolution calculation in Fig. 7a is:

$$0 \times 1 + 1 \times 2 + 2 \times 4 + 3 \times 5 + 1 = 26$$

The calculation methods for other situations in Fig. 2.7 are similar and will not be repeated here.

3. **Stride**

In the example above, we carried out convolution by sliding 1 pixel to the right or down, and the number of pixels slid here is actually the stride. In fact, the stride can be set to a value greater than 1, which can reduce the size of the feature map in the convolution neural network (CNN) calculation, effectively reducing the network's computing load.

Figure 2.8 shows the calculation process of convolution when the stride is 2. The image size is 4×4, and the stride is 2. That is to say: each time it slides right or down by 2 positions, and the final feature map size is 2×2.

4. **Padding**

In the example shown in Fig. 2.7, the size of the input image is 3×3, and the size of the output image is 2×2. If we carry out another convolution calculation on the output image with the size of 2×2, the size of the image will become 1×1. That is to say, every time a convolution calculation is carried out, the size of the output image will decrease. In order to solve this problem, the padding technique was proposed.

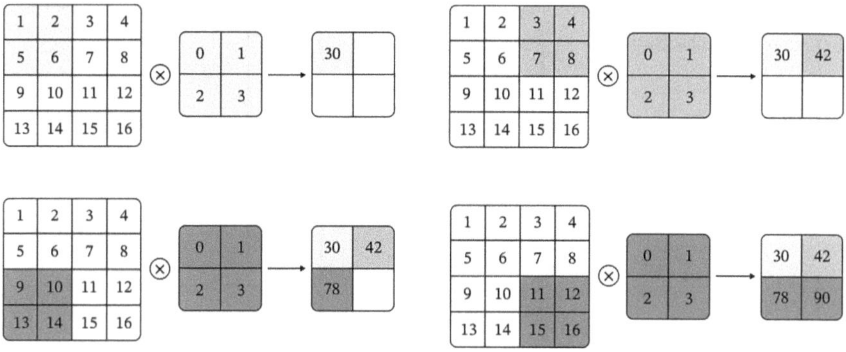

Fig. 2.8 Convolution calculation with stride of 2

Specifically, padding refers to supplementing the edge positions of the input image with "0" pixel values, so that the edge pixels of the input image can also participate in convolution computation as centered points. As shown in Fig. 2.9, after padding "0" in the four directions of the original image with the size of 3 × 3, the size of the image becomes 5 × 5. If the stride is set to 1, the feature map with the size of 3 × 3 can be obtained after calculation using the convolution kernel with the size of 3 × 3, which is the same size as that of the original input image.

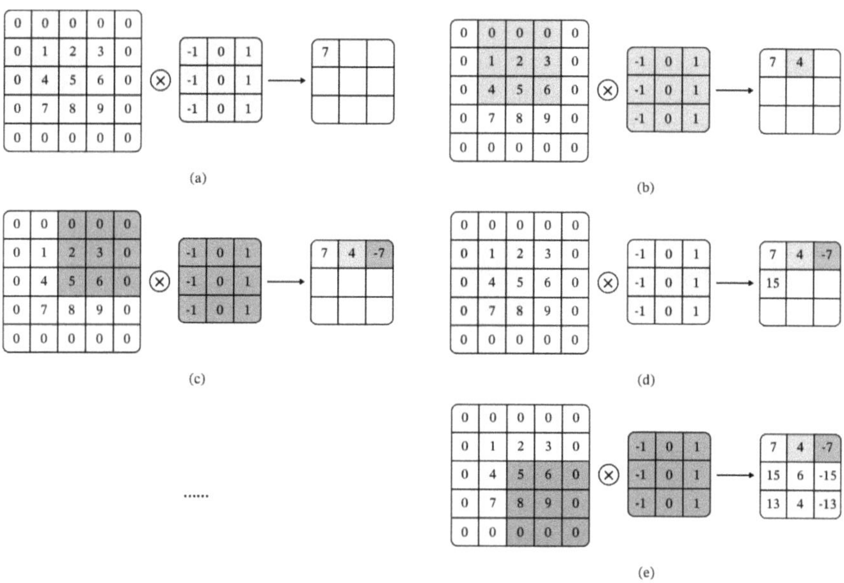

Fig. 2.9 Convolution calculation with padding and a stride of 1

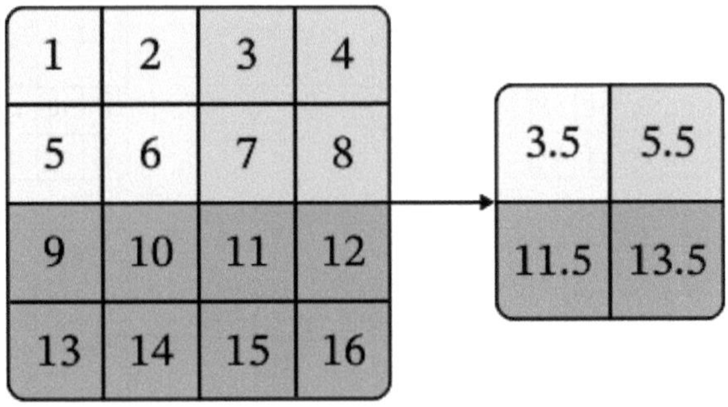

Fig. 2.10 Example of average pooling calculation

2.4.2 Pooling

In the convolution introduced above, the convolution kernel slides over the image, and the convolution kernel and the values of corresponding image blocks are weighted and summed for each slide. Then the feature map after convolution can be obtained. Pooling has a similar effect as convolution. It is achieved by setting a pooling window and using the slide of the window on the image to perform pixel calculation on the selected image block. According to different calculation methods, pooling methods are divided into different types, among which the commonly used pooling methods are average pooling and max pooling.

1. **average pooling**

The average pooling method calculates the mean of all the values within the feature map block covered by the pooling window and outputs the result. Figure 2.10 shows the result of average pooling using a pooling window with the size of 2×2 to carry out the average pooling of a feature map with the size of 4×4. It should be noted that the stride moved each time is 2. In convolution neural networks, pooling windows with the size of 2×2 and a stride of 2 are usually used.

2. **max pooling**

The max pooling method is to select the maximum value in the feature map block covered by the pooling window for output. Figure 2.11 shows the result of max pooling on the feature map with the size of 2×2 using the pooling window with the size of 4×4. It should be noted that the stride for each movement is also 2.

2.4.3 Convolution Neural Network

A complete convolution neural network is usually stacked by convolution layers, pooling layers, and fully connected layers alternately. Figure 2.12 shows a typical convolution neural network, which includes 3 convolution blocks and 1 fully connected layer. Each convolution block includes 1 convolution layer, 1 activation

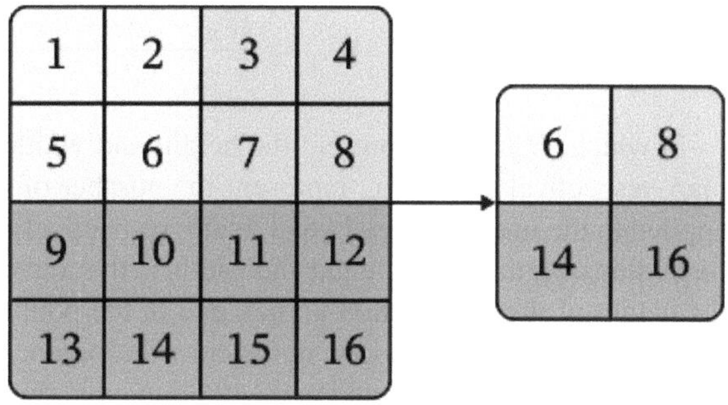

Fig. 2.11 Example of max-pooling calculation

layer, and 1 pooling layer. The activation layer usually uses the ReLU activation function. After the input image is processed by multiple convolution blocks, we can obtain a feature map that can better express the image. Then it is stretched to become a one-dimensional vector and finally used as the input of the fully connected layer to complete specific image processing tasks, such as image classification.

According to the previous introduction, it is not difficult to understand that the feature map after convolution and pooling becomes smaller, leading to a smaller calculation scale. Therefore, the example given in Fig. 2.12 also reflects this continuous shrinking process.

The size change of the image after convolution and pooling can be calculated as follows:

$$H_{out} = \frac{H_{in} + 2p_h - k_h}{s_h} + 1 \tag{2.11}$$

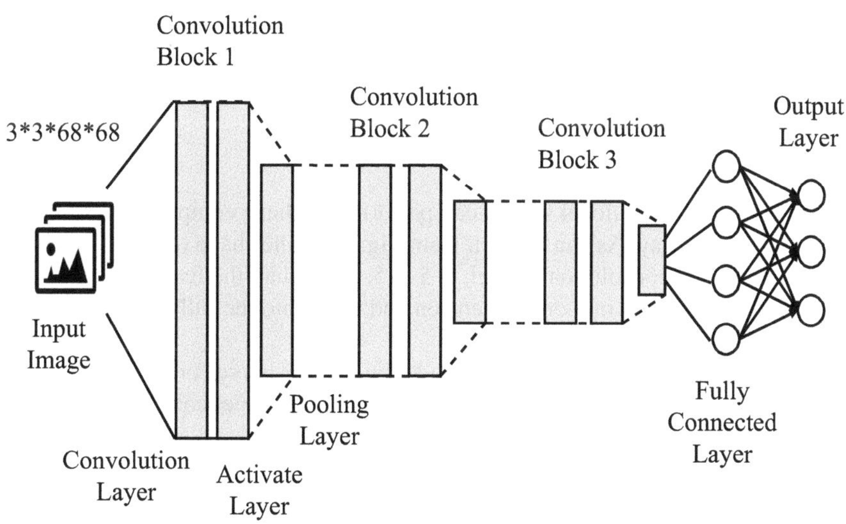

Fig. 2.12 Typical convolution neural network example

$$W_{out} = \frac{W_{in} + 2p_w - k_w}{s_w} + 1 \tag{2.12}$$

In which, $H$and W represent the height and width of the input image or feature map, respectively; $p_h$andp_w represent the number of rows and columns of padding applied to the image on height and width, respectively; s_h and s_w represent the stride for sliding over the image height and width; in the convolution operation, k_h represents the height of the convolution kernel, and in the pooling operation, it represents the height of the pooling window; in the convolution operation, k_w represents the width of the convolution kernel, and in the pooling operation, it represents the width of the pooling window.

In Fig. 2.12, assuming that an image with the size of 68 × 68 is input (RGB is not considered for the time being), in the convolution operation, the convolution kernel used with the size of 5 × 5 has a stride of 1 and no padding; in the pooling operation, the shape of the pooling window is set to 2 × 2, with the stride is 2 and no padding. Then, according to Formulae (2.11) and (2.12), the shape of the feature map after convolution in the first convolution block can be obtained as:64 × 64.

$$H_{out} = \frac{68 + 2 \times 0 - 5}{1} + 1 = 64$$

$$W_{out} = \frac{68 + 2 \times 0 - 5}{1} + 1 = 64$$

Since the activation layer does not change the size of the image, the shape of the feature map remains the same: 64 × 64. Next, the feature map will pass through the pooling layer, and its shape will be further reduced to 32 × 32.

$$H_{out} = \frac{64 + 2 \times 0 - 2}{2} + 1 = 32$$

$$W_{out} = \frac{64 + 2 \times 0 - 2}{2} + 1 = 32$$

Subsequent convolutional blocks can perform the shape computation of feature map in the same way. As can be seen from Fig. 2.12, the shape of the feature map output by the last convolutional block is 5 × 5. After that, the feature map of each image can be stretched into one dimension and input into the fully connected layer for further processing.

It should be noted that in the actual modeling process, more convolutional blocks and fully connected layers can be stacked to construct a deeper convolutional neural network to further enhance the learning ability of the model.

2.5 Recurrent Neural Network

In natural language, both speech and text sentences can be seen as sequential signals, and there is a sequential relationship between signals (Chinese characters). Changing the order may directly change the meaning of the sentence. Please see the following example:

(1) Xiaoming gave Xiaohong a harmonica as a birthday gift.
(2) Xiaohong gave Xiaoming a harmonica as a birthday gift.

In the example above, changing the order of names led to the change of the relationship between the implementer and the recipient, and the meaning of the sentence before and after is completely different. So how to model this kind of data with sequential relationships? In recent years, recurrent neural networks have been widely used. Next sections will introduce simple recurrent neural networks, long-short-term memory, and gated recurrent unit.

2.5.1 Simple Recurrent Neural Network

1. **The Working Principle of RNN**

RNN is a very classic model for sequence processing tasks, which can model natural language sentences or other sequential signals. In RNN, the processing unit (we usually call it RNN cell) unfolds in chronological order. It receives both the current input $\boldsymbol{x}_t$ and the output $\boldsymbol{h}_{t-1}$ from the previous moment at each moment t, calculates the output $\boldsymbol{h}_t$ at the current moment, and sends it to the RNN cell at the next moment.

As shown in Fig. 2.13, assuming there is a sentence: “I love artificial intelligence.” After word segmentation, it becomes 4 words “I love artificial intelligence.” RNN cell processes these four words in turn according to the sequential relationship (writing order from left to right). At the first moment, the word “I” is processed, then the output $\boldsymbol{h}_1$ is generated after calculating “I” and the initial state $\boldsymbol{h}_0$. And then enter the next moment. At the second moment, $\boldsymbol{h}_1$ and the word “love” is simultaneously input, and the output $\boldsymbol{h}_2$ generated after calculation is sent to the next RNN cell. In the same manner, such processing procedures will be carried out until the last word is processed.

It can be seen from examples that when RNN “reads” a sentence from left to right, it constantly uses a same RNN cell to process sequential information. When “reading” each word, RNN always combines the word $\boldsymbol{x}_t$ at the current moment t with the state vector $\boldsymbol{h}_{t-1}$ of the internal memory of the model to form a state vector $\boldsymbol{h}_t$ with the latest memory. When RNN reads the last word, it has read the whole sentence. Generally, it can be considered that the state vector output by the last word can represent the semantic information of the whole sentence, because it “accumulates” the information of all the words before, which can be seen as the semantic vector of the whole sentence.

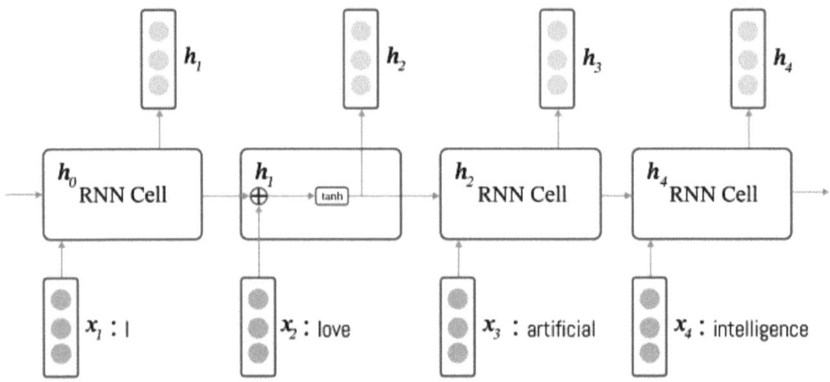

Fig. 2.13 RNN network structure

2. **Understanding RNN from the Perspective of Formula**

Assuming the vector $\boldsymbol{x}_t \in \mathbb{R}^M$ is the network input at time t, and $\boldsymbol{h}_{t-1} \in \mathbb{R}^D$ is the output value at time $t-1$, then the output $\boldsymbol{h}_t$ at the current time can be simply expressed as:

$$\boldsymbol{h}_t = f(\boldsymbol{W}\boldsymbol{x}_t + \boldsymbol{V}\boldsymbol{h}_{t-1} + \boldsymbol{b}) \tag{2.13}$$

In which, $\boldsymbol{W} \in \mathbb{R}^{D \times M}$ is the weight matrix for the current time state input, V $\in \mathbb{R}^{D \times D}$ is the weight matrix from the previous time state to the current time state, and b is the bias. Then at time t, the RNN cell performs a linear transformation on the two inputs x_t and h_{t-1}. After that, it obtains the output h_t at the current time using the f function. Usually, f uses the tanh function.

The tanh function is a function with a value range of $(-1,1)$, which can maintain $\boldsymbol{h}$ values within a fixed numerical range for a long period of time, preventing numerical explosion caused by multiple iterations and updates. At the same time, the derivative of tanh is a smooth function, making neural network training easier.

RNN looks like a perfect sequential data processing model, but there is a significant flaw in the real task training process. That is, when reading a long sequence, the internal information of the network will gradually become more and more complex, exceeding the memory capacity of the network, making the final output information chaotic and useless. Take for example the following sentence:

> I think the dishes in this restaurant are very good. The roast duck is very authentic, the buns are also good, and the beef in soy sauce is very chewy. However, the waiter's attitude is too bad. We waited for 50 minutes at the door to line up a place, and finally managed to get one. But the table wasn't cleared for a long time, and the whole environment was very noisy. My child was even scared to tears. I won't bring friends here next time.

Obviously, this is a relatively long text sequence. When the RNN reads this sentence, it may be able to accurately calculate the vector representation of the sentence (which we temporarily consider as "semantic representation") in the first half of the sentence. But it may become completely confused when reading the second half of the sentence, and cannot accurately calculate the semantic representation of the sentence. That is, it cannot maintain the dependency between long-term information.

Therefore, improved models based on RNN have been proposed for solving this problem, and LSTM is one of the classic models.

2.5.2 *Long-Short Term Memory, LSTM*

1. **LSTM working principle**

LSTM is a type of recurrent neural network that introduces a gated mechanism to address the inherent flaws of RNNs. It uses three gates: input gate, forget gate, and output gate to control which information the LSTM cell needs to remember and which to forget, thus maintaining longer information dependency and expressing the semantics of the whole sentence more accurately. As with RNNs. LSTM processing units (LSTM cells) process sequential data in the chronological order, as shown in Fig. 2.14.

In Fig. 2.14, we can see that the LSTM cell receives three types of data at each moment t: the word $\boldsymbol{x}_t$ (word embedding) at the current time, the state vector $\boldsymbol{c}_{t-1}$ at the previous time, and the hidden state vector $\boldsymbol{h}_{t-1}$ at the previous time. The output of the LSTM unit at the time t is the state vector $\boldsymbol{c}_t$ and the hidden state vector $\boldsymbol{h}_t$.

Table 2.1 shows the input and output of LSTM at each moment when processing the sentence "I love artificial intelligence." We can see that at the first time, the input of the model is the word embedding of the word "I," the initial state vector $\boldsymbol{c}_0$ and the initial hidden state vector $\boldsymbol{h}_0$. The output of the model is the state vector $\boldsymbol{c}_1$ and the hidden state vector h_1; at the second time, the input of the model is the word embedding of the word "love," the state vector $\boldsymbol{c}_1$ and the hidden state vector $\boldsymbol{h}_1$ at

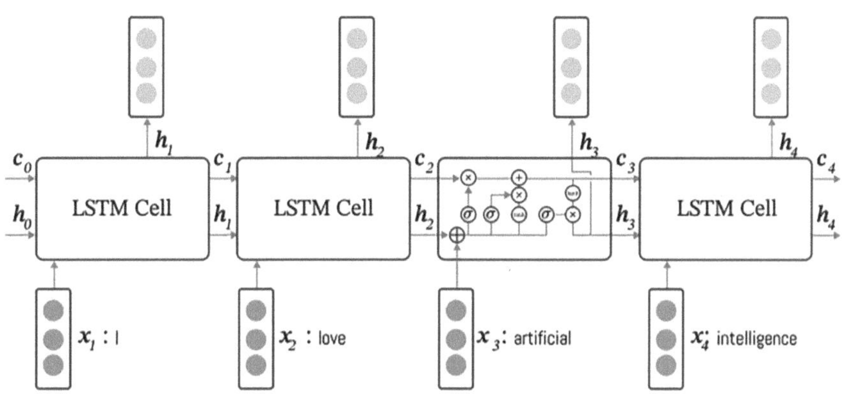

Fig. 2.14 LSTM network structure

the first time, and the output of the model is the state vector $\boldsymbol{c}_2$ and the hidden state vector $\boldsymbol{h}_2$. The subsequent time is deduced in turn.

It should be noted here that there are two state vectors $\boldsymbol{c}_t$ and $\boldsymbol{h}_t$ in LSTM. Generally speaking, we consider the first one as the state that represents the information that can be read by the current LSTM unit, while the second one is the output state of the current LSTM unit to the outside world. It is the actual working state vector, which is generally used to perform specific tasks.

2. **Understanding LSTM from the Formula Perspective**

First, let's analyze the computing logic inside the LSTM unit to see how it maintains long-term data dependency.

Figure 2.15 shows the internal structure of the LSTM unit, which includes the following components:

(1) State vector c_t: It controls the state or memory of the entire LSTM unit. It is updated based on the input at each moment, thus maintaining the memory of the LSTM unit in real time.

Table 2.1 LSTM data processing sample table

	Time step 1	Time step 2	Time step 3	Time step 4
Input	×1: I	×2: love	×3: artificial	×4: intelligence
	c0	*c1*	*c2*	*c3*
	h0	*h1*	*h2*	*h3*
Output	*c1*	*c2*	*c3*	*c4*
	h1	*h2*	*h3*	*h4*

ci: Cell state vector at time step i, hi: Hidden state vector at time step i

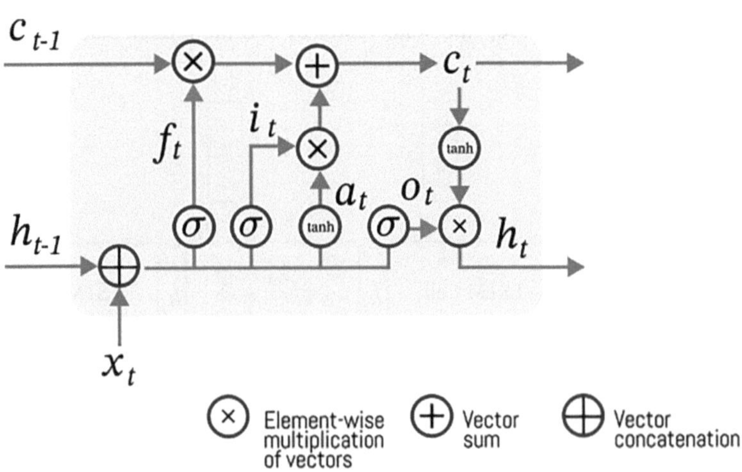

Fig. 2.15 The internal structure of the LSTM unit

(2) Hidden state vector h_t: It is the external output state of the current LSTM unit, which is the actual working state vector h_t. It is generally used to perform specific tasks.
(3) Input gate i_t: It controls which information from the current input should be sent into the state vector c_t. For example, when the input information consists of words without actual meanings, such as "的," the model may prevent such information from flowing into the state vector, thus maintaining the semantic representation of the model.
(4) Forget gate f_t: It controls which information from the previous state vector c_{t-1} needs to be masked or forgotten. For example, if the input is a spoken sentence which has been corrected: "I went to climb the Great Wall yesterday, oh, no, it was the day before yesterday." When the model processes "oh, no, it was the day before yesterday," it needs to forget the previous word "yesterday."
(5) Output gate o_t: It controls which information from the current state vector c_t needs to be output to the outside. The final output is h_t

The calculation formulas for three gates at time t is:

$$\boldsymbol{i}_t = \text{Sigmoid}(\boldsymbol{W}_i\boldsymbol{x}_t + \boldsymbol{U}_i\boldsymbol{h}_{t-1} + \boldsymbol{b}_i) \tag{2.14}$$

$$\boldsymbol{f}_t = \text{Sigmoid}(\boldsymbol{W}_f\boldsymbol{x}_t + \boldsymbol{U}_f\boldsymbol{h}_{t-1} + \boldsymbol{b}_f) \tag{2.15}$$

$$\boldsymbol{o}_t = \text{Sigmoid}(\boldsymbol{W}_o\boldsymbol{x}_t + \boldsymbol{U}_o\boldsymbol{h}_{t-1} + \boldsymbol{b}_o) \tag{2.16}$$

When calculating the values of three gates, the input data $\boldsymbol{x}_t$and$\boldsymbol{h}_{t-1}$ at time t is used, and after a linear transformation, the result is passed to the S igmoid function. Because the value range of the Sigmoid function is (0,1), it can map the data to this fixed range, thus controlling the flow of information.

The formula for calculating the candidate state information of the LSTM unit at time t is:

$$\boldsymbol{a}_t = \tanh(\boldsymbol{W}_a\boldsymbol{x}_t + \boldsymbol{U}_a\boldsymbol{h}_{t-1} + \boldsymbol{b}_a) \tag{2.17}$$

When calculating $\boldsymbol{a}_t$, the input data $\boldsymbol{x}_t$and$\boldsymbol{h}_{t-1}$ are also sent to linear transformation, and then the transformation result is passed to the tanh function. The final result is the information that needs to be input into the state vector $\boldsymbol{c}_t$ of the current LSTM unit.

With the above components, the current LSTM cell state vector $\boldsymbol{c}_t$ can be updated at current time t, and the calculation formula is:

$$\boldsymbol{c}_t = \boldsymbol{f}_t \cdot \boldsymbol{c}_{t-1} + \boldsymbol{i}_t \cdot \boldsymbol{a}_t \tag{2.18}$$

Clearly, the LSTM cell state $\boldsymbol{c}_t$'s update is a selective forgetting of the previous moment's state $\boldsymbol{c}_{t-1}$, and a selective input of the current moment's candidate state information $\boldsymbol{a}_t$, and finally adds the results of the two. This means that the current LSTM unit is integrated into the previous state information $\boldsymbol{c}_{t-1}$, while also inputting the latest current information $\boldsymbol{a}_t$. After calculating the current moment's state vector $\boldsymbol{c}_t$, it can be output according to this state vector. The calculation formula is:

$$\boldsymbol{h}_t = \boldsymbol{o}_t \cdot \tanh(\boldsymbol{c}_t) \tag{2.19}$$

The current state information $\boldsymbol{c}_t$ is selectively output through the output gate.

2.5.3 Gated Recurrent Unit

1. **GRU principle**

GRU is another classic recurrent neural network. Compared with LSTM, GRU has been optimized to some extent in terms of model complexity. Thus, the model has a faster speed while achieving comparable performance to LSTM models.

According to the previous introduction, in the LSTM model, the input gate, output gate, and forget gate control the input and filtering of information. The information is transmitted between adjacent moments through state vectors c and hidden state vectors h. In the structure of the GRU model, there are only two gates: reset gate and update gate, and the GRU cell between adjacent moments only transmits information through hidden states, as shown in Fig. 2.16.

At time t, the GRU cell receives two types of input: the word $\boldsymbol{x}_t$(word embedding) and the hidden state vector $\boldsymbol{h}_{t-1}$ at the previous time. The output of the GRU cell is the hidden state vector $\boldsymbol{h}_t$ at the current time t. For example, when processing the sentence "I love artificial intelligence," the input at the second time is the second word "love" and the hidden state vector $\boldsymbol{h}_1$ at the first time. After internal calculations by the GRU cell, the hidden state $\boldsymbol{h}_2$ is output.

2. **Anatomy of GRU from the Perspective of Formula**

The internal structure of the GRU cell is shown in Fig. 2.17.

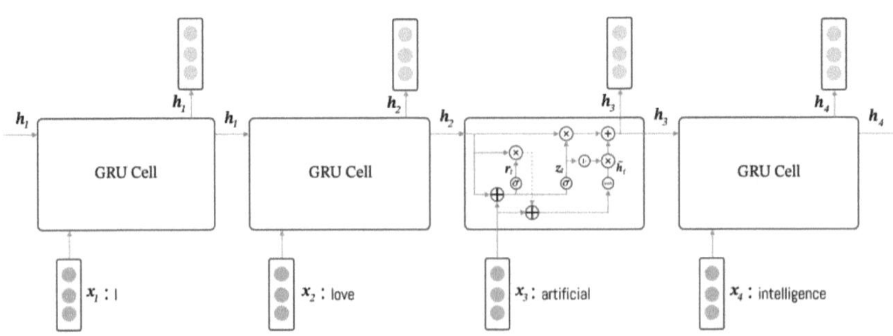

Fig. 2.16 GRU network structure

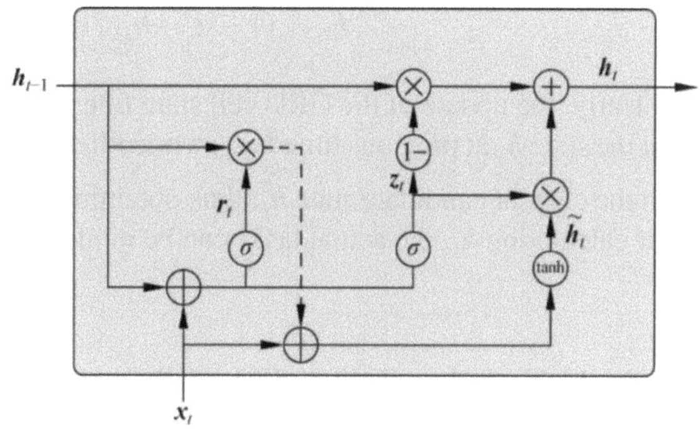

Fig. 2.17 The internal structure of GRU cell

As can be seen from Fig. 2.17, the internal structure of the GRU cell consists of the following components:

(1) Hidden state vector $\boldsymbol{h}_t$: It is the state vector of the current GRU cell and also the actual working state vector. It is generally used for specific tasks.
(2) Reset gate $\boldsymbol{r}_t$: It controls how the state vector $\boldsymbol{h}_{t-1}$ from the previous time flows into the candidate state vectors at the current time.
(3) Update gate z_t: It controls which information from the previous time's state vector $\boldsymbol{h}_{t-1}$ needs to be masked or forgotten. For example, for the spoken sentence "I went to climb the Great Wall yesterday, oh, no, it was the day before yesterday." When the model processes "oh, no, it was the day before yesterday," it needs to update the previous word "yesterday."

The calculation formula for resetting the gate and updating the gate at time t is:

$$\boldsymbol{r}_t = \text{Sigmoid}(\boldsymbol{W}_r\boldsymbol{x}_t + \boldsymbol{U}_r\boldsymbol{h}_{t-1} + \boldsymbol{b}_r) \tag{2.20}$$

$$z_t = \text{Sigmoid}(\boldsymbol{W}_z\boldsymbol{x}_t + \boldsymbol{U}_z\boldsymbol{h}_{t-1} + \boldsymbol{b}_z) \tag{2.21}$$

The calculation process of reset gate and update gate is the same as the gate calculation in LSTM, both using input data $\boldsymbol{x}_t$ and $\boldsymbol{h}_{t-1}$ and then passing the result to the Sigmoid function after linear transformation, thus controlling the flow of information.

Next, use the following calculation formula to calculate the candidate state information of the GRU cell at the current time:

$$\tilde{\boldsymbol{h}}_t = \tanh[\boldsymbol{W}_h\boldsymbol{x}_t + \boldsymbol{U}_h(\boldsymbol{r}_t \cdot \boldsymbol{h}_{t-1}) + \boldsymbol{b}_h] \tag{2.22}$$

Similarly, the input data $\boldsymbol{x}_t$and $\boldsymbol{h}_{t-1}$ are firstly linearly transformed and then passed to the tanh function, and the final result is the information to be input into the state vector $\boldsymbol{h}_t$ of the current GRU cell. After obtaining the above components, the state vector $\boldsymbol{h}_t$ of the current GRU cell can be updated:

$$h_t = (1 - z_t) \cdot h_{t-1} + z_t \cdot \tilde{h}_t \tag{2.23}$$

Clearly, the update of the GRU cell state filters out some unvaluable information from the state h_t at previous time h_{t-1}, while filtering out some valuable information from the current candidate state $\tilde{h}_t$, thus obtaining the current time's state vector h_t. After calculating h_t, the actual tasks can be modeled according to $.h_t$

2.5.4 Knowledge Expansion of Recurrent Neural Network

1. **Stacked Recurrent Neural Network, SRNN**

The stacked recurrent neural network (SRNN) is a deeper network formed by stacking single-layer recurrent neural networks. Figure 2.18 shows a three-layer stacked recurrent neural network. The input of each recurrent neural network is the output of the next recurrent neural network and the output of the recurrent neuron in the previous moment of the same layer. For the top-level recurrent neural network, its output will be the final output of the stacked recurrent neural network.

Assuming that the simple recurrent neural network is used in Fig. 2.18, we may as well take the second moment as an example to see how it works. From the first layer to the last layer of the recurrent neural network, its calculation formulas are as follows:

$$h_2^{(1)} = f(W^{(1)}x_2 + U^{(1)}h_1^{(1)} + b^{(1)}) \tag{2.24}$$

$$h_2^{(2)} = f(W^{(2)}h_2^{(1)} + U^{(2)}h_1^{(2)} + b^{(2)}) \tag{2.25}$$

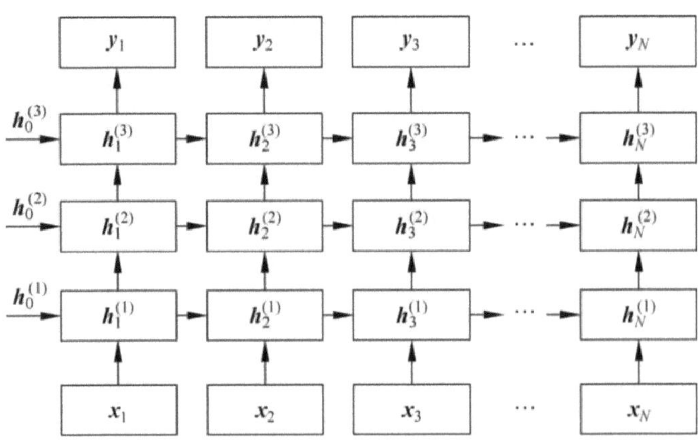

Fig. 2.18 Stacked recurrent neural network

$$h_2^{(3)} = f(W^{(3)} h_2^{(2)} + U^{(3)} h_1^{(3)} + b^{(3)}) \tag{2.26}$$

In these calculation formulas, $f(\cdot)$ is the activation function; $h_j^{(i)}$ represents the output vector of the recurrent neural network at layer i and at time j; $W^{(i)}$,$b^{(i)}$ and $U^{(i)}$ represent the weight and bias of the layer i recurrent neural network, respectively. Through layer-by-layer calculation, the final output $y_2 = h_2^{(3)}$ of the stacked recurrent neural network can be obtained. The calculation methods at other moments are similar, which will not be repeated here.

2. **Bidirectional Recurrent Neural Network**

The recurrent neural network introduced earlier is a network that processes sequential information from left to right. However, in practical situations, sometimes the output of the recurrent neural network at a certain moment is not only related to past information but also to future information. Therefore, the bidirectional recurrent neural network (BRNN) was proposed. Figure 2.19 shows the structure of the bidirectional recurrent neural network.

From Fig. 2.19, we can see that the bidirectional recurrent neural network passes the sequential information from the front to the back (from left to right) and from the back to the front (from right to left), respectively, so each time corresponds to two output hidden state vectors. BRNN concatenates these two vectors as the corresponding output y. Such a bidirectional structure allows the vector output at each time to contain both the information of the left sequence and the right sequence. Compared with a unidirectional LSTM, BRNN can obtain more comprehensive semantic information.

Next, we will take the second moment as an example to introduce the working principle of a bidirectional recurrent neural network. The calculation formulas from the front to the back and from the back to the front are as follows:

$$\overrightarrow{h}_2 = f(\overrightarrow{W} x_2 + \overrightarrow{U} \overrightarrow{h}_1 + \overrightarrow{b}) \tag{2.27}$$

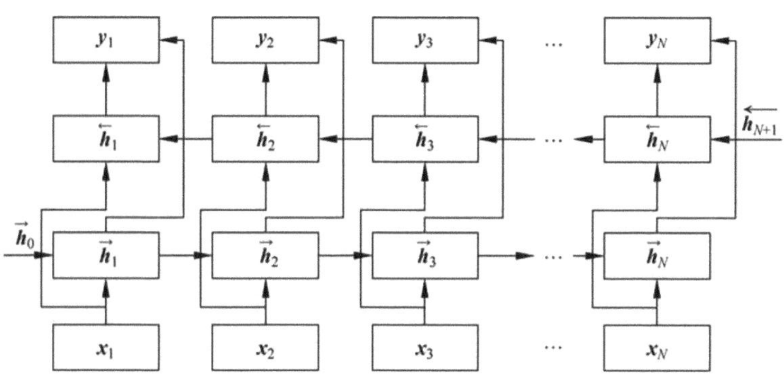

Fig. 2.19 Bidirectional recurrent neural network

$$\overleftarrow{\boldsymbol{h}_2} = f(\overleftarrow{\boldsymbol{W}}\boldsymbol{x}_2 + \overleftarrow{\boldsymbol{U}}\overleftarrow{\boldsymbol{h}_3} + \overleftarrow{\boldsymbol{b}}) \tag{2.28}$$

In which, f is the activation function. $\overrightarrow{\boldsymbol{h}_i}$ represents the output of the forward recurrent neural network at time i. $\overrightarrow{W}$, $\overrightarrow{U}$ and $\overrightarrow{b}$ respectively represent the weight and bias of the forward recurrent neural network. The variable with a left arrow represents the relevant vectors of the backward recurrent neural network, with the same meaning as above. In this way, the forward and backward output vectors can be obtained at each time. For a bidirectional recurrent neural network, the two output vectors are generally concatenated as the final output of the bidirectional recurrent neural network. For example, the final output at the second time is:

$$\boldsymbol{y}_2 = \text{concat}(\overrightarrow{\boldsymbol{h}_2}, \overleftarrow{\boldsymbol{h}_2}) \tag{2.29}$$

The function concat completes the concatenation operation of two vectors. The calculation and processing methods at other times are similar and will not be elaborated here.

Stacked recurrent neural networks and bidirectional recurrent neural networks can be constructed based on different types of recurrent neural networks, for example, Bi-LSTM or Bi-GRU.

For a detailed introduction to the basic theory of neural networks, readers can refer to (Qiu Xipeng, 2020) and Chapter 2 "Deep Networks: Modern Practices" of (Zhao Shenjian et al., 2017).

Chapter 3
Distributed Representation

Abstract This chapter introduces distributed representation as a foundational technique in natural language processing that maps text into low-dimensional, dense vectors to capture semantic meaning, improving upon traditional one-hot encoding methods that fail to represent relationships between words. The text first covers word-level representations, differentiating between static word vectors, such as Word2Vec and GloVe, which have fixed representations, and dynamic word vectors, like ELMo, that adapt to context to address polysemy. It then progresses to the representation of phrases, discussing compositional methods like the bag-of-words model and more advanced techniques that consider word order. Finally, the chapter addresses sentence representation, outlining universal and task-related methods that leverage recurrent neural networks and pre-trained language models like BERT to encode the meaning of entire sentences.

Text representation is the foundation of natural language processing. In order to help machine learning algorithms process text data, it is particularly important to effectively represent the text in an appropriate computable form. The method based on statistical learning adopts one-hot representation, ignoring the semantic relationships between words. In recent years, the distributed representation method maps text data into low-dimensional dense vectors, significantly improving the performance of NLP tasks and has become a widely used text representation method.

This chapter introduces distributed representation methods for words. Based on this, distributed representation methods for phrases and sentences will also be introduced.

C. Zong et al., *Natural Language Processing and Large Language Models*,
https://doi.org/10.1007/978-981-92-0682-7_3

3.1 Distributed Representation of Words

A word is the smallest language unit with independent meanings and is the fundamental unit of phrases, sentences, and documents. There are two main ways to represent words: one-hot encoding and distributed representation, as shown in Table 3.1.

One-hot encoding is a discrete representation method, where the current word's position in the vocabulary corresponds to one dimension, while the values in other dimensions are all 0. In this representation, the Euclidean distance between any two word vectors is equal, and in the representation space, any two vectors are orthogonal, making it difficult to accurately capture semantic information between words. Moreover, when the dictionary is large, the one-hot encoding vector dimension of each word is high, which brings trouble to subsequent calculations.

Distributed semantic representation assumes that the meaning of words can be represented by their contextual distribution. If we have the contextual information of a word, then we have mastered its semantics. Based on this assumption, in distributed representation, words are represented as low dimensional, dense, and real vectors. Each word can be represented by a point in space, and semantically similar words are also adjacent in the real vector space. Word embedding is a typical distributed representation method. Figure 3.1 shows two-dimensional representation.

Table 3.1 Methods for word representation

Word	One-hot representation	Distributed representation
A	[1 0 0 0 0 0 0 0]	[0.34 0.5]
B	[0 0 0 1 0 0 0 0]	[0.7 0.9]

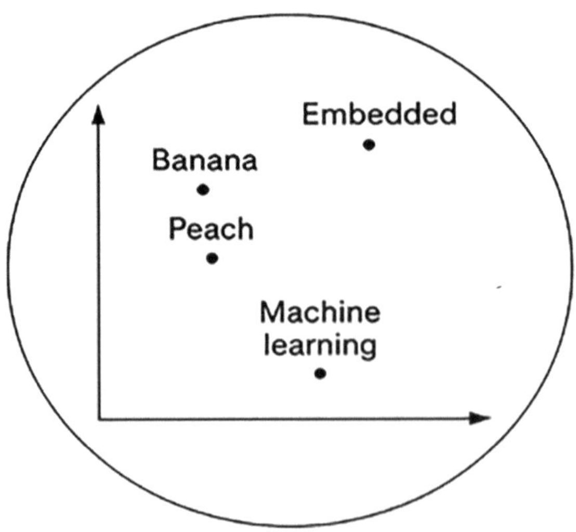

Fig. 3.1 Illustration of word embedding

In Fig. 3.1, the three words "Bananas, pineapples, peaches" are all fruits with similar semantics and commonly used in the same context. Therefore, the distance between them is also relatively close. The semantics expressed by the words "machine learning" and "embedded space" are far apart, and there are also significant differences from the meanings of fruit vocabulary, so their positions in the graph are relatively scattered.

According to whether words automatically adjust semantic information based on context, the distributed representation of words can be divided into two categories: static word vectors and dynamic word vectors. Static word vectors refer to situations where the distributed representation of a word is fixed and does not change with its context, making it unable to express polysemy. Dynamic word vectors can dynamically adjust the distributed representation of words based on their contextual changes, thus accurately representing situations where a word is polysemous. For example, in the following two sentences, the meaning of "apple" is completely different:

"This apple is really sweet."

"This year's apple launch event will be held on September 16th."

If static word vectors are used for representation, "apple" has only one representation and cannot accurately express semantic information in different contexts, while dynamic vectors can do this.

3.1.1 Static Word Vectors

Common static word vector algorithms include Word2Vec, GloVe, FastText, etc. Word2Vec was proposed by Thomas Mikolov, etc. in 2013 and includes two models: Skip-gram and CBOW (Mikolov et al. 2013a, b). Its implementation involves unsupervised learning from local corpus within a sliding window. The Word2Vec model can preserve the linear relationships between words, and words with similar semantics can also obtain similar word vectors from the Word2Vec model. Vector computation to some extent reflects the semantic relationships between words. For example, $E(King) - E(Man) + E(Woman) \approx E(Queen)$, here it refers to the word embedding vector obtained through Word2Vec.

GloVe was proposed by Stanford University in 2014, which utilizes co-occurrence information between words and trains it using global corpus (Pennington et al. 2014). Due to the use of comprehensive global and local contextual information, GloVe performs better in word vector learning compared to Word2Vec.

FastText is a tool used for word vector computing and text classification, open-sourced by Facebook in 2017. Its model structure is similar to Word2Vec, and its performance of shallow network is comparable to deep networks, with faster training and inference speed (Joulin et al. 2017).

Fig. 3.2 The structure of Skip-gram network

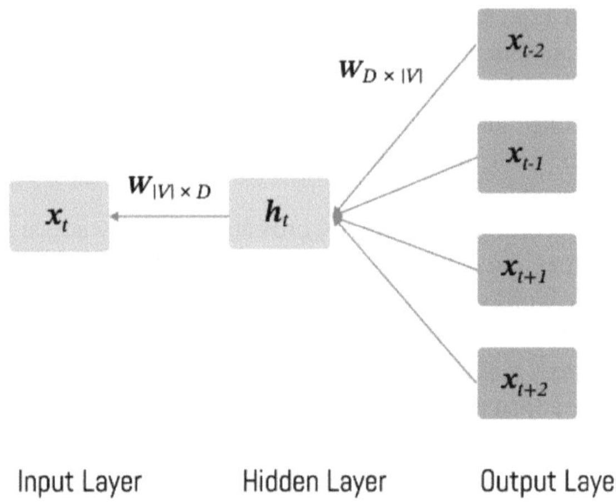

(a) **Skip-gram Model**

Given a text sequence $\boldsymbol{w} = \{w_0, w_1, w_2, \cdots, w_i\}$, each word in the text is sequentially used as the central word. Assuming that the central word at time t is w_t, the probability of the occurrence of context $\boldsymbol{C}$ within a certain window range is predicted. The vector representation of w_t is optimized through the objective function to ensure the maximum probability of $\boldsymbol{C}$, and the vector representation of w_t is ultimately learned. For example, if the window size is 5, then the current word w_t has two words on each side as the context, that is, $\boldsymbol{C} = \{w_{t-2}, w_{t-1}, w_{t+1}, w_{t+2}\}$.

As shown in Fig. 3.2, the network structure of the Skip-gram model is a feed-forward neural network consisting of input layers, hidden layers, and output layers, where the linear layer has no bias.

Input layer: The input of the model is the one-hot encoding $x_t \in \mathbb{R}^{|V|}$ of the central word w_t. $|V|$ is the length of the vocabulary.

Hidden layer: The embedding word vector $h_t \in \mathbb{R}^D$ of the hidden layer is obtained by multiplying the one-hot encoding of the input word with the weight matrix $W_{|V| \times D}$. D is the dimension of the word vector.

Output layer: The embedded word vector is multiplied by the weight matrix $W_{D \times |V|}$ to obtain the output vector, and the numerical value of each dimension in the output vector represents the score of the corresponding word. By using the Softmax function to normalize these scores, the contextual probability of predicting the central word can be obtained.

During the model training process, each contextual word w_i forms a training sample (w_t, w_i) with the central word w_t. And then input the training samples into the model in the form of batch for training, with the training parameters being the weight matrix $\boldsymbol{W}_{|V| \times D}$ and $\boldsymbol{W}_{D \times |V|}$. The loss function is the cross-entropy loss. After the model training is completed, the vector of each word in the vocabulary can be obtained from the matrix $\boldsymbol{W}_{|V| \times D}$, and each row vector in the matrix is the corresponding word vector.

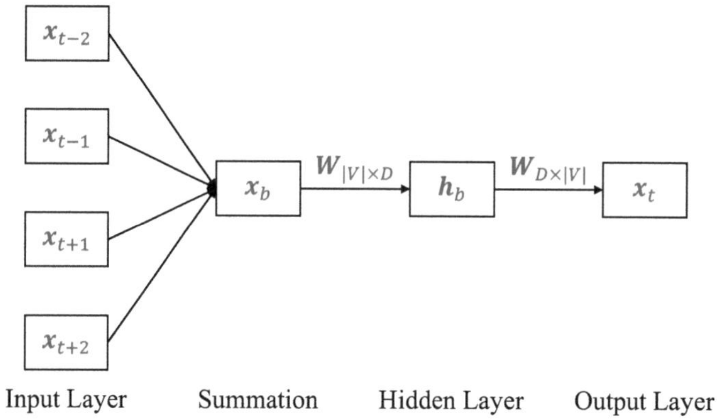

Fig. 3.3 The structure of CBOW model

(b) **CBOW Model**

The model structure of CBOW is similar to Skip-gram, which includes input layers, hidden layers, and output layers. The main difference between the two models is that CBOW predicts the probability of the centered word's appearance based on the words in the context within a specified range (window). The structure of the CBOW model is shown in Fig. 3.3. $\boldsymbol{x}_b$ is the result of summing the contextual word vectors of the central word w_t, and the output w_t is the probability of predicting the central word.

Experiments have shown that the Skip-gram model performs slightly worse than CBOW in syntactic analysis tasks, but performs better than CBOW in semantic analysis tasks. The training speed of the CBOW model is faster than that of Skip-gram.

Whether it is the Skip-gram model or the CBOW model, the output layer is a vector with dimensions equal to the length of the vocabulary. When the vocabulary is large, normalizing high-dimensional vectors is an extremely resource-consuming task that has a significant impact on computation efficiency. Here are two approximate calculation methods to solve this problem: negative sampling and hierarchical Softmax.

(a) Negative Sampling

The process of selecting negative samples corresponding to positive samples from a dictionary is called negative sampling. During the training process, two words are given each time to determine if they are a pair of centered words and contextual words. If they are a pair of centered words and contextual words, they are positive samples; otherwise, they are negative samples. This problem can be simplified as a binary classification problem.

Taking the Skip-gram model as an example, the contextual word of the centered word w_{t-2}, $w_{\mathrm{t}-1}$, w_{t+1}, w_{t+2} is a positive sample w_b. The negative sampling process involves randomly w_b selecting N negative samples $w_{2\prime}$, ..., $w_{N\prime}$ from the vocabulary to participate in training. The negative sampling method updates weights only through positive and negative samples, greatly reducing computing load and improving training efficiency.

(b) Hierarchical Softmax

Hierarchical Softmax is implemented using a binary tree, in which each leaf node in the tree represents a word in the vocabulary, which is a vector. Each non-leaf node is also a vector, and the distance from high-frequency words to the root node is smaller than that from low-frequency words to the root node. In the path from the root node to each word, each non-leaf node either selects the left or right child node, and the probability is determined based on the Sigmoid function. The calculation formula for the hierarchical Softmax is:

$$P(w|w_I) = \prod_{j=1}^{L(w)-1} \sigma([[n(w, j+1) = ch(n(w, j))]].v\prime_{n(w,j)}^{T} v_{w_I}), \tag{3.1}$$

In this formula, $n(w, j)$ is the node j in the path from the root node to the word w. $L(w)$ is the length of the path, and $ch(n)$ is any determined child node of the node n. If x is true, $[[x]]$ is 1, otherwise, it is -1. σ is the Sigmoid function.

Assuming there are only 5 words in the dictionary, including *Pineapples*, *are*, *spiked*, *and*, and *yellow*, Fig. 3.4 shows an example of a binary tree built based on the frequency of each word.

Based on the probability in the graph, the final predicted probability of the word "spiked" is: $0.7 \times 0.6 \times 0.8 \times 0.6 = 0.2016$

In some open-source implementations of Word2Vec, negative sampling is adopted by default. The subsequent chapters of this book will specifically introduce the implementation of negative sampling.

3.1.2 *Dynamic Word Vectors*

As mentioned earlier, dynamic word vectors can generate word vectors based on the context in which the word is located, hence also known as contextualized word embedding. There is a close relationship between dynamic word vectors and pre-training language models. We will only introduce one dynamic word vector method, ELMo. The other dynamic word vectors and pre-training language models will be introduced in Chap. 6. ELMo (embeddings from language models) was proposed by Peters et al. in 2018 and it is a model based on recurrent neural networks (Peters et al. 2018). The training task of the ELMo model is to calculate the probability of a text sequence, in which word vector is the function of the entire input. Unlike Word2Vec, the vector representation of ELMo is multi-level, and the model structure is shown in Fig. 3.5.

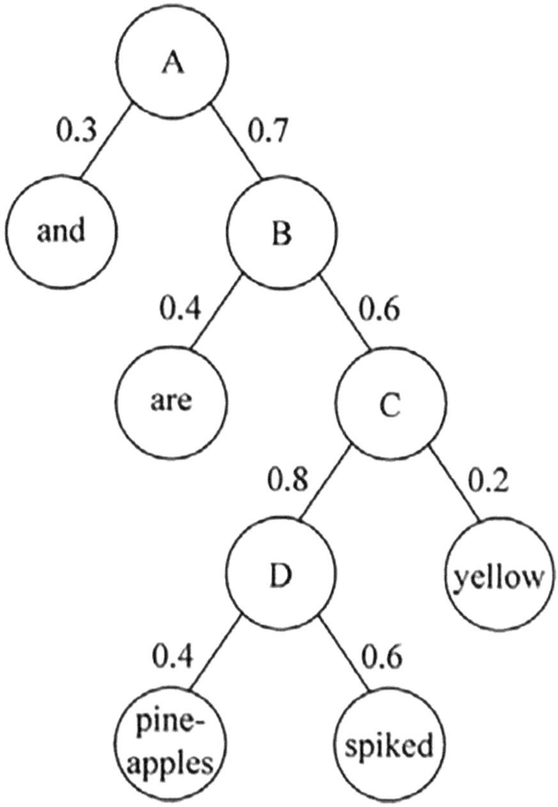

Fig. 3.4 Example of predicting the probability of word vectors based on binary tree

Training ELMo requires a large amount of corpus for semi-supervised learning. After the training is completed, when a new sentence is input, each word in the sentence can obtain three levels of representations, corresponding to the contextual features of the word.

- Embedding layer: It can obtain the features of words, and convert the input word (letter) into a vector without contextual information.
- LSTM layer: It is composed of two LSTM modules in different directions. Each layer generates vectors in two directions, corresponding to the contextual features of words.
- Softmax layer: It calculates the probability of the corresponding next word.

In ELMo, for each input word, a layer L of bidirectional language model can calculate $2L + 1$ representations, that is, the LSTM model of each layer generates representations in two directions, plus the representations obtained by the embedding layer. For downstream tasks, ELMo integrates all the representations obtained into a vector through linear combination.

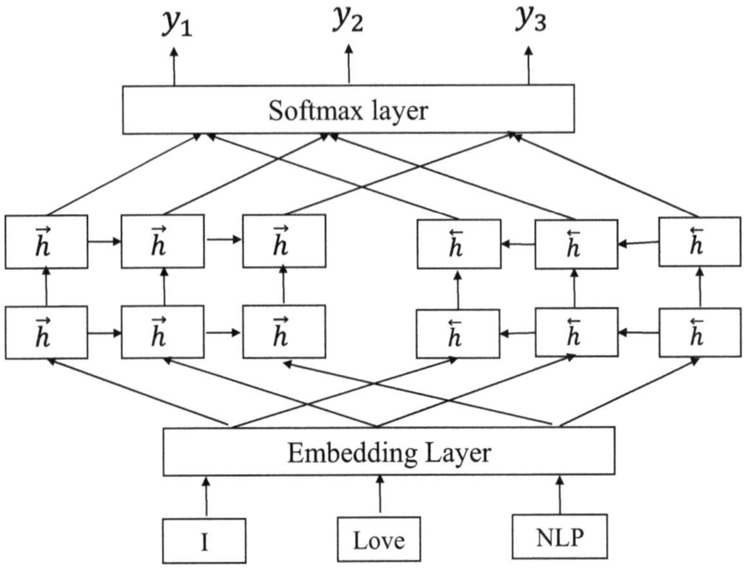

Fig. 3.5 The structure of ELMo model

The vector representation of each word obtained by the ELMo model is related to sentences in which the words are located. Compared with static word vector training, dynamic word vectors require a complete calculation process (forward and backward computing). Experiments showed that ELMo achieved good results in multiple NLP tasks, including question answering systems and sentiment analysis.

3.2 Distributed Representation of Phrases

In natural language processing, phrases generally refer to continuous word strings, not just noun phrases, verb phrases, and prepositional phrases in linguistic sense. There are two methods for learning the distributed representation of phrases. One considers phrases as indivisible and independent semantic units and then learns the semantic vector representation of phrases based on the distributional hypothesis. The other approach holds that the semantics of a phrase are composed of word combinations, and the key to learning the distributed representation of a phrase is to learn the semantic combinations between words.

The bag-of-words (BOW) model is the simplest text representation method, which treats texts as a set of corresponding words. This method is also used to represent phrases, and the commonly used semantic combination method is to average the word vectors, as shown in Eq. (3.2):

$ph_N = \frac{1}{N}\sum_{k=1}^{N}\boldsymbol{x}(w_k)$, (3.2)

In this formula, N represents the number of words in a sentence in a phrase, and $\boldsymbol{x}(w_k)$ represents the word vector w_k of the word. Or we can take the maximum value for each dimension of the word vector, as shown in Formula (3.3):

$ph_N = max_{k=1}^{d}(\boldsymbol{x}(w_1)_k, \boldsymbol{x}(w_2)_k, ..., \boldsymbol{x}(w_N)_k)$ (3.3)

In this formula, $\boldsymbol{x}(w_1)_k$ represents the k element of $\boldsymbol{x}(w_1)$.

Sometimes, words' weight information is added on top of word vector averaging:

$ph_N = \frac{1}{N}\sum_{k=1}^{N} v_k \boldsymbol{x}(w_k)$, (3.4)

In this formula, v_k can be information such as term frequency–inverse document frequency (TF-IDF) corresponding to the word w_k.

Phrase vectors represented by the bag-of-words model cannot capture word order information in phrases, and in practical applications, phrases with different word orders often have completely different semantics. So Socher and others proposed a recursive autoencoder (RAE) to solve word order problems. As the name suggests, a recursive autoencoder recursively merges the vector representations of two child nodes from bottom to top until it obtains the final vector representation of the phrase (Socher et al. 2011). Although RAE can solve word order problems, it is difficult to obtain semantic information. Therefore, Zhang and others proposed a bilingual constrained phrase representation method for machine translation. The basic assumption of this method is that two mutually translated phrases have the same semantics, so they can share the same vector representation. Based on this assumption, co-training can be used to simultaneously learn phrase vector representations of two languages, thereby enabling phrases to obtain richer semantic information (Zhang et al. 2014).

3.3 Distributed Representation of Sentences

For natural language processing tasks such as sentence-level text classification and sentiment analysis, the granularity of word and phrase representation is too small, requiring distributed representation vectors for sentences. There are two main types of sentence representation methods: one is universal, and the other is task related. The universal sentence representation is almost always based on unsupervised learning methods and neural network-based models designed for training on large-scale corpora. The specific methods include distributed representation based on the bag-of-words model and the Skip-Thought model. Task-related sentence representations are optimized based on specific task performance indicators. For example, in sentiment classification tasks, the vector representation of a sentence is ultimately input into the classification model to predict the sentiment category expressed by the sentence. Specific methods include sentence vector representation based on recursive neural networks, convolutional neural networks, recurrent neural networks, attention models, and pre-training language models.

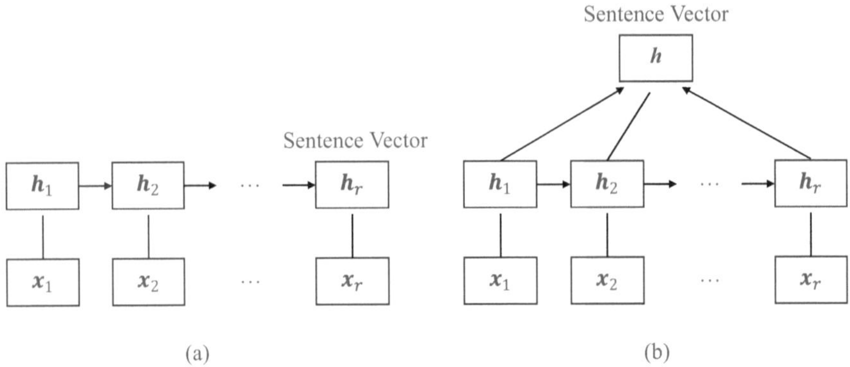

Fig. 3.6 Recurrent neural network representation for sentence **a** Normal mode; **b** Average sampling mode by time

3.3.1 Representation Methods Based on Recurrent Neural Networks

In vector representation of sentences, the order of words is crucial for the final semantic expression. In the following two sentences, the words used are exactly the same, but the semantics are completely opposite.

"I defeated him in this badminton match."

"He defeated me in this badminton match"

In Chap. 2, a detailed introduction was given to recurrent neural networks. When using this network to model sequential information, each sequential unit receives the output of the previous unit and the input at the current moment to generate the output. The last sequential unit obtains the information of all previous units in this sequence, so the output of this unit can be seen as a vector representation of the entire sentence, as shown in Fig. 6a. In addition, the average output at each moment can also be taken as the word vector for the entire sentence, as shown in Fig. 6b.

3.3.2 Representation Methods Based on Pre-Trained Language Models

Pre-trained language models refer to initial language models trained on large-scale language datasets, which record the general syntax and semantic knowledge learned from datasets. In downstream tasks, transfer learning is performed based on these kinds of initial models, transferring knowledge learned from open domains to downstream tasks, and fine-tuning to obtain specific models related to the task. In 2018, Google was the first to propose the pre-trained model BERT (bidirectional encoding representation from transformers) for natural language processing (Devlin et al. 2019). Subsequently, a series of pre-trained language models were proposed and widely applied into many areas, such as the GPT (generative pre-training) model series (Radford et al. 2018, 2019; Brown et al. 2020), RoBERTA (Liu et al. 2019), XLNet (Zhang et al. 2019), Albert (Lan et al. 2020), and ERNI. E (Sun et al. 2019).

When representing a sentence based on BERT, a special word "[CLS]" is first inserted before the input sentence. "[CLS]" has no specific semantic information, and its function is to integrate the semantic information of other words in the sentence. The output vector corresponding to "[CLS]" is usually used as the semantic representation or the sentence vector of the entire input sentence, which will be used in subsequent downstream tasks, such as text classification and sentiment analysis.

BERT and other pre-trained language models will be further introduced in Chap. 6 and will not be further elaborated here. For more sentence and document representation methods, this book will not introduce them one by one. Readers who need to know more can refer to references such as (Zong Chengqing et al. 2022).

For a detailed introduction to text representation methods, please refer to Chap. 3 of (Zong Chengqing et al. 2022) and other related papers.

Chapter 4
Sequence Generation Models

Abstract This chapter introduces sequence generation models, beginning with the foundational sequence-to-sequence (Seq2Seq) framework, which uses an encoder-decoder architecture for tasks like machine translation. It first explains the implementation of Seq2Seq models using Recurrent Neural Networks (RNNs) and discusses common decoding methods such as greedy decoding and beam search strategies. The text then introduces the attention mechanism as a method to improve model focus and efficiency. Subsequently, the chapter details the Transformer architecture, which leverages a self-attention mechanism to overcome the limitations of RNNs, such as difficulties with parallelization and long-distance dependencies. The core components of the Transformer, including its encoder and decoder structures, multi-head attention, and positional encoding, are explained. Finally, the chapter provides an overview of modern models derived from the Transformer's structure, categorizing them into autoencoding (AE) models like BERT and autoregressive (AR) models like GPT.

In many natural language processing tasks, both input and output are sequences, such as machine translation, automatic summarization, and question-answering systems. To complete these tasks, we not only need to encode the input sequences, but also decode the input representation to complete the generation of the output sequences, that is, to model the conversion relationship between the input sequences and the output sequences. This model is what we usually call the sequence-to-sequence transformative generative model (Seq2Seq), referred to as the sequence generative model. Taking machine translation as an example, assuming the input is a word sequence $\boldsymbol{x} = [x_1, x_2, \ldots, x_M]$ with length M, and the output $\boldsymbol{y} = [y_1, y_2, \ldots, y_N]$ is a translation sequence with length N corresponding to x. N and M are not necessarily equal. Then, the key to Seq2Seq modeling is to learn the mapping relationship between the input sequences and the output sequences:

$$\mathrm{y} = f(x) \tag{4.1}$$

C. Zong et al., *Natural Language Processing and Large Language Models*,
https://doi.org/10.1007/978-981-92-0682-7_4

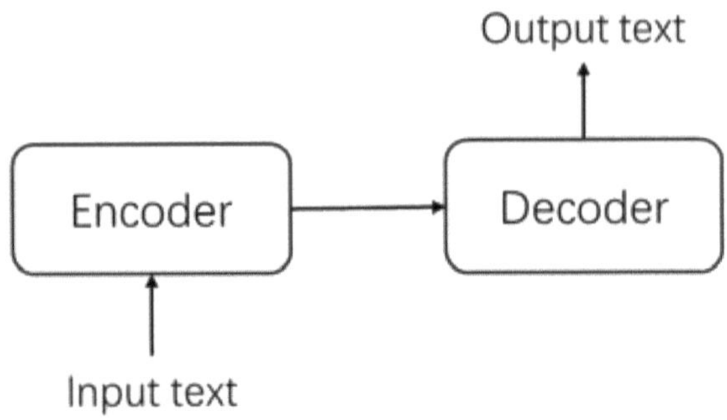

Fig. 4.1 The network structure of the Seq2Seq model

The network structure of the Seq2Seq model is shown in Fig. 4.1, which is an encoder-decoder structure. The encoder is responsible for processing the information of input sequence, encoding it, and then passing the encoded vector to the decoder. The decoder decodes the encoded vector and generate the output sequence.

At present, the common Seq2Seq models are mainly based on recurrent neural network and Transformer.

4.1 Seq2Seq Model Based on Recurrent Neural Network

Chap. 2 introduces the classic Recurrent Neural Network (RNN) and its variant Long Short-Term Memory (LSTM) used for sequence processing tasks. We will now introduce how to implement a Seq2Seq model based on a recurrent neural network, and address the long-distance dependence problem of recurrent neural networks by introducing a self-attention mechanism.

4.1.1 The Working Principle of the RNN-Based Seq2Seq Model

Taking machine translation as an example, the working principle of the Seq2Seq model based on recurrent neural network is shown in Fig. 4.2. The input sequence of the model is the source language phrase "Deep Learning," and the output sequence is the target language phrase "深(shen)度(du) 学(xue)习 (xi)."

The execution process of the model mainly includes two steps:

Step 1: The encoder encodes the input sequence. Then the encoder processes the input text sequence from left to right, using the hidden state vector of the RNN at the last time step as the encoded vector $\boldsymbol{h}_M^e$ of the input sequence, in which M refers to the last moment. The encoder is composed of multiple RNN cells, which can be LSTM or GRU.

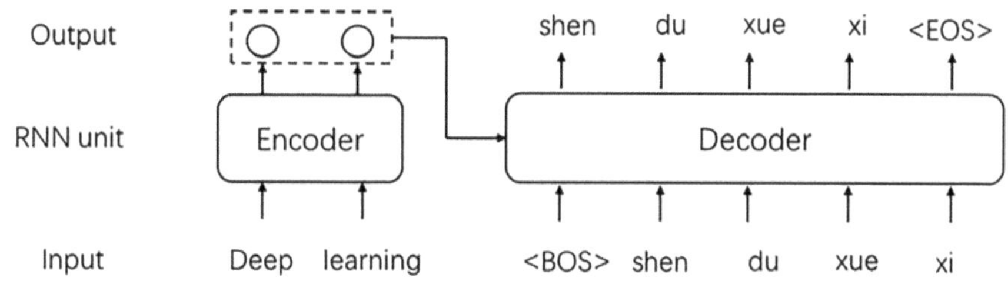

Fig. 4.2 The working principle of Seq2Seq machine translation based on recurrent neural network

Step 2: The decoder generates the target language sentence based on the vector $\boldsymbol{h}_M^e$. The decoder is also composed of multiple RNN cells. The decoding process is carried out word by word from the beginning to the end, that is, after generating the previous word, the next word is generated. There are two special characters <BOS> and < EOS> in the target language sentence, representing the start and end symbol of the decoding sequence, respectively. In the example given in Fig. 4.2, if the unit is a character (Chinese character), at the first moment, the decoder will generate the first Chinese character "深" in the translation sequence based on the <BOS> symbol and the semantic vector. Then it will pass "深" to the next step (second moment) as input to decode the second Chinese character "度," and so on. When decoder generates the character <EOS> or reaches the preset maximum length, the decoding process is terminated, and the output sequence of the Seq2Seq model is finally obtained.

4.1.2 Decoding Strategies

How does the model decode the word at the moment t in the testing phase? Generally speaking, there are two most commonly used decoding strategies: greedy decoding strategy and beam search strategy.

1. Greedy Decoding Strategy.

Assuming at the moment t of the decoder, the output vector of the recurrent neural network is $\boldsymbol{h}_t \in \mathbb{R}^D$, in which D is the dimension of the vector. Through the mapping of the linear layer, the corresponding word probability is obtained:

$$p_t = \text{softmax}(\boldsymbol{W}\boldsymbol{h}_t + \boldsymbol{b}) \tag{4.2}$$

In the formula above, $\boldsymbol{W} \in \mathbb{R}^{V \times D}$ and $\boldsymbol{b}$ are parameters which can be updated, and V is the number of words in the vocabulary.

The calculation method of greedy decoding strategy is relatively simple. At each moment t, the predicted word with the largest probability value is selected, and the formula is:

$$y_t = \text{argmax}(p_t) \tag{4.3}$$

It can be seen that the greedy decoding strategy is a depth-first-search, which only decodes one "optimal" path.

2. Beam Search Strategy.

The greedy decoding strategy only considers the optimal situation at the current moment and ignores the relationship between the words decoded at the $t-1$ moment and the words decoded at the t moment. Therefore, the final decoding sequence is not necessarily the optimal solution. To address this, the exhaustive search strategy was proposed, which lists all possible prediction candidates one by one at each decoding moment, and then selects the most probable one as the basis for decoding at the next moment until the final decoding is completed. Although this decoding strategy can ensure that the output sequence is the global optimal solution, the number of extended paths is huge, and the search efficiency is low.

The beam search strategy adopts a compromise solution. It does not only select the candidate word (or character) with the highest probability at each moment, nor does it retain all candidates. Instead, it retains the top K most probable candidates and sequentially expands only the retained candidates. Here, K is a hyperparameter, called beam size. Compared with the greedy search strategy, the beam search strategy is more likely to find the optimal solution. Compared with the exhaustive search strategy, the beam search strategy is more computationally efficient.

Figure 4.3 shows the calculation process of the beam search strategy, in which the vocabulary $V = [w_1, w_2, w_3, w_4]$, beam size is 2, and the solid arrows show the decoding sequence of the two highest-scoring sequences decoded at each moment t. The calculation process is:

Step 1: At time t = 1, select the two words $[w_1]$ and $[w_3]$ with the highest probability values.

Step 2: At time t = 2, select the top 2 possible sequences $[w_1, w_2]$ and $[w_3, w_4]$ from the candidate outputs.

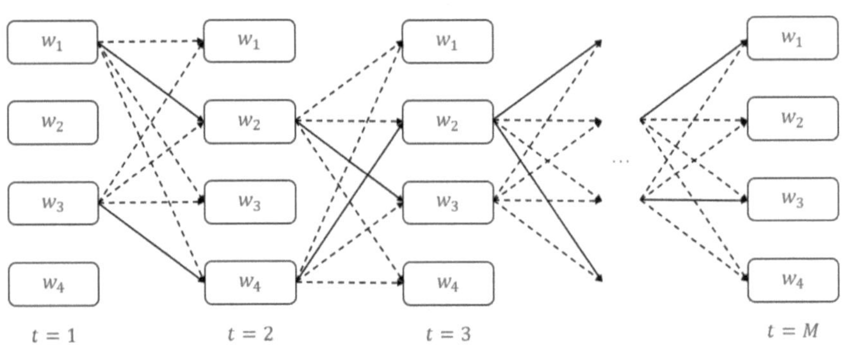

Fig. 4.3 The calculation process of beam search

Step 3: At time $t = 3$, select the top 2 possible sequences $[w_1, w_2, w_3]$ and $[w_3, w_4, w_2]$ from the candidate outputs.

Step 4: Repeat the process until the decoding is completed. After decoding is complete, select the sequence with the highest probability as the final output.

4.2 The Seq2Seq Model with Attention Mechanism

In sequence generation tasks, the information needed at different times is often different. The attention mechanism can allow the model to focus on more important information when processing data, thereby improving computing efficiency. The attention mechanism can be used independently or as a component combined with other networks.

Taking machine translation as an example, Fig. 4.4 shows the working principle of the Seq2Seq model with attention mechanism.

The specific implementation process is as follows:

Step 1: Perform attention calculation between the output vector $\boldsymbol{q}_i$ (i = 2) of the decoder at the current moment and the output vector $\boldsymbol{h} = [\boldsymbol{h}_1, \boldsymbol{h}_2, \ldots, \boldsymbol{h}_M]$ (M =4) of the encoder to obtain the attention distribution at the current moment, represented by α_{ij}. α_{ij} is a normalized numerical value, indicating the importance of the output vector $\boldsymbol{h}_j$ when decoding the moment i. The attention calculation formula is:

$$\alpha_{ij} = \frac{\exp\left(s\left(\boldsymbol{q}_i, \boldsymbol{h}_j\right)\right)}{\sum_{k=1}^{M} \exp\left(s\left(\boldsymbol{q}_i, \boldsymbol{h}_k\right)\right)} \tag{4.4}$$

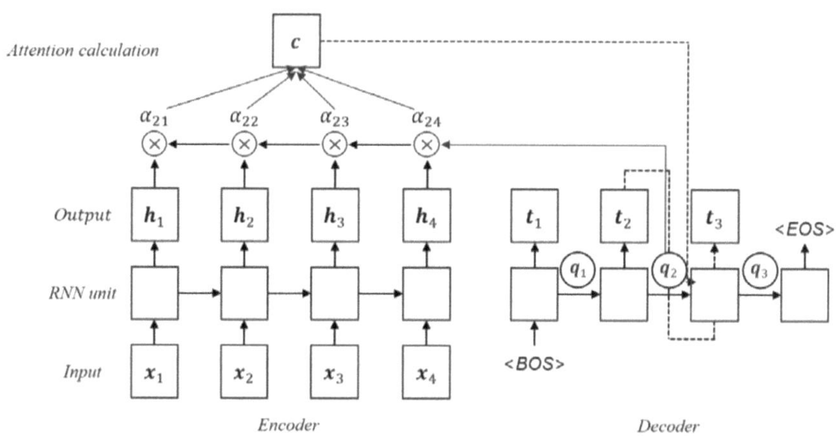

Fig. 4.4 The working principle of the Seq2Seq model with attention mechanism

In this formula, $s(\boldsymbol{q}_i, \boldsymbol{h}_j)$ is the attention scoring function. Higher scores indicate higher correlation between the vectors $\boldsymbol{q}_i$ and $\boldsymbol{h}_j$, and the calculation formula is:

$$s(\boldsymbol{h}, \boldsymbol{q}) = \boldsymbol{v}^{\mathrm{T}} \tanh(\boldsymbol{W}\boldsymbol{h} + \boldsymbol{U}\boldsymbol{q}) \tag{4.5}$$

In this formula, $\boldsymbol{v}$, $\boldsymbol{W}$ and U are parameters which can be updated. $\boldsymbol{v}$ is a vector. $\boldsymbol{W}$ and $\boldsymbol{U}$ are matrices.

Step 2: Obtain the attention vector $\boldsymbol{c}_i$ at the current time i by getting the weighted average and then summing the results together. The calculation formula is:

$$\boldsymbol{c}_i = \sum_{j=1}^{M} \alpha_{ij} \cdot \boldsymbol{h}_j \tag{4.6}$$

Step 3: Concatenate $\mathbf{c}_i$ and the vector output from the decoder at the previous moment as the input of the RNN cell of the decoder at the current moment. Follow the above steps at subsequent moments until the decoding is complete.

4.3 Transformer-Based Seq2Seq Models

The recurrent neural network is a serial modeling method, meaning the unit state of the next moment t depends on the input of the unit state at the previous moment $t - 1$. This design has two bottleneck questions in practical applications:

1. The calculation process is difficult to parallelize. When the model is large, parallel computing techniques cannot be used to accelerate the training speed.
2. When the sequence is long, there is a problem of long-distance dependency (exploding gradient or vanishing gradient). In theory, recurrent neural networks can establish dependencies between states with long time intervals, but due to long-distance dependencies problem, important information from the earlier part of the sequence may be lost during subsequent calculations in the sequence.

In order to solve the problems mentioned above, Vaswani et al. proposed the Transformer model in 2017, which was successfully applied to machine translation tasks, achieving significant improvements in both translation quality and computing efficiency (Vaswani et al., 2017). The success of the Transformer model is largely attributed to its core component: the self-attention model. This mechanism allows the Transformer to avoid the serial computation inherent in recurrent neural networks. It enables the model to focus selectively on the most relevant information in the input sequence at any given time, thus effectively alleviating the issue of long-distance dependencies in modeling long sequences.

4.3.1 Self-Attention Models

The self-attention model adopts query-key-value (QKV) pattern. Assuming the input sequence is $\boldsymbol{X} = [\boldsymbol{x}_1, \ldots, \boldsymbol{x}_L] \in \mathbb{R}^{L \times D}$, after linear transformation, the query vector $\boldsymbol{q}_i \in \mathbb{R}^{D_k}$, key vector $\boldsymbol{k}_i \in \mathbb{R}^{D_k}$ and value vector $\boldsymbol{v}_i \in \mathbb{R}^{D_v}$ corresponding to $\boldsymbol{x}_i$ can be obtained. For the entire input sequence $\boldsymbol{X}$, the process of linear transformation can be abbreviated as:

$$\boldsymbol{Q} = \boldsymbol{X}\boldsymbol{W}^Q \epsilon \mathbb{R}^{L\times D_k}, \tag{4.7}$$

$$\boldsymbol{K} = \boldsymbol{X}\boldsymbol{W}^K \epsilon \mathbb{R}^{L\times D_k}, \tag{4.8}$$

$$\boldsymbol{V} = \boldsymbol{X}\boldsymbol{W}^V \epsilon \mathbb{R}^{L\times D_v}, \tag{4.9}$$

In the formulas above, $\boldsymbol{W}^Q \in \mathbb{R}^{D\times D_k}$, $\boldsymbol{W}^K \in \mathbb{R}^{D\times D_k}$ and $\boldsymbol{W}^V \in \mathbb{R}^{D\times D_k}$ are mapping matrices which can be updated. By default, the dimensions of the mapped feature vectors Q, K and V can be set to the same, all equal to D.

The working principle of the self-attention model is shown in Fig. 4.5. When the input is ["I," "love," "deep," "learning"], the self-attention calculation process at the position of "love" mainly consists of the following three steps:

Step 1: Perform a linear transformation on the input vector $[\boldsymbol{x}_1, \boldsymbol{x}_2, \boldsymbol{x}_3, \boldsymbol{x}_4]$ to obtain the query vector Q, key vector K, and value vector V.

Step 2: The query vector $\boldsymbol{q}_2$ performs dot product calculations with $[\boldsymbol{k}_1, \boldsymbol{k}_2, \boldsymbol{k}_3, \boldsymbol{k}_4]$, and the results are then divided by $\sqrt{D}$ to obtain the attention scores of other

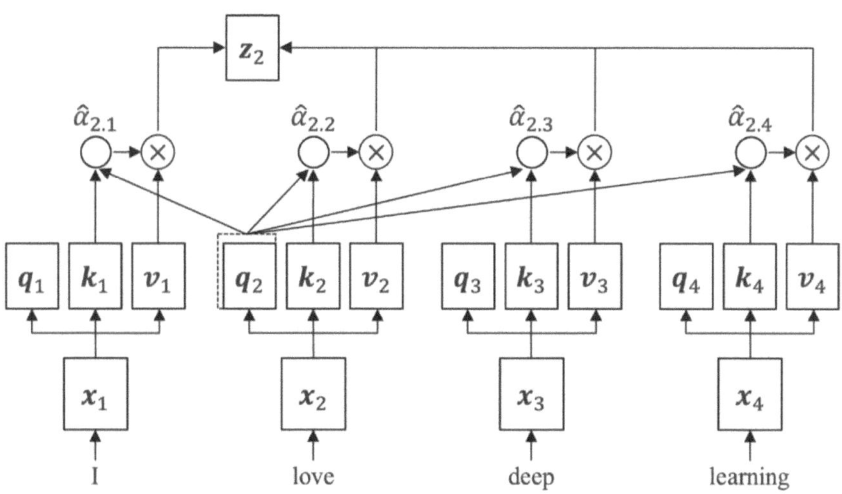

Fig. 4.5 The architecture of self-attention model

words for "love." These scores are subsequently normalized using the softmax function to yield the attention distribution of other words at the current word position, $\left[\hat{\alpha}_{2,1}, \hat{\alpha}_{2,2}, \hat{\alpha}_{2,3}, \hat{\alpha}_{2,4}\right]$.

Step 3: Calculate the output vector of the current position "love" by computing the weighted average and then summing these weighted results. The calculation formula is:

$$z_2 = \sum_{i=1}^{4} \hat{\alpha}_{2,j} v_j \tag{4.10}$$

Similarly, we can obtain the output vectors z_1, z_3 and z_4 for other positions. In order to improve the computing efficiency, in practical applications, the self-attention output vectors at all positions can be calculated by matrix calculation:

$$\boldsymbol{Z} = \text{attention}(\boldsymbol{Q}, \boldsymbol{K}, \boldsymbol{V}) = \text{softmax}\left(\frac{\boldsymbol{Q}\boldsymbol{K}^T}{\sqrt{D}}\right)\boldsymbol{V} \tag{4.11}$$

In this formula, $\boldsymbol{Z} \in \mathbb{R}^{L \times D_v}$. In the self-attention calculation process, to prevent a large variance in the attention distribution, a smaller gradient of the softmax is created which is not conducive to the convergence of the model. Thus, the self-attention calculation mechanism is divided by $\sqrt{D}$ during the calculation process. It can be seen that self-attention can effectively focus on important information in the model, alleviating the issues caused by the parallel computing of recurrent neural networks.

4.3.2 Transformer's Network Architecture

The Transformer network consists of an encoder and a decoder, both of which are stacked by N layers of networks, as shown in Fig. 4.6. This multi-layer architecture can enhance the Transformer's ability to extract semantic features, thereby improving its performance in modeling natural language processing tasks.

In both the encoder and the decoder, the output of each layer is used as the input of the next layer, extracting deep semantic information layer by layer. The output vector of the last layer of the encoder is considered to contain the semantic information of the source text and is passed to each layer of the decoder for use.

4.3.3 The Encoder of Transformer

The network architecture of the Transformer's encoder includes an embedding layer and N stacked encoding layers (multi-head self-attention layers and feedforward

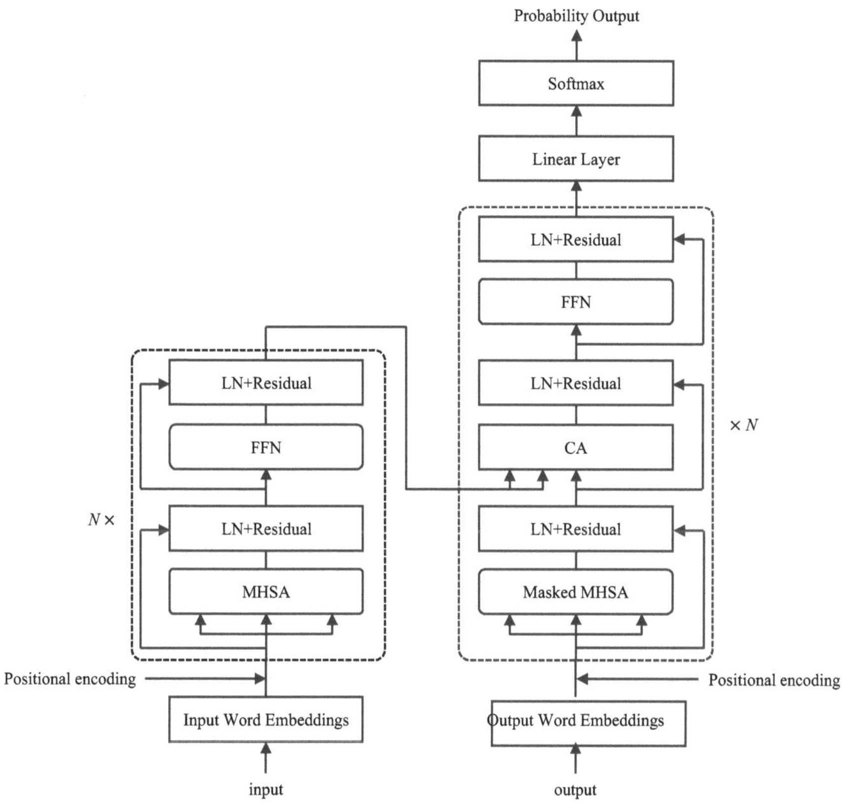

Fig. 4.6 The architecture of Transformer model

layers, with cross-layer connections and layer normalization). Its working principle is shown in Fig. 4.7.

1. **Embedding Layer.**

The role of the embedding layer is to convert the input text into its corresponding vector representation. The input text sequence is first segmented into words or sub-words to obtain a sequence of words or sub-words, such as [I, love, machine, learning]. Then the semantic vectors $\boldsymbol{x}_1, \boldsymbol{x}_2, \boldsymbol{x}_3, \boldsymbol{x}_4$ and position vectors $\boldsymbol{p}_1, \boldsymbol{p}_2, \boldsymbol{p}_3, \boldsymbol{p}_4$ corresponding to each word (sub-word) are obtained. Finally, the two are added according to position to get the feature representation of each word.

The Transformer's encoder uses self-attention mechanism to process data, which can improve the parallel computing ability of the model but ignores the positional relationship between words. It is necessary to introduce positional embedding (PE) to capture the order information between words. This order information is crucial for correctly encoding the semantics of the input sequence. A commonly used method to generate positional embedding is trigonometric functions (sine or cosine).

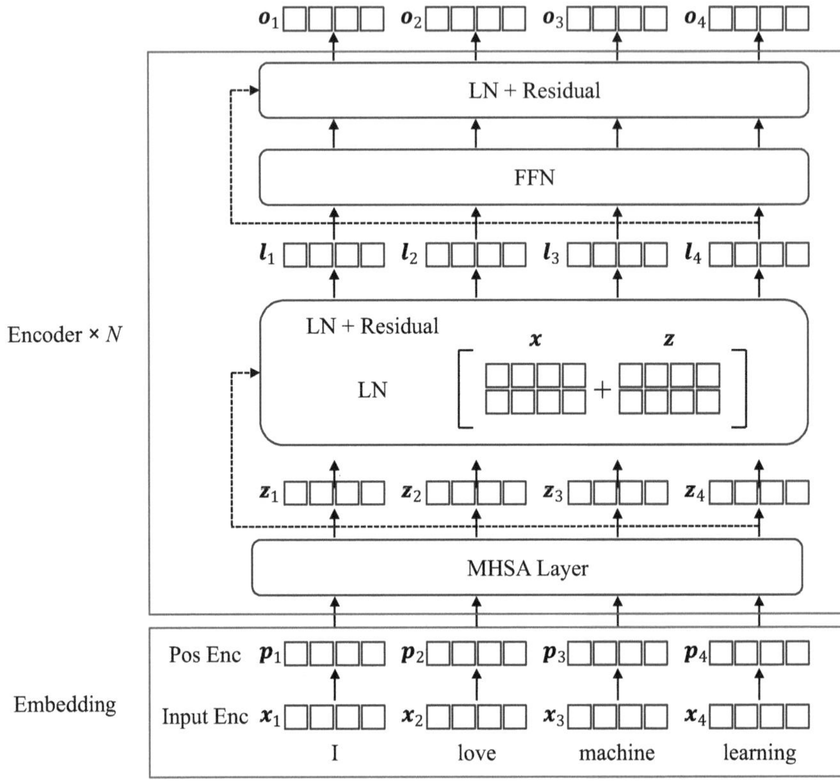

Fig. 4.7 The encoder of Transformer

Assuming the dimension of positional embedding is D, the calculation formula for each dimension value is:

$$PE(pos, 2i) = \sin\left(\frac{t}{10\,000^{2i/D}}\right) \tag{4.12}$$

$$PE(pos, 2i+1) = \cos\left(\frac{t}{10\,000^{2i/D}}\right) \tag{4.13}$$

In the formulas above, t indicates the position of the current word in the text sequence. $0 \leq i \leq \frac{D}{2}$ is the dimension of the encoded vector. Sinusoidal encoding is used for even-numbered dimensions; cosine encoding is used for odd-numbered dimensions.

2. **Multi-head Self-attention Layer.**

The essence of multi-head self-attention (MHSA) is the combination of multiple sets of self-attention computations, in which each set of self-attention is called a "head."

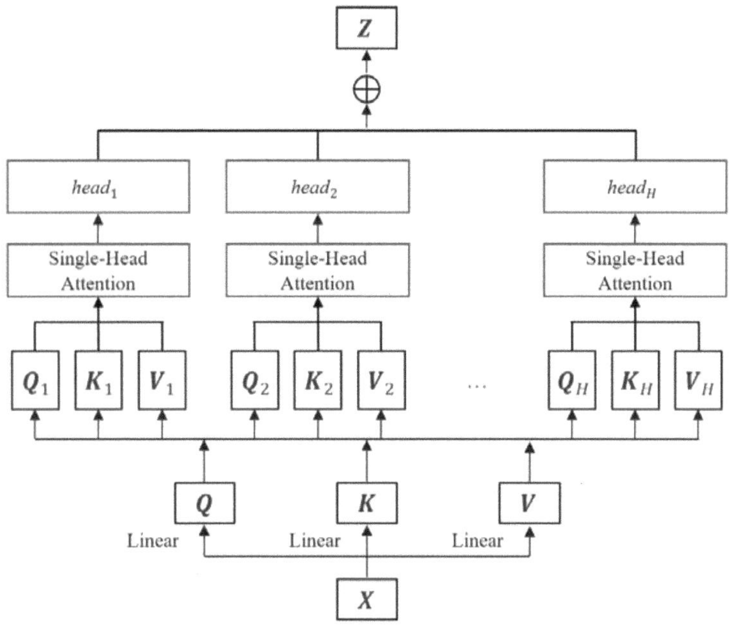

Fig. 4.8 Multi-head attention calculation

The computing processes of different sets of self-attention are independent of each other, and its working principle is illustrated in Fig. 4.8.

The calculation method of multi-head self-attention is completed through the following three steps:

Step 1: Assuming the input sequence is $\boldsymbol{X} = [\boldsymbol{x}_1, \ldots, \boldsymbol{x}_L] \in \mathbb{R}^{L \times D}$, after linear transformation, we obtain the query vector $\boldsymbol{q}_i \in \mathbb{R}^D$, key vector $\boldsymbol{k}_i \in \mathbb{R}^D$, and value vector $\boldsymbol{v}_i \in \mathbb{R}^D$ corresponding to $\boldsymbol{x}_i$. Next, the Q, K, V vectors are evenly split into H parts along their last dimension of, according to the number of heads. Finally we can get $Q_i \in \mathbb{R}^{L \times D_h}$, $K_i \in \mathbb{R}^{L \times D_h}$, $V_i \in \mathbb{R}^{L \times D_h} i = 1, 2, \ldots, H$, in which $D_h = \frac{D_v}{H}$.

Step 2: Calculate the self-attention of each head separately. For the j-th head, denoted as $head_j$, the formula is:

$$head_j = \text{attention}\left(\boldsymbol{Q}_j, \boldsymbol{K}_j, \boldsymbol{V}_j\right) \in \mathbb{R}^{L \times D_h} \tag{4.14}$$

Step 3: Combine the self-attention results of multiple heads to obtain the final output vector $\boldsymbol{Z}$.

$$\boldsymbol{Z} = \text{MultiHeadAttention}(\boldsymbol{X}) \triangleq (head_1 \oplus head_2 \oplus \cdots \oplus head_H)\boldsymbol{W} \tag{4.15}$$

In the calculation formula, $\oplus$ denotes the vector concatenation operation. The dimension of the concatenated vector may not be equal to the original vector dimension D, so a matrix $\boldsymbol{W} \in \mathbb{R}^{(H \times D_h) \times D}$ is multiplied here to map the vector back to the original input dimension D.

3. **Cross-layer Connection and Layer Normalization.**

As shown in Fig. 4.7, each Transformer's encoding layer employs cross-layer connection and layer normalization operations. The purpose of these operations is to enhance the stability of network training and improve convergence by incorporating two components: residual connections and layer normalization. Assuming that the input and output of multi-head self-attention are $\boldsymbol{X} \in \mathbb{R}^{L \times D}$ and $\boldsymbol{Z} \in \mathbb{R}^{L \times D}$ respectively, then cross-layer connection and layer normalization can be represented as:

$$\boldsymbol{L} = \text{LayerNorm}\,(\boldsymbol{X} + \boldsymbol{Z}) \tag{4.16}$$

In this formula, LayerNorm represents layer normalization. Next, the vector $\boldsymbol{L}$ will pass through a position-wise feedforward layer, and will also undergo cross-layer connections and layer normalization. Finally, the output vector $\boldsymbol{O} \in \mathbb{R}^{L \times D}$ of a Transformer's encoding layer can be obtained. At the same time, the output vector of this encoding layer serves as the input vector of the next encoding layer and continue perform the aforementioned computations until the last encoding layer.

4.3.4 The Decoder of Transformer

The Transformer's decoder also uses a self-attention-based model, with the following two key differences compared to the encoder:

1. The decoder uses a masked self-attention mechanism, and its working principle is shown in Fig. 4.9. The main difference between masked self-attention and regular self-attention is that masked self-attention only calculates attention weights based on the current state and the previous state, rather than using the current state and all states. In other words, the decoder can only predict the current word based on the words that have already been generated and cannot use future words to predict the current word.

2. The cross-attention mechanism is adopted, and its working principle is shown in Fig. 4.10. The decoder takes the input vector at the current moment as $\boldsymbol{q}_i$, and calculates it with $\boldsymbol{k}_i$ and $\boldsymbol{v}_i$ in the encoder respectively according to the self-attention calculation formula, thereby introducing the global information generated by the encoder into the decoding process.

For a more detailed introduction to the Transformer's framework and sequence generation models, please refer to (Vaswani et al., 2017; Raffel et al., 2020) and other references.

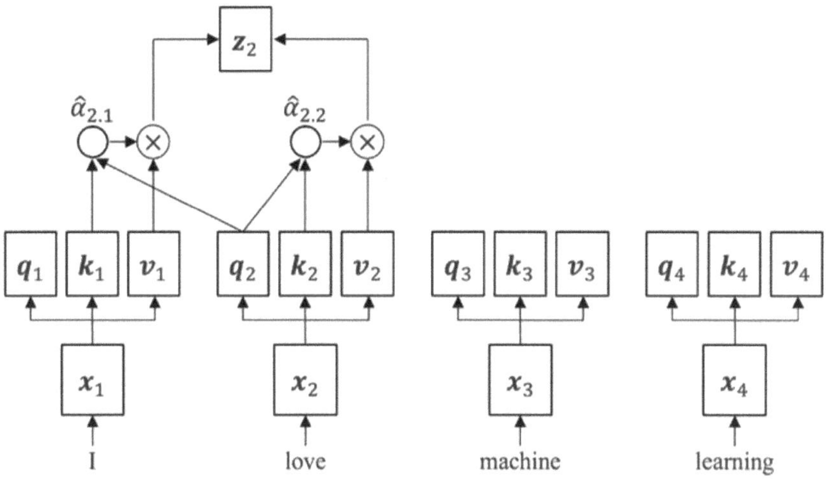

Fig. 4.9 The working principle of masked self-attention

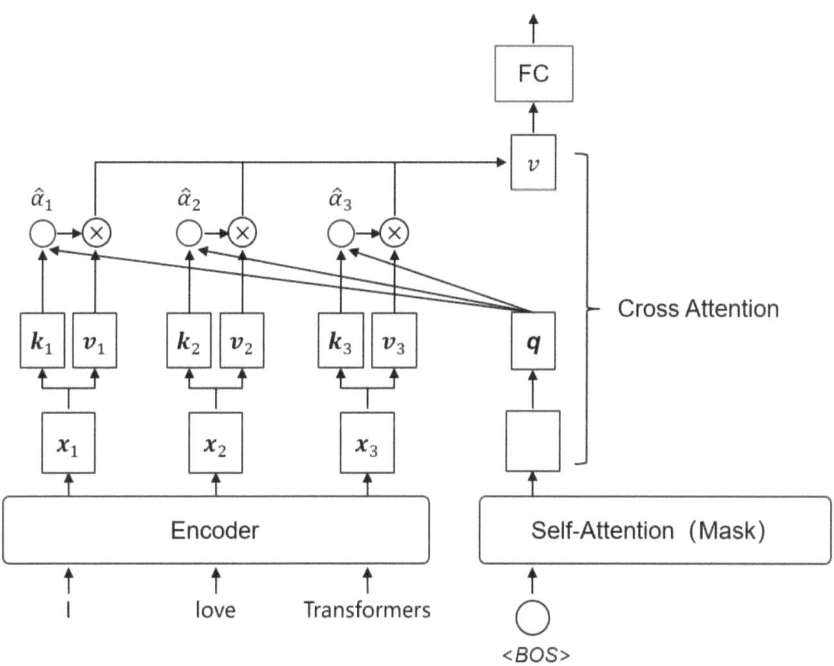

Fig. 4.10 The architecture of cross attention network

4.3.5 Knowledge Extension: Models Based on Transformer's Structure

In recent years, models based on the Transformer's architecture have gained significant attention, especially with the proposal of the pre-training model Bidirectional Encoder Representation from Transformers (BERT) in 2018, which has made breakthrough progress in various natural language processing tasks, marking a milestone. Since then, both academia and industry have seen a boom in research and applications based on Transformer pre-training models, which are widely used in many areas from natural language processing to computer vision, speech, multimodal information processing.

Models based on the Transformer' structure are mainly divided into two categories: autoencoding model (AE) and autoregressive model (AR). Autoencoding models are designed based on the working principle of the encoder, while autoregressive models are designed based on the working principle of the decoder.

1. **Autoencoding Model, AE.**

The autoencoding model first adds different noises to the input text, and then uses the input text sequence to reconstruct it, hoping to retrieve the original text. For example, BERT reconstructs the original sequence by predicting the words at the masked positions. Its advantage is that it can capture the information from both before and after a word when predicting it, but its disadvantage is that the masked language model (MLM) is used in the pre-training process, masking a word in the sentence. However, the fine-tuning stage is trained with the complete sentences, resulting in differences between the pre-training and fine-tuning stages. In addition, during the training process, the predictions of different masked words are independent of each other. Assuming that the masked words in the sequence are $\boldsymbol{w} \in \boldsymbol{W}_m$, and the unmasked words are $\boldsymbol{w} \in \boldsymbol{W}_n$, then the corresponding probability calculation formula is:

$$p(x) = \prod_{\boldsymbol{w} \in \boldsymbol{W}_m} p(\boldsymbol{w} | \boldsymbol{W}_n) \tag{4.17}$$

Currently, the main autoencoding models based on Transformer's architecture include BERT, ERNIE, RoBERTa, XLNet. Autoencoding models are very suitable for handling tasks such as text classification and text matching, and their performance on these tasks is constantly improving.

2. **Autoregressive Model, AR.**

The autoregressive model perform model by estimating the generative probability of a text sequence, calculating the sequence probabilities in either a forward or backward order. However, no matter which order, it is unidirectional. When predicting a word, it cannot simultaneously access the information from both before and after the word's position. Assuming a given text sequence $\boldsymbol{X} = (\boldsymbol{x}_1, \boldsymbol{x}_2, \ldots, \boldsymbol{x}_N)$, its left-to-right sequence generation probability can be formulated as:

$$p(\boldsymbol{X}) = \prod_{t=1}^{N} p(\boldsymbol{x}_t|\boldsymbol{x}_{<t}) \tag{4.18}$$

In the formula above, $\boldsymbol{x}_{<t}$ are the words before the t-th position.

Currently, autoregressive models based on Transformer's structure mainly include GPT-1, GPT-2, GPT-3, GPT-4, BART, and T5. Autoregressive models are very suitable for handling natural language generation tasks. With the proposal of various improvement strategies, the performance of natural language generation tasks has been continuously improved.

Chapter 5
Basic Language Models

Abstract This chapter provides an introduction to basic language models, which calculate the probability of a sentence or predict the next word in a sequence. It first covers statistical language models, specifically the n-gram model, which calculates a word's probability based on the preceding n-1 words and uses data smoothing to address the issue of data sparsity. The chapter then introduces neural network language models, including feedforward and recurrent network-based models, as a solution to the limitations of statistical models, such as their inability to learn long-distance semantic relationships. Finally, the text explores the close relationship between language models and word vectors, explaining how models like Word2Vec and ELMo are trained using a language model approach to generate static and dynamic word vectors, respectively.

A language model (LM) is one of the classic models in the field of natural language processing, used to calculate the probability of a natural language sentence or word sequence, or to predict the next possible word under given the context. An ideal language model should calculate a higher probability for a grammatically correct natural language sentence and a lower probability for sentences that do not conform to grammar.

The language models generally include statistical language models and neural language models. Statistical language models are also called n-gram models. Neural language models refer to language models based on neural network models, including language models based on feedforward neural networks and recurrent neural networks, and pre-training language models. This chapter briefly introduces these models one by one.

C. Zong et al., *Natural Language Processing and Large Language Models*,
https://doi.org/10.1007/978-981-92-0682-7_5

5.1 Statistical Language Model

5.1.1 *Model's Definition*

Given a sentence $s = w_1, w_2, \ldots, w_m$ composed of m words, a reasonable probability calculation formula is:

$$p(s) = p(w_1)p(w_2|w_1)p(w_3|w_1w_2)\cdots p(w_{\mathrm{m}}|w_1\ldots w_{m-1}) \\ = \prod_{i=1}^{m} p(w_i|w_1\cdots w_{i-1}). \quad (5.1)$$

Assuming the premise of Formula (5.1) is: the probability of No. i word is determined by the previous $i-1$ words. For example, after the word segmentation of the sentence "The weather is really nice today," it is "The/weather/is/really/nice/today" (in which the slash "/" represents the word separator), and the probability of this sentence is:

$$\begin{aligned} p(\text{The/weather/is/really/nice/today}) = {} & p(\text{The}) \times p(\text{weather}|\text{The}) \\ & \times p(\text{is}|\text{The, weather}) \\ & \times p(\text{really}|\text{The, weather, is}) \\ & \times p(\text{nice}|\text{The, weather, is, really}) \\ & \times p(\text{today}|\text{The, weather, is, really, nice}). \end{aligned}$$

In the formula above, each conditional probability can be calculated through maximum likelihood estimation on the train set, for example:

$$p(\text{is }|\text{The, weather}) = \frac{C(\text{is}|\text{The, weather})}{C(\text{The, weather})},$$

in which, $C(\cdot)$ is the number of times the corresponding word or word sequence appears in the train set (also known as frequency).

Ideally, we can always calculate the conditional probability of each word's appearance, and then calculate the probability of the sentence according to Formula (5.1). However, as the length of the sentence increases, the frequency of the sequence $w_1w_2\cdots w_i$ after position i in the corpus becomes less, or even 0, resulting in the inability to calculate the probability of the sentence. Even if it can be calculated theoretically, it cannot be trained in practice because of the large number of historical parameters $w_1w_2\cdots w_{i-1}$. To solve this problem, the n-gram model uses the Markov chain hypothesis: the probability of a word's appearance is only related to $n-1$ words that appear before it. Therefore, Formula (5.1) is rewritten as:

$$p(s) = \prod_{i=1}^{m} p(w_i|w_{i-n+1}^{i-1}) \quad (5.2)$$

At time $n = 1$, the probability of the next word has nothing to do with the previous context, the model is called unigram model. At time $n = 2$, the probability of the next word only relates to the previous word, and the model is called bigram model. At time $n = 3$, the probability of the next word relates to the previous two words, and it is called trigram model.

In the n-gram model, smaller values of n lead to smaller context range considered by the model, and the model is simpler, and vice versa. Broader context range considered by the model leads to more complex model calculation. As the value of n increases, the occurrence of unregistered words is inevitable. In fact, due to the dynamic development of language, new words are constantly being generated, and the phenomenon of unfamiliar words always exists. Once unfamiliar words appear, it will lead to zero probability events, making it impossible to estimate the probability of sentences and predict the emergence of new words. This situation is called data sparsity. To solve the data sparsity problem, data smoothing must be carried out.

5.1.2 Data Smoothing Methods

The simplest data smoothing method is by adding 1 to the count. This method assumes that the frequency of all n-gram sequences to be counted is one more than the actual frequency in the corpus, that is, the probability calculation formula is corrected by adding 1 to the count. Taking the unigram model as an example, the probability calculation formula after smoothing is:

$$p(w_i) = \frac{C(w_i) + 1}{\sum_{k=1}^{|V|}(C(w_k) + 1)} = \frac{C(w_i) + 1}{\sum_{k=1}^{|V|} C(w_k) + |V|}. \tag{5.3}$$

In the formula, V is the vocabulary, and |V| represents the number of words in the vocabulary.

For the bigram model, the formula for calculating the probability of bigram after smoothing is:

$$p(w_i|w_{i-1}) = \frac{C(w_{i-1}w_i) + 1}{\sum_{k=1}^{|V|}(C(w_{i-1}w_k) + 1)} = \frac{C(w_{i-1}, w_i) + 1}{C(w_{i-1}) + |V|}. \tag{5.4}$$

It can be seen that the smoothing method of adding 1 can well solve the problem of n-gram probability being 0. When the corpus size is large, the probability change caused by adding 1 to the count can be neglected.

In the application of statistical language models, many data smoothing methods have been proposed, such as Good–Turing method, Katz smoothing method, absolute decrement method, linear decrement method, and interpolation smoothing method. We will not elaborate on these methods here, but readers interested in these can refer to (Zong Chengqing, 2013).

5.1.3 Language Model Evaluation

In general, perplexity is used as an indicator to evaluate the learning quality of n-gram language models. For a sequence of text, the lower the perplexity value, the higher the language rationality of the sequence, and the greater the probability value of the language model, and vice versa. Assuming that a sentence $s = w_1 w_2 \cdots w_N$ is composed of N words, the calculation formula of its perplexity is:

$$\text{perplexity}(s) = p(w_1 w_2 \cdots w_N)^{-\frac{1}{N}} = (\frac{1}{\prod_{i=1}^{N} p(w_i|w_{i-n+1})})^{\frac{1}{N}}. \tag{5.5}$$

In practice, in order to prevent the numerical overflow caused by multiple consecutive multiplications, logarithmic form is often used:

$$\log(\text{perplexity}(s)) = -\frac{1}{N}\sum_{i=1}^{N} \log p(w_i|w_{i-\text{n}+1}). \tag{5.6}$$

Usually, the perplexity of Chinese texts ranges from 50 to 1000. More colloquial and simpler word choices and sentence structures lead to lower perplexity. More complex sentence structures and more obscure word choices lead to higher perplexity. The higher the perplexity of a sentence, the lower the possibility of accurately predicting words.

5.1.3.1 Neural Network Language Model

Although the n-gram model provides a feasible way to calculate language models and plays a significant role in practical applications, it has two obvious shortcomings. First, the model is susceptible to the data sparsity problem, which can lead to the calculated probability of a sentence being 0, forcing the use of data smoothing methods to smooth out data and use approximate probabilities to solve data sparsity problems. Second, the n-gram model can only count the co-occurrence frequency between words within a limited sequence range, and cannot learn the semantic relationships between contexts at longer distances, resulting in insufficient generalization ability of the model. Moreover, the n-gram model counts the conditional probabilities between discrete symbols (with "word" as the statistical unit), even if two words with very similar semantics cannot replace each other in the same context, which to some extent contributes to the existence of data sparsity problem.

The neural network language model (NNLM) effectively solves the data sparsity problem. In this section, we briefly introduce neural language models, taking feedforward neural network language models and recurrent neural network language models as examples.

5.1.4 Feedforward Neural Network Language Model

Yoshua Bengio et al. (2003) proposed a language model based on feedforward neural network. Similar to n-gram models, feedforward neural network language model predicts the No.n word by using the previous $n - 1$ words. But the introduction of word vectors greatly solves the data sparsity problem of n-gram models. Its modeling formula is as follows:

$$p(s) = \prod_{i=1}^{n} p(w_i | w_{i-n+1} \cdots w_{i-1}). \tag{5.7}$$

The network structure of the feedforward neural network language model is composed of input layers, hidden layers, and output layers, as shown in Fig. 5.1.

(1) Input Layer: The input to the model is a sequence of $n - 1$ words $w_{i-n+1} w_{i-n+2} \cdots w_{i-1}$. Through the word embedding matrix $C \in \mathbb{R}^{|V| \times m}$, we can obtain the word embedding for each preceding word. Here, $|V|$ denotes the number of words in the vocabulary, and m represents the dimension of the corresponding word embedding, meaning that each row represents a word embedding. After obtaining the embeddings for all preceding $n - 1$ words, these embeddings are concated to obtain the representation vector $\boldsymbol{x} \in \mathbb{R}^{m(n-1)}$ of the historical context. The embedding concatenation formula is:

$$\boldsymbol{x} = \boldsymbol{C}(w_{i-n+1}) \oplus \boldsymbol{C}(w_{i-n+2}) \oplus \cdots \oplus \boldsymbol{C}(w_{i-1}). \tag{5.8}$$

(2) Hidden layer: Perform linear transformation on the input vector $\boldsymbol{x}$, the formula is:

$$\boldsymbol{h} = \tanh(\boldsymbol{Hx} + \boldsymbol{b}). \tag{5.9}$$

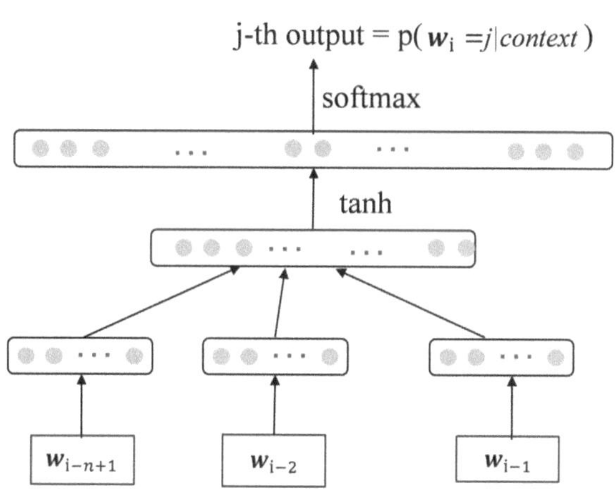

Fig. 5.1 The network structure of the feedforward neural network language model

In the formula above, $\boldsymbol{H} \in \mathbb{R}^{d \times m(n-1)}$ is the weight matrix and $\boldsymbol{b} \in \mathbb{R}^{d}$ is the bias term.

(3) Output layer: After linear transformation, the input vector $\boldsymbol{x}$ is mapped to the dimension $|\boldsymbol{V}|$, and the value on each dimension is the score corresponding to the predicted word. Then, the predicted value is normalized using the softmax function to obtain the predicted probability of each word in the vocabulary.

$$\boldsymbol{y} = \text{softmax}(\boldsymbol{W}x + \boldsymbol{U}h + \boldsymbol{b}), \tag{5.10}$$

in which $\boldsymbol{W} \in \mathbb{R}^{|V| \times (n-1)m}$, $\boldsymbol{U} \in \mathbb{R}^{|V| \times d}$, $\boldsymbol{b} \in \mathbb{R}^{|V|}$.

5.1.5 Recurrent Neural Network Language Model

In feedforward neural network language models, the length of the preceding sequence is still limited (n-1 words), which also affects the modeling ability of the model. The recurrent neural network introduced in Chapter 2 can effectively handle such long sequences. Therefore, Mikolov et al. proposed a language model based on recurrent neural networks (Mikolov et al., 2010).

As introduced in Sect. 2.5, the recurrent neural network is a computing model that unfolds in temporal sequences. The output of the previous moment will be used as the input of the current moment, and the calculation will be repeated until the entire sequence is processed. Therefore, the state vector h_{i-1} output at moment $i-1$ can be regarded as the semantic vector generated after the model reads the previous text $w_1 w_2 ... w_{i-1}$. Based on this semantic vector, the next word w_i can be directly predicted. According to this idea, the following formula is given:

$$p(w_i | w_1, w_2, \cdots, w_{i-1}) = \text{RNN}(w_i | h_i). \tag{5.11}$$

The network structure of the recurrent neural network language model is shown in Fig. 5.2, which consists of an input layer, an RNN layer, and an output layer.

(1) Input layer: The input is the word vector $x_1, x_2, \cdots, x_{i-1}$ at moment $i-1$, in which $\boldsymbol{x} \in \mathbb{R}^m$. The vector will be fed into the RNN cell in turn.

(2) RNN layer: After receiving the input x_{i-1} at time $i-1$ and the output h_{i-2} at the previous time, the RNN cell performs a transformation to obtain the output vector h_{i-1} at time $i-1$. The transformation formula is:

$$h_{i-1} = \tanh(\boldsymbol{W}x_{i-1} + \boldsymbol{U}h_{i-2} + \boldsymbol{b}), \tag{5.12}$$

in which $\boldsymbol{W} \in \mathbb{R}^{d \times m}$, $\boldsymbol{U} \in \mathbb{R}^{d \times d}$, $\boldsymbol{b} \in \mathbb{R}^{d}$ are all training parameters of the RNN layer.

(3) Output layer: The state vector at time $i-1$ is mapped to the vocabulary dimension through linear transformation, and finally normalized using the softmax function

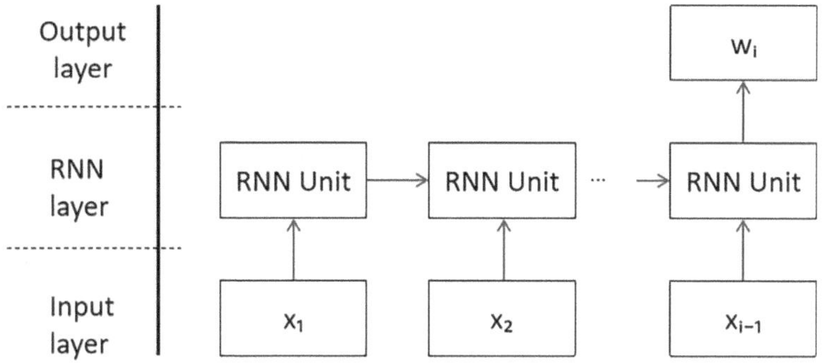

Fig. 5.2 The structure of a simple recurrent neural network

to obtain the probability distribution of the predicted word.

$$y_i = \text{Softmax}(\boldsymbol{W}h_{i-1} + \boldsymbol{b}), \tag{5.13}$$

in which $\boldsymbol{W} \in \mathbb{R}^{|V| \times d}and\boldsymbol{b} \in \mathbb{R}^{|V|}$ are the training parameters of the output layer.

The language model introduced above is based on a simple recurrent network. Although in theory a simple recurrent network can model sequences of infinite length, its ability to consider long-distance relationships in semantics between words or phrases when modeling is also limited due to the problem of gradient vanishing or exploding. Therefore, in practical cases, the gated recurrent unit (GRU) or long short-term memory network (LSTM) is usually chosen to solve the problem of gradient vanishing or exploding. No matter which recurrent neural network is chosen, the modeling idea of its language model is the same as the one introduced above.

5.1.6 *Language Models and Word Vectors*

Whether it is a feedforward neural network language model or a recurrent neural network language model, a word vector matrix is used to map words to word vectors. The word vectors are initialized at the beginning of the model training, and the word vector matrix is constantly updated during the training process. The word vector representation gradually converges to a relatively stable value. After the language model training is completed, it can be considered that each row vector in the matrix can accurately express the semantic information of the corresponding word.

In Sect. 3.1.1, we introduced two methods to obtain static word vectors, in which the Word2Vec model is actually a kind of language model. It learns language knowledge by modeling the relationship between context and the central word, and then learns the corresponding word vectors. However, after the training of the Word2Vec model is completed, the word vectors will not change with the context (such word vectors are called static word vectors). To solve this kind of problem, some methods for training dynamic word vectors gradually emerged, such as the ELMo model introduced in Sub-Sect. 3.1.2. The ELMo model is also trained through a language model approach, and its model structure is designed based on LSTM. Therefore, the output at each moment can be considered as a vector that combines the semantic information in the context. By combining these vectors with contextual information, the corresponding dynamic word vectors are generated. After the model training is completed, the dynamic word vectors will be used in downstream natural language processing tasks, and they can often achieve better results.

Whether it is static word vectors or dynamic word vectors, language models and word vectors are two closely related concepts. If the language model works well, the corresponding word vectors will be more accurate for their semantic expression. Conversely, if the word vectors perform well during the training process, it can also promote the language model to learn more accurate language knowledge.

Chapter 6
Pre-trained Large Models

Abstract This chapter provides an overview of pre-trained large models, which have established a dominant "pre-training + fine-tuning" paradigm in natural language processing. It examines two main categories: autoregressive language models like GPT, which utilize a Transformer decoder structure to predict the next word, and autoencoder models such as BERT, which use a Transformer encoder and are trained with Masked Language Model (MLM) and Next Sentence Prediction (NSP) tasks. The text also introduces knowledge-enhanced models like ERNIE, which improves upon BERT by implementing phrase-level and entity-level knowledge masking strategies. The chapter culminates in a discussion of large language models (LLMs), noting their massive parameter scale and emergent abilities. It concludes by outlining the modern three-stage process for building advanced models like ChatGPT: training a foundational large model, followed by instruction tuning and reinforcement learning from human feedback (RLHF).

After the proposal of Transformer network, pre-training language models represented by GPT and BERT have initiated a new training paradigm for natural language processing tasks: pre-training + fine-tuning, and continually set new performance records for various natural language processing tasks. Pre-training language models have become the mainstream models at present, driving natural language processing research into a new historical stage. At the same time, pre-training models with the integration of image, speech, and multimodal information have shown strong development potential in the entire information processing field. Coupled with the simplicity of expression, pre-training models seem to be used more frequently than pre-training language models. The subtle disappearance of the word "language" in this context also indicates the strong scalability and wide application scenarios of this kind of models. In the latter part of this book, pre-training models are considered synonymous with pre-training language models.

Pre-training language models can be roughly divided into two categories: autoregressive language models represented by GPT and autoencoder language models represented by BERT. The autoregressive language model uses the classic modeling

C. Zong et al., *Natural Language Processing and Large Language Models*,
https://doi.org/10.1007/978-981-92-0682-7_6

method of language models introduced in Sect. 5.2; the autoencoder language model uses a method similar to "fill in the blank" to predict masked words.

6.1 Generative Pre-training Language Models

The GPT model, proposed by OpenAI in 2018 (Radford et al., 2018), is a generative language model. Its model structure is shown in Fig. 6.1. It adopts the Transformer decoder network structure, which is stacked by decoding layers. Since there is no need to receive output information from the encoder, the cross-attention layer is removed, and it only retains masked multi-head attention layers and feedforward layers.

6.1.1 The Pre-Training of GPT Models

The pre-training task of GPT language models is similar to that of n-gram models. Given the input text sequence $s = w_1, w_2, ..., w_n$, slide through the sequence using a window with the length of n. Then use the $n-1$ words in the window to predict the No.n word, and use the maximum likelihood estimation method for modeling. The likelihood function is:

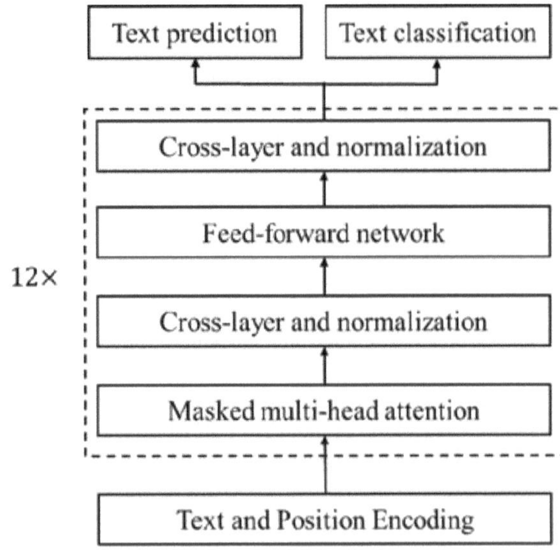

Fig. 6.1 The structure of GPT's network

$$L_p(s) = \sum_i \log p(w_i | w_{i-n+1}, ..., w_{i-1}) \tag{6.1}$$

Specifically, the input sequence $\hat{s} = w_{i-n+1}, ..., w_{i-2}, w_{i-1}$ can be converted into the corresponding one-hot vector $\boldsymbol{U} = (u_{i-n+1}, ..., u_{i-1})$ by the word table. Then the word vector matrix $\boldsymbol{W}_e \in \mathbb{R}^{|V| \times m}$ is added to the corresponding position matrix $\boldsymbol{W}_p \in \mathbb{R}^{K \times m}$ of the input sequence to obtain the input vector $\boldsymbol{H}_0$ of the model, with the formula as follows:

$$\boldsymbol{H}_0 = \boldsymbol{U}\boldsymbol{W}_e + \boldsymbol{W}_p. \tag{6.2}$$

In the formula, $|V|$ denotes the number of vocabulary words, and m denotes the dimension of the vector.

Next, $\boldsymbol{H}_0$ will be sent into each layer (which can be denoted as Transformer_block) of the generative pre-training model. Its computing process is same as Transformer, and the output of the lower layer will be used as the input of the upper layer, with the formula as follows:

$$\boldsymbol{H}_l = \text{transformer\_block}(\boldsymbol{H}_{l-1}), \ \forall l \in [1, L]. \tag{6.3}$$

The output vector $\boldsymbol{h}_n$ of the last layer $\boldsymbol{H}_L$ is multiplied by the transpose of the word embedding matrix, and the probability distribution of the predicted word w_i is obtained through the softmax layer:

$$p(w_i) = \text{softmax}\left(\boldsymbol{h}_n \boldsymbol{W}_e^T\right). \tag{6.4}$$

6.1.2 Application of GPT in Downstream Tasks

During the pre-training stage, the GPT model learns extensive general linguistic knowledge from large-scale corpus, but it lacks knowledge directly related to downstream tasks. If it is directly used for the prediction of downstream tasks, it may not achieve the expected results. Therefore, further fine-tuning is needed to achieve better results. Taking the text classification task as an example, given a text sequence $x = x_1, x_2, \cdots, x_n$, the corresponding output label is y. First, the GPT model is used to calculate x, and the output vector h_n of the last layer is the semantic vector that reads the entire sequence. Then, the vector h_n is multiplied by the weight matrix $\boldsymbol{W}_y$ and the corresponding prediction label y is obtained through softmax:

$$p(y | x_1, x_2, \cdots, x_n) = \text{softmax}\left(\boldsymbol{h}_n \boldsymbol{W}_y\right). \tag{6.5}$$

The corresponding objective function is:

$$L_F(C) = \sum_{(x,y)\in C} \log p(y|x_1, x_2, \cdots, x_n). \tag{6.6}$$

In this formula, C denotes the entire dataset.

In classification tasks, the parameters of the GPT models are updated in real time. Thus, general linguistic knowledge contained in the GPT models are largely forgotten. To avoid this situation, the modeling task of the language model can be introduced at the same time in the fine-tuning process to remind the model not to forget linguistic knowledge. Specifically, when fine-tuning the downstream task, the task objective L_P of the language model combined with the loss L_F of the downstream task should both be considered, and the formula is:

$$L(C) = L_F(C) + \lambda L_P(C), \tag{6.7}$$

where λ represents the weight in the formula above, which is used to balance with value ranging from 0 to 1. When $\lambda = 0$, the loss function does not consider the language models' tasks; when $\lambda = 1$, the language models' tasks and the fine-tuning tasks are both important in the loss function. Generally speaking, for different downstream tasks, the optimal value of λ is also different, which should be determined according to specific tasks in practice.

Figure 6.2 shows the GPT models applied to different downstream tasks, including text classification, text entailment, text similarity judgment, and multiple-choice tasks. The input samples have added a starting token < Start > and an ending token < Extract >, and different text segments are separated by < Delim >.

(1) Text Classification Tasks

The text classification task specifically refers to classifying a single input sentence, that is, the input is a text sequence, and the output is the classification label of this sequence, which is the simplest GPT models' application tasks. For example, there are sentences as follows:

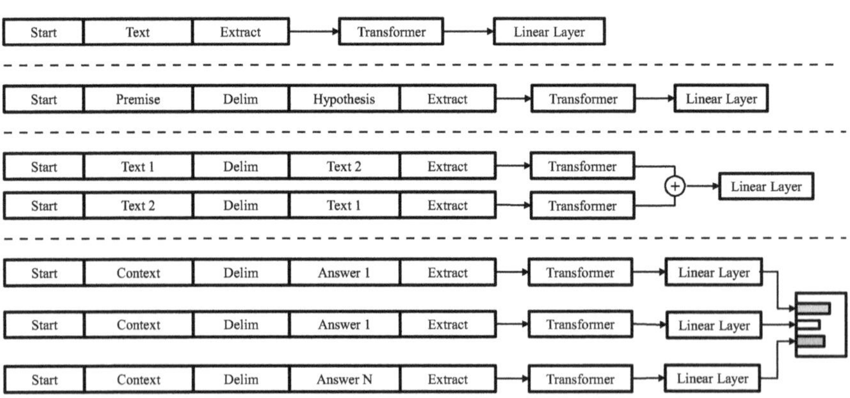

Fig. 6.2 GPT applied to different downstream tasks

<Start> The weather is really nice today. <Extract>

If we classify the sentiment polarity of this sentence, the output label should be positive.

(2) Text Entailment Tasks

The text entailment task is to infer the entailment relationship between a hypothesis and a given premise. The input of this task consists of two sentences: the premise and the hypothesis, which are in a specific order. The former is the premise while the latter is the hypothesis. A special separator $ (as shown as <Delim> in Fig. 6.2) is usually used to separate the two sentences. For example, the following input:

<Start> Xiao Ming left his football at my house yesterday.$ Xiao Ming came to my house yesterday.<Extract>

The judgment result should be "true."

(3) Text Similarity Judgment Tasks

The text similarity judgment task is to let the model judge whether the two given sentences are similar in meaning. If they are similar, the output is "1" or "yes," otherwise, the output is "0" or "no." Unlike the text entailment task, there is no order relationship between the two sentences in this task. To eliminate the interference caused by the order, the GPT model treats two sentences as two separate statements in different orders, and concatenates them in different orders to generate the input text. For example:

<Start> How do I set a profile picture on WeChat?$ How to set a profile picture on WeChat.<Extract>

<Start> How to set a profile picture on WeChat.$ How do I set a profile picture on WeChat?<Extract>

Use the GPT model to process these two sentences separately, obtain the corresponding output vectors, then add these two vectors element by element and send them to the linear layer for classification.

(4) Multiple-Choice Tasks

Multiple-choice tasks involve selecting the most suitable candidate from a given set of multiple candidates for an input text. Common multiple-choice tasks include multiple-choice answer tasks and sentence inference tasks. In multiple-choice answer tasks, a context document, a question, and a set of possible answers are given, requiring the model to select the correct answer from these candidates. The context document may not be provided. In sentence inference tasks, a text description sentence is given, and the most reasonable inference result is selected from multiple candidate inferences. Here is an example of a sentence inference task:

- <Start> What's the weather like today?$ Let's play ball today. <Extract>
- <Start> What's the weather like today?$The weather will be very good tomorrow. <Extract>
- <Start> What's the weather like today?$It's going to rain today. <Extract>
- <Start> What's the weather like today?$ I'm going to my friend's house today. <Extract>

The GPT model processes the above four sentence sequences (sequences composed of text descriptions and candidate inferences) independently and generates the corresponding output vectors. Then, after mapping the respective output vectors through a linear layer and softmax, the prediction probability distributions of multiple candidate answers can be obtained. Finally, the inference results are derived based on the prediction probability distributions of all candidate answers.

6.2 BERT Language Models

The BERT language model was proposed by Devlin et al. in 2019 (Devlin et al., 2019). The basic idea is to adopt the "pre-training + fine-tuning" processing paradigm, and it has achieved breakthrough progress in various natural language processing tasks, marking a significant milestone.

The BERT model adopts the network structure of the Transformer encoder, but the model is different from the Transformer encoder at the input layer. The embedding layer of the BERT model adds segment embedding, meaning the input vector of BERT consists of three parts: token embeddings, segment embeddings, and positional embeddings. The network structure of the BERT model is shown in Fig. 6.3.

The BERT model supports one or multiple sentences as input to complete different downstream tasks, but the input text data needs to follow certain rules. In the two examples given in Fig. 6.4, there are rules for inputting one sentence and two sentences respectively. In these two examples, special characters [CLS] and [SEP] are added at the beginning and end of the sentence, and [SEP] is used to separate different sentences. The output vector at the position of [CLS] can learn the semantics of the entire sentence through the self-attention mechanism, so this vector can be regarded as the semantic vector of the entire sentence. Based on this vector, different

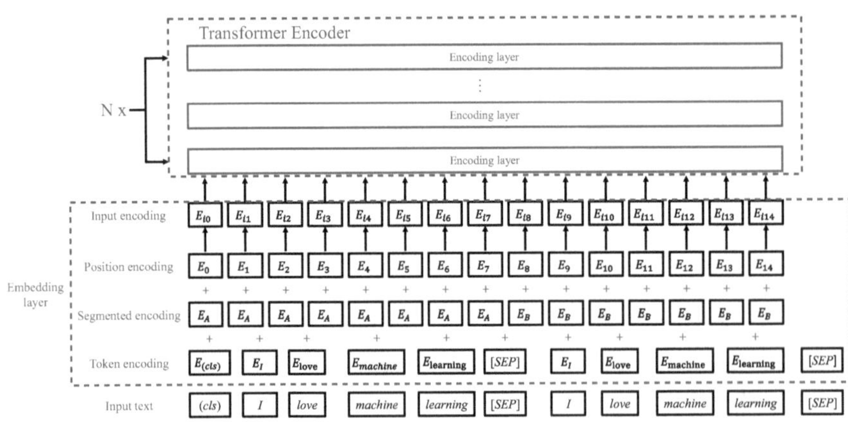

Fig. 6.3 The structure of BERT's network

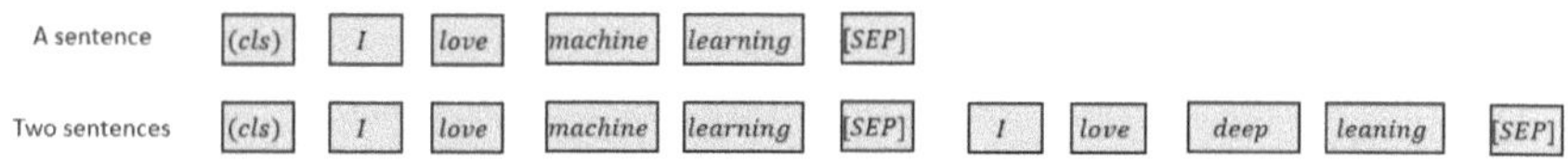

Fig. 6.4 The rules of BERT model's input

downstream tasks can be carried out, such as text classification, sentiment, or emotion classification.

As mentioned earlier, due to the addition of segmented embeddings in the embedding layer, the BERT model introduces more embedding information.

Token embedding: After the input sentences are compressed by word segmentation or sub-words, each word (sub-word) (denoted as Token) is mapped to the corresponding vector. It can be considered that the Token vector contains semantic information of each word.

- Positional embedding: By introducing positional embedding, the corresponding positional information is recorded for each input token. It should be noted that the initialization of positional embedding in the BERT model is not performed using trigonometric functions. Instead, the positional embedding is obtained as parameters which can be updated, just like token embeddings.
- Segment embedding: BERT supports the input of multiple sentences simultaneously. Segment embedding is used to remember which sentence each token in the input sequence belongs to, such as A for the first sentence and B for the second sentence. After adding segment encoding, segment encoding E_A can be used to represent sentence A, E_B represents sentence B, etc.

The final input vector to the model can be obtained by adding the token embedding, positional embedding, and segment embedding. This vector contains three parts of information and can help the BERT model better model the input sentence or sentence pair (for example, in tasks such as sentence similarity classification and sentence implication inference, two sentences need to be concatenated).

6.2.1 *Pre-Training Tasks of the BERT Models*

The modeling methods of statistical language models (n-gram model) and GPT models both use the previous words to predict the current word (target word) by default, and only use unidirectional sentence information. While the BERT model can predict the target word using left–right bidirectional information. In the pre-training stage, the BERT model uses two pre-training tasks: masked language model (MLM) and next sentence prediction (NSP).

Input text

The pre-training model is a milestone in natural language processing.

Inference Results of the BERT Model:

The pre-training[MASK] model is a milestone in natural language processing.	Predictive answers:model"
The pre-training model is a [MASK] milestone in natural language processing.	Predictive answers:milestone"
The pre-training model is a milestone in [MASK] natural language processing.	Predictive answers:natural"

Fig. 6.5 An example of masked language model task

6.3 Masked Language Models

A masked language model (MLM) is a model that masks some tokens in a sentence and asks the model to predict these masked tokens during the training process. Specifically, some of the tokens are selected from the training corpus and replaced with special characters [MASK]. The masked language model uses the contextual information of the sentence to predict the masked tokens to help the model learn language knowledge, like the "fill in the blank" task. Figure 6.5 is an example of a masked language model task, in which the model predicts the masked tokens based on the contextual information before and after [MASK].

During the pre-training phase of the BERT model, there is a problem when replacing tokens in the training corpus with [MASK]: the real text does not contain [MASK], which leads to a mismatch between the pre-training process and the fine-tuning and prediction of downstream tasks. To solve this problem, the BERT model adopts the following strategies:

- replace the words in the sentence with special characters [MASK] with a probability of 80%;
- replace the correct words in the sentence with other tokens randomly in the vocabulary with a probability of 10%;
- keep the sentence tokens unchanged with a probability of 10%.

6.4 Sentence Prediction

For the convenience of expression, we abbreviate next sentence prediction as NSP. The task of sentence prediction is to determine whether the two input sentences have a neighboring relationship in the original document. Figure 6.6 shows an example of sentence prediction. It can be seen that "The pre-training model is a milestone" and "It gets a lot of breakthroughs in natural language processing" are sentences next to each other in the original document, so there is a contextual proximity relationship in semantics between them. However, sentences "The pre-training model is a milestone" and "We had an unforgettable weekend" are not adjacent and are completely unrelated in terms of semantics.

The pre-training model is a milestone.	It gets a lot of breakthroughs in natural language processing.	Yes
The pre-training model is a milestone.	We had an unforgettable weekend.	No

Fig. 6.6 An example of next sentence prediction task

The common practice in sentence prediction is to concatenate the two sentences X and Y next to each other in the corpus as follows: "[CLS] X [SEP] Y [SEP]," input the concatenated sequence into the BERT model. The hidden state h_{cls} of the special character "[CLS]" is extracted and then passed into the softmax layer. If the prediction label is "Yes," it indicates that X and Y are two sentences next to each other. Otherwise, if the prediction label is "No," it means that X and Y are not two sentences next to each other.

Using the two pre-training tasks of masked language model (MLM) and sentence prediction to traverse massive unlabeled text data and optimize the network parameters in BERT can help the BERT models learn richer linguistic knowledge.

6.4.1 Application of BERT in Downstream Tasks

Practice has proved that fine-tuning a pre-training BERT model for downstream tasks can help achieve better results. Typical application scenarios include single-sentence automatic classification, sentence-pair automatic classification, automatic question-answering, and sequence labeling. The model structures for different application tasks are shown in Fig. 6.7.

(1) Automatic Single-Sentence Classification

The automatic single-sentence classification task means that the model automatically categorizes a given sentence according to the preset categories. Common single-sentence classification tasks include news category classification, sentiment classification, emotion classification, and spam classification.

During the fine-tuning stage of the single-sentence classification task, sentences, and corresponding labels can be directly used as training data to fine-tune all the parameters in BERT.

(2) Automatic Sentence-Pair Classification

The automatic sentence-pair classification task refers to the model's ability to predict the relationship between two given sentences. Common sentence classification tasks include text similarity and text entailment. Text similarity refers to judging whether the meanings of two sentences are the same, while text entailment refers to judging whether one sentence can infer the other given two sentences.

In the fine-tuning phase of the sentence-pair classification tasks, two sentences are first concatenated following the sentence prediction task during the training phase.

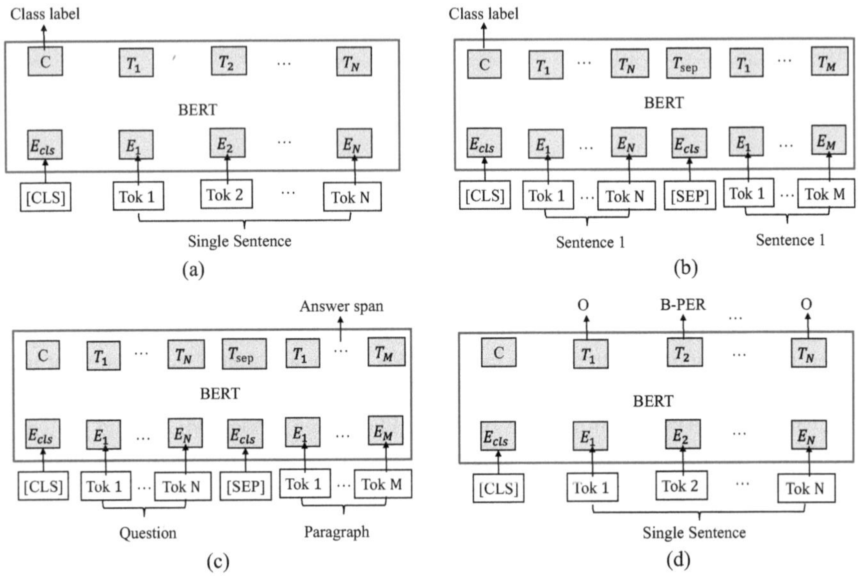

Fig. 6.7 Mainstream modeling methods based on BERT

The concatenated sentence sequences and corresponding labels (Yes/No) are used as training data to fine-tune all parameters in BERT.

(3) Automatic Question-Answering

The automatic question-answering task is to extract the corresponding answers from the given question and relevant documents and locate the start and end positions of the answers. A typical automatic question-answering task is extractive reading comprehension.

In the fine-tuning stage of the automatic question-answering task, the question and the given document are first concatenated, and the sequence after concatenation together with the corresponding answers is used as training data to fine-tune all parameters of BERT.

(4) Sequence Labeling

Sequence labeling task refers to the model's prediction of the label corresponding to each word or character in a given text sequence. Common sequence labeling tasks include named entity recognition, automatic word segmentation, and part-of-speech tagging.

The fine-tuning method for sequence-labeling tasks involves using the sequences to be labeled and the corresponding label sequences together as training data to directly fine-tune the parameters of BERT.

6.5 ERNIE Language Models

ERNIE is a knowledge-enhanced pre-training language model released by Baidu. Its backbone architecture is still the encoder of Transformer, but by introducing external knowledge, it implements three levels of knowledge masking strategies to help the model better learn language knowledge, and its performance on multiple tasks exceeds BERT.

In Sect. 6.2, BERT uses a masked language model (MLM) for pre-training to learn language knowledge from the training corpus. However, considering that the training corpus contains a large amount of language knowledge such as phrases and entities, it is very challenging for the model to automatically and effectively learn this complex language knowledge, often leading to poor results. Therefore, based on the masked language model, ERNIE proposes a knowledge masking strategy, which includes three levels of masking methods: token-level, phrase-level, and entity-level. The main functions of each masking method are introduced as follows:

Token-level: The masking method at token level is the same as in BERT. During the pre-training process, some tokens (e.g., "is") are randomly masked, and then the model predicts these masked tokens;

Phrase-level: Mask phrases in the sentence, and then directly predict the masked phrases based on the context. Phrases in the sentence can be identified using existing chunk identification methods and other shallow syntactic analysis methods.

Entity-level: Mask named entities in sentences, and then directly predict the masked entities based on the context, such as "Natural Language Processing". Entities in sentences can be identified using existing identification methods.

ERNIE can learn more knowledge about words, phrases, and entities through the combination of the above three masking strategies, thus achieving the goal of improving text representation ability. Figure 6.8 provides a rough comparison of the masking strategies at the three levels mentioned above and the masking strategy of BERT.

In the downstream application tasks, the fine-tuning methods of ERNIE and BERT are the same, so we will not repeat them here. For a detailed introduction of pre-training language models, please refer to (Zong Chengqing et al., 2022).

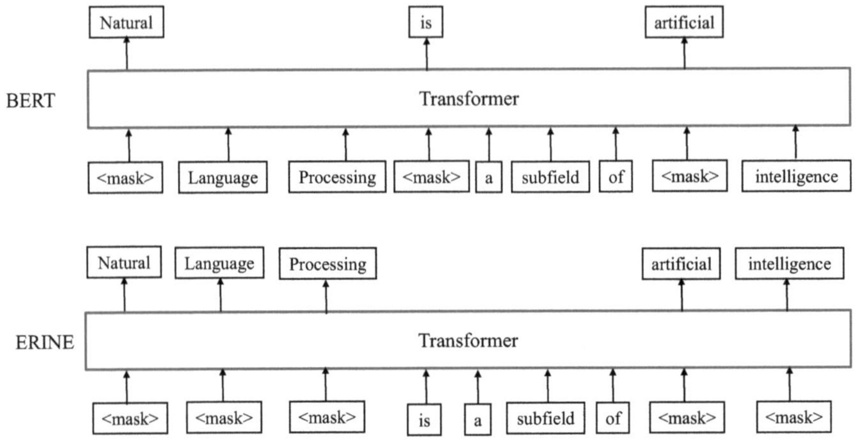

Fig. 6.8 A comparison of masking strategies between ERNIE (Sun et al., 2019) and BERT (Devlin et al., 2019)

6.6 Pre-training Large Models

Based on pre-training language models, researchers have found through experiments that when the amount of data used in the pre-training stage is large and the parameters of the pre-training language models exceed a certain threshold, the pre-training language models can not only significantly improve the performance on multiple natural language processing tasks, such as classification, information extraction, reasoning, and dialogue. Additionally, the pre-training language models show some special abilities that do not exist in small-scale language models, such as in-context learning ability (Brown et al., 2020), emergent abilities (Wei et al., 2022a), and chain of thought ability (Wei et al., 2022b).

To highlight the importance of the increase of parameter size, researchers have proposed the concept of "large language models" (LLMs) or "pre-training large models" to represent language models with large parameter sizes. It should be noted that there is no consensus on what size of parameters is required to be considered a large language model. One viewpoint suggests that a parameter scale exceeding 10 billion (10B) can be called a large language model (Zhao et al., 2023).

Recently, both academia and industry have been vigorously promoting the research of large language models (LLMs), especially after OpenAI's ChatGPT was released in 2022, which attracted widespread attention from all sectors of society. Users can interact with ChatGPT through natural language and complete different tasks at the same time, such as question-answering, chatting, classification, summarization, translation, conversation, and story continuation. The model has achieved good results in many natural language understanding and generation tasks. Nowadays, many internet companies and research institutions have built their own large-scale language models, such as Google's PaLM (Thoppilan et al., 2022), LaMDA (Thoppilan et al., 2022), Meta's OPT (Zhang et al., 2022), LLaMA (Touvron et al.,

2023), and Baidu's ERNIE. Table 6.1 shows the number of parameters, release time, representative papers or websites of some common large language models.

Different large language models (LLMs) vary in terms of data usage, model architecture, training methods, and technical routes adopted. This chapter mainly takes InstructGPT and ChatGPT as examples to introduce the main construction process of InstructGPT and ChatGPT. Overall, the entire implementation process includes three steps:

(1) Training the foundational large model: In this stage, we primarily use a large amount of text data and code data, and adopt an autoregressive language model to train the foundational large model.
(2) Instruction tuning: Describe multiple natural language processing tasks as instructions and use manual labeling of answers to fine-tune models' parameters through supervised learning, so that they have the ability to complete various tasks.
(3) Reinforcement learning from human feedback (RLHF): Collect human's feedback data and use reinforcement learning methods to make the model's outputs better align with human needs.

It should be noted that as of the time of writing this book, many technical details of ChatGPT have not been fully disclosed, and at the same time, a large number of

Table 6.1 Some common pre-training large language models

Model	Is it open source	Release time	The number of parameter	Paper or website
T5	YES	October 2019	11 billion	(Raffel et al., 2019)
T0	YES	October 2021	11 billion	(Sanh et al., 2022)
OPT	YES	May 2022	175 billion	(Zhang et al., 2022)
BLOOM	YES	October 2022	176 billion	(Scao et al., 2022)
LLaMA	YES	February 2023	65 billion	(Touvron et al., 2022)
GPT-3	NO	May 2020	175 billion	(Brown et al., 2020)
FLAN	NO	September 2021	137 billion	(Wei et al., 2021)
LaMDA	NO	November 2022	137 billion	(Thoppilan et al., 2022)
InstructGPT	NO	May 2022	175 billion	(Ouyang et al., 2022)
ChatGPT	NO	November 2022	–	http://openai.com/blog/chatgpt
PaLM	NO	April 2022	540 billion	(Chowdhery et al., 2022)
ERNIE	NO	April 2023	–	http://liyan.baidu.com/
GPT-4	NO	April 2023	–	(OpenAI.2023)

technologies are still rapidly evolving and changing. Therefore, some of the details discussed below are based on InstructGPT (Ouyang et al., 2022) and related papers published by OpenAI.

6.6.1 Foundational Large Models

The foundational large model stage primarily involves training with a large amount of text data and code data. OpenAI calls the large model trained on text data and code data as GPT-3.5. OpenAI has not published the training process and implementation details of GPT-3.5. Here, we mainly take GPT-3 as an example and introduce it from two aspects: data training and model framework.

In terms of training data, GPT-3's training data includes four datasets: the CommonCrawl dataset, the WebText dataset, a curated book dataset, and the English Wikipedia dataset. Among them, the largest dataset is the CommonCrawl dataset. GPT-3 used web data from 2016 to 2019, with the original data amounting to 45 TB. After filtering, 570 GB of data was retained. The biggest difference in training data between GPT-3 and GPT-3.5 is that GPT-3.5 added a large amount of code data. The size of the code data added by GPT-3.5 has not yet been disclosed. Here, we can refer to the Codex model proposed by OpenAI in July 2021 (Chen et al., 2021), which collected 179 GB of Python code from GitHub and retained 159 GB after filtering. From the perspective of training data, the foundational large model learns rich language knowledge and certain facts from massive text data on one hand and learns certain logical reasoning abilities from massive code data on the other hand.

In terms of model framework, GPT-3 adopts a network structure similar to GPT-1 and GPT-2, utilizing the Transformer decoder. Its training method adopts an autoregressive language model. Unlike previous models, GPT-3 has a significant increase in model parameters, reaching a total of 175 billion parameters. A comparison of specific parameters is shown in Table 6.2. It should be noted that although GPT-3 has not published the codes and corresponding models, (Zhang et al., 2022) has implemented and open-sourced the OPT (open pre-trained Transformer language models) system based on the experimental details of GPT-3.

Table 6.2 A comparison of parameters of GPT-1, GPT-2, and GPT-3

Model	The number of parameter	Number of layers	The dimension of hidden layer vectors	The number of attention heads	The vector dimension of each head
GPT-1	125 M	12	768	12	64
GPT-2	1.5B	48	1600	25	64
GPT-3	175B	96	12,288	96	128

6.6.2 Instruction Tuning

Traditional pre-training language models use parameter fine-tuning, which involves adjusting the parameters of the pre-training language model using a small amount of labeled data from downstream tasks. The above method is generally suitable for pre-training models with smaller parameter scales. However, when the model parameters are large, fine-tuning for each downstream task can lead to significant computational overhead and high space requirements. Additionally, when the labeled data for downstream tasks is limited, there is a risk of parameter overfitting.

To address the problems existing in parameter tuning, researchers proposed a method called prompting learning. This method first converts the downstream tasks into a form that is easier for pre-training language models to process by adding prompts, so that the pre-training tasks and the downstream tasks can be as consistent as possible.

Unlike parameter tuning and prompting learning, ChatGPT uses the method of instruction tuning (Wei et al., 2021), which first describes different tasks in the form of instructions, and then uses instruction data to fine-tune the model. The instructions for machine translation tasks are given below.

The instruction data for machine translation is:

Translate this sentence to English: 今天天气不错 Today is a nice day.

"Translate this sentence to English" is the task-related description (instruction), "今天天气不错" is the source Chinese language, and "Today is a nice day." is the standard generated translation. ChatGPT optimizes the pre-training model by collecting instruction data from real users, writing standard answers manually, and using the labeled instruction data in a manner of supervised learning.

A large amount of current research work shows that after instruction tuning, pre-training large models not only perform well on fine-tuned tasks but also show significant performance improvements on many tasks that are not included in the instruction tuning dataset. This demonstrates good instruction generalization and emergent capabilities.

6.6.3 Reinforcement Learning from Human Feedback, RLHF

Pre-trained models that have undergone instruction tuning are already capable of following user instructions. However, creating standard answers for each instruction requires significant human effort and incurs substantial costs. On the basis of instruction tuning, pre-training large language models use reinforcement learning from human feedback to further align model-generated content with real user preferences and requirements, enhancing alignment with human expectations.

(1) Train a reward model that aligns with user preferences

The input of this function is the text sequence output by the pre-training large models, and the output of the model is the reward value of this text sequence. The reward value can reflect whether the result output by the pre-training model conforms to user preferences. A high reward value means that the text sequence output by the pre-training model conforms to user preferences more, and vice versa. According to the real user instructions, the pre-training model generates different output results through sampling. Evaluators compare and evaluate the output results of the pre-training model and sort them. After collecting a large dataset of instructions, model outputs, and corresponding sorting results, the reward model learns user preferences using the above dataset. InstructGPT uses the GPT-3 model with a parameter size of 6 billion as the initialized model of the reward model and obtains the final reward model through instruction tuning.

(2) Use deep reinforcement learning to optimize parameters of pre-training models

In the process of optimizing pre-training language models using deep reinforcement learning, the agent is the pre-training language model, the environment is the real user, and the reward is the score output by the reward model. The model optimizes the expected cumulative rewards through deep reinforcement learning methods, increasing the probability of generating samples with larger reward values. Specifically, ChatGPT uses the proximal policy optimization (PPO) method in the policy gradient method to update parameters (Schulman et al., 2017).

Chapter 7
Word Segmentation and Tokenization

Abstract This chapter details two distinct approaches to word segmentation and tokenization for Natural Language Processing (NLP). The first method addresses Chinese Word Segmentation (CWS) by treating it as a character-based sequence-labeling task, where each character is tagged based on its position within a word (e.g., beginning, middle, end, or single). It provides a step-by-step guide for implementing this task by fine-tuning a pre-trained BERT model, covering everything from data processing and model construction to training and evaluation. The second method introduces sub-word segmentation using the Byte Pair Encoding (BPE) algorithm, a technique designed to handle out-of-vocabulary (OOV) words and reduce vocabulary size by breaking words into smaller, more frequent sub-word units. This section explains how to build a BPE vocabulary by iteratively merging the most frequent adjacent character pairs and then use that vocabulary to segment new text.

Word is the smallest unit of language that can be used independently. In some NLP tasks, words are often the most basic processing unit of texts. Western languages (such as English, French, and German) have word separators such as spaces, so word segmentation is not necessary. However, Chinese and some agglutinative languages (such as Japanese, Vietnamese, and Tibetan) do not have any symbol indicating word boundary, so when analyzing and processing Chinese, word segmentation should be firstly finished. This process is called Chinese word segmentation (CWS). Simply speaking, automatic Chinese word segmentation is to let the computer automatically add space or other boundary marker between two adjacent words in Chinese texts. There is a lot of research work for Chinese word segmentation, including dictionary-based word segmentation methods (such as maximum matching method, and shortest-path word segmentation method), the statistical method based on n-gram models, the character-based sequence-labeling method. In addition, methods based on neural networks have emerged as a prominent approach in recent years.

In the neural network-based NLP method, if words are used as the basic granularity, there are the following disadvantages: words in the corpus conform to the "long tail" distribution, and there are a large number of low-frequency words and

C. Zong et al., *Natural Language Processing and Large Language Models*,
https://doi.org/10.1007/978-981-92-0682-7_7

out-of-vocabulary (OOV) words. The semantic representation learning of these low-frequency words and OOV words is often inaccurate. In addition, if the number of words retained in the vocabulary is large, the computing complexity of the model will significantly increase. For the above reasons, researchers have proposed a sub-word segmentation method. This is done by compressing two letters (or Chinese characters) next to each other to form a granular representation between characters (or Chinese characters) and words, called sub-words, to alleviate the computing burden caused by long-tail words and too many words. The algorithm based on byte pair encoding (BPE) is the most commonly used sub-word segmentation method (Sennrich et al. 2016).

This chapter introduces two methods for implementing word segmentation: (1) The Chinese word segmentation based on the pre-training model BERT; (2) The sub-word segmentation based on the BPE algorithm.

7.1 Character-Based Chinese Word Segmentation Based on BERT

7.1.1 Task Objectives

The Character-based tagging method for Chinese word segmentation was proposed by Xue and Converse (2002). In this method, a Chinese sentence can be viewed as a sequence of words. Furthermore, each character in a sentence (including Chinese characters, punctuation, numbers, and any other letters) can only take four positions within the word: at the beginning, at the end, in the middle, or as a single word. This experiment uses BMESO tags to label these four possibilities. Among these, B represents that the character is at the beginning of the word; E indicates that the character is at the end of the word; M means that the character is in the middle of the word; S indicates that the character is a single-character word, and O indicates the character is a special character (such as padding). For a correct labeling, B and E should appear in pairs. When using neural networks to model Chinese word segmentation tasks, the Chinese word segmentation task is converted into a sequence-labeling task to predict each word's sequence of the word position in the sentence. Based on that, the word segmentation result can be obtained. As shown in Fig. 7.1, the "B" labels are located to "计," "语","Beijing," "held in" separately, and "BME" label is located to "linguistics," and the "S" label is located to "on" because it is a single-character word.

This section introduces the sequence labeling method of modeling the Chinese word segmentation task based on the pre-training model BERT. BERT is a model based on a large amount of corpus after pre-training. It has strong language prediction and generation capabilities, and only needs fine-tuning on downstream tasks to achieve good results.

Source	计算语言学大会在北京召开										
Cut	计	算	语	言	学	大	会	在	北	京	召	开
Tag	B	E	B	M	E	B	E	S	B	E	B	E

Fig. 7.1 An example of character-based Chinese word segmentation of sentence "计算语言学大会在北京召开 (The Conference on Computational Linguistics is being held in Beijing)"

7.1.2 Implementation Ideas and Processes

The idea of implementing Chinese word segmentation based on BERT is shown in Fig. 7.2. The input of the model is a text sentence, and the output is the Chinese word segmentation result. During the modeling process, the input text to be segmented is first processed, and then BERT is employed to encode the text sequence to obtain the semantic vector representation of the text. Then, the linear layer is used to obtain the scores of word position tagging for each word in the input text, and then decoding is performed to obtain the corresponding word segmentation result of the text.

PaddleNLP is a simple and powerful NLP development base with multiple pre-training models, which provides an API system that can help users efficiently model related tasks in the NLP field. The following six steps show the implementation process of Chinese word segmentation by using PaddlePaddle:

(1) Data processing: processing the input text data according to the data format required by the BERT model, including text truncation and padding, word2id, and so on. And then put the data into a minibatch format and pass it into the BERT model.
(2) Model construction: use the BERT model to encode the input text data and obtain the corresponding vector representation.
(3) Training configuration: instantiate the model, select the computing resources (CPU or GPU) for the model, and specify the optimization algorithms for model upgrading.
(4) Model training: train the parameters of the model to achieve optimal results.
(5) Model testing: evaluate and test the trained model to observe the changes in accuracy and loss function.
(6) Model inference: select any Chinese text and obtain the word segmentation results of the input text.

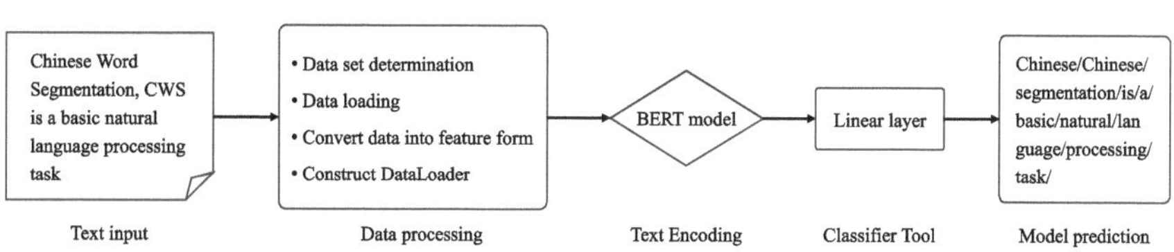

Fig. 7.2 A flowchart of implementing Chinese word segmentation task

The following sub-sections describe the specific implementation processes of each step above.

7.1.3 *Data Processing*

Data processing includes steps such as determining the dataset, loading the data into memory, converting the data into the feature form, constructing a DataLoader, and finally processing the same batch of data into an equal-length feature sequence. The DataLoader is used to pass the data of each round of upgrading into the BERT model batch by batch.

(1) **Determination of the Dataset**

This practice uses the ICWB2 dataset for training Chinese word segmentation models.[1] ICWB2 was jointly provided by Academia Sinica of Taiwan, Hong Kong City University, Peking University, and Microsoft Research Asia. The former two use traditional Chinese characters and the latter two use simplified Chinese characters. The following practice uses the dataset provided by Peking University. Due to the lack of a validation set in the original dataset, we divided 1000 samples (sentences) from the train set as the validation set. And finally the train set has 18,056 samples (traditionally "sentences" are referred to as "pieces"), with a validation set of 1000 samples, and a test set of 1945 samples.

To facilitate readers' learning and use, we have processed the data ahead and converted the original Chinese word segmentation data into the sequence-labeling format. Below are three sample data sets from the train set. Each row is a sample, and each sample contains "text" (characters) and character "labels," separated by spaces.

{"text": "黄 河 的 臂 弯, 秦 岭 的 胸 膛[2] "label": "B E S B E S B E S B E"}.
{"text": "黄 土 地 上 的 蒲 公 英[3]", "label": "B M E S S B M E"}.
{"text": "瞩望过秦宫汉阙的兴衰④[4]", "label": "B E S B E B E S B E"}.

2. **Data Loading**

The purpose of this task is to read the training data, validation data, and test data into memory. In this book, unless special conditions, the datasets are loaded by Huggingface by default, an open-source tool. The implementation codes are shown as follows:

[1] For detailed introduction of the icwb2 ICWB2 dataset, please refer to http://sighan.cs.uchicago.edu/bakeoff2005/.

[2] The bend of the Yellow River, the chest of the Qinling Mountains.

[3] Dandelions on the Loess Plateau.

[4] Gazing on the rise and fall of Qin's palaces and Han's imperial towers.

```
# Install dataset

%pip install datasets
```

```
# Load PaddleNLP
import paddlenlp
from datasets import load_dataset


# Use datasets to load the train set, validation set, and test set. Here, we use the path of the datasets of the projects in the AI Studio. Readers should conform to reality.

train_path = "data/data174918/train.json"
dev_path = "data/data174918/dev.json"
test_path = "data/data174918/test.json"
dataset = load_dataset("json", data_files={"train":train_path, "dev":dev_path, "test":test_path})
```

```


# Print the first three data in the train set
print(dataset["test"][:3])
```

The output results are:

{'text': ['共 同 创 造 美 好 的 新 世 纪 — — 二 ○ ○ 一 年 新 年 贺 词', '(二 ○ ○ ○ 年 十 二 月 三 十 一 日) (附 图 片 1 张)[5]', '女 士 们, 先 生 们, 同 志 们, 朋 友 们[6]:'], 'label': ['B E B E B E S S B E B E B M M M E B E B E', 'S B M M M E B M E B M M E S S S B E S S S', 'B E S S B E S S B E S S B E S S']}.

From the output results, we can see that the data structure consists of text data "text" and corresponding labels "label," which contain the first three data of the training set.

1. **Convert Data into the Feature Form**

The loaded text data is in the form of a string, and the BERT model cannot read it directly. It is necessary to convert the text data into corresponding feature data. In the BERT network structure introduced in Sect. 6.2, we learned that the input data of the BERT model needs to follow a fixed format, including input_ids, token encoding, token_type_ids, position_ids, and others. This practice needs to convert text data into feature data that meets the requirements of the BERT format.

PaddleNLP provides two Tokenizer tactics to help us automatically convert text data into the input features required by the BERT model: normal Tokenizer and AutoTokenizer.

---

[5] Let's create a beautiful new century together—New Year's message for 2001', '(December 31, 2000) (with 1 photo).

[6] Ladies, gentlemen, comrades, friends.

Table 7.1 Commonly used pre-training models and their tokenizers

Model	BERT	ERNIE	RoBERTa	GPT	BART
Tokenizer	BertTokenizer	ErineTokenizer	RobertaTokenizer	GPTTokenizer	BartTokenizer

(1) Method 1: Use Tokenizer to convert data

In PaddleNLP, each pre-training model corresponds to its respective Tokenizer, as shown in Table 7.1. The name of the Tokenizer by default is composed of the "model's name" + "Tokenizer."

According to the Tokenizer rules shown in Table 7.1, the following uses the BertTokenizer to convert text data into feature form. Then codes can be implemented as follows:

```
from paddlenlp.transformers import BertTokenizer
# Initialize tokenizer
tokenizer = BertTokenizer.from_pretrained('bert-base-chinese')

# Use the tokenizer to convert data into the corresponding feature form
```

```
inputs = tokenizer(text='中文分词是一项重要的自然语言处理领域任务 ①')
print(inputs)
```

See Footnote. [7]

From the output results, by default, the tokenizer will return input_ids and token_type_ids in the form of a list. In addition, the tokenizer also provides more functions, with the following commonly used keyword parameters:

- text: sequences to be processed.
- text_pair: another specified sequence forms a pair with sequences.
- is_split_into_words: indicate whether the input text data has been segmented into words.
- max_seq_len: the maximum sequence length of the sequence after word segmentation.
- return_position_ids: return positional embedding.
- return_attention_mask: return the attention_mask value.
- return_length: whether to return the length of sequences.
- return_tensors: set the form of the returned data. If it is set to pd, then return a paddle.Tensor object; if it is set to np, then return a np.ndarray object.

[7] Chinese word segmentation is an important task in natural language processing.

Examples of codes are as follows:

```
inputs = tokenizer('中文分词是一项重要的自然语言处理领域任务',
return_length=True, max_seq_len=128, return_position_ids=True,
return_offsets_mapping=True, return_attention_mask=True)
print(inputs)
```

The output results are:

```
{'input_ids': [101, 704, 3152, 1146, 6404, 3221, 671, 7555, 7028,
6206, 4638, 5632, 4197, 6427, 6241, 1905, 4415, 7566, 1818, 818,
1218, 102], 'token_type_ids': [0, 0, 0, 0, 0, 0, 0, 0, 0, 0, 0, 0,
0, 0, 0, 0, 0, 0, 0, 0, 0, 0], 'offset_mapping': [(0, 0), (0, 1), (1,
2), (2, 3), (3, 4), (4, 5), (5, 6), (6, 7), (7, 8), (8, 9), (9, 10),
(10, 11), (11, 12), (12, 13), (13, 14), (14, 15), (15, 16), (16, 17),
(17, 18), (18, 19), (19, 20), (0, 0)], 'position_ids': [0, 1, 2, 3, 4,
5, 6, 7, 8, 9, 10, 11, 12, 13, 14, 15, 16, 17, 18, 19, 20, 21], 'atten-
tion_mask': [1, 1, 1, 1, 1, 1, 1, 1, 1, 1, 1, 1, 1, 1, 1, 1, 1, 1, 1, 1,
1, 1], 'length': 22, 'seq_len': 22}.
```

The following shows some other commonly used methods of Tokenizer to implement codes:

```
# Implement the word segmentation of text sequences
tokens = tokenizer.tokenize('中文分词是一项重要的自然语言处理领域任务')
print(tokens)

# Convert token into id
input_ids = tokenizer.convert_tokens_to_ids(tokens)
print(input_ids)
# Convert id into token
tokens = tokenizer.convert_ids_to_tokens(input_ids)
print(tokens)

# Convert token into strings of characters
text = tokenizer.convert_tokens_to_string(tokens)
print(text)
```

The output results are:

```
['中', '文', '分', '词', '是', '一', '项', '重', '要', '的', '自', '然', '语
', '言', '处', '理', '领', '域', '任', '务']
[704, 3152, 1146, 6404, 3221, 671, 7555, 7028, 6206, 4638, 5632, 4197, 642
7, 6241, 1905, 4415, 7566, 1818, 818, 1218]
['中', '文', '分', '词', '是', '一', '项', '重', '要', '的', '自', '然', '语
', '言', '处', '理', '领', '域', '任', '务']
中 文 分 词 是 一 项 重 要 的 自 然 语 言 处 理 领 域 任 务
```

(2) Method 2: Convert data using AutoTokenizer (Recommended).

Due to the fact that different pre-training models correspond to different tokenizers, to simplify operations, PaddleNLP provides the AutoTokenizer function, which automatically loads the corresponding tokenizer by specifying the model's name. For example, by specifying the model's name as 'bert-base-chinese', the corresponding tokenizer can be automatically loaded. Then codes can be implemented as follows:

```
from paddlenlp.transformers import AutoTokenizer

# Initialize tokenizer
tokenizer = AutoTokenizer.from_pretrained('bert-base-chinese')
print(tokenizer.__class__)


# Use tokenizer to convert data into the corresponding feature form
inputs = tokenizer(text='中文分词是一项重要的自然语言处理领域任务')
print(inputs)
```

The output results are:

```
{'input_ids': [101, 704, 3152, 1146, 6404, 3221, 671,
7555, 7028, 6206, 4638, 5632, 4197, 6427, 6241, 1905,
4415, 7566, 1818, 818, 1218, 102], 'token_type_ids':
[0, 0, 0, 0, 0, 0, 0, 0, 0, 0, 0, 0, 0, 0, 0, 0, 0, 0,
0, 0, 0, 0]}.
```

From the output results, the class name of the Tokenizer is BertTokenizer, which is the same as the output results of Method 1.

Note

In the subsequent Chapters of this book, AutoTokenizer is used by default.

The following defines the convert_example_to_feature function and converts the input text data by loading the tokenizer into feature data. In the loaded dataset, the labeled data "label" is a word position labeling sequence BEMSO. For the ease of processing, it is necessary to define a word-list label2id and convert it into a numerical sequence. The mapping rules for word position labeling are: "O" maps to 0, "B" maps to 1, "M" maps to 2, "E" maps to 3, and "S" maps to 4. Among them, "O" is mainly used to map some special tokens to 0, such as "[CLS]," "[SEP]," and "[PAD]."

The implementation of converting data in batches is as follows:

```
from paddlenlp.transformers import AutoTokenizer

def convert_example_to_feature(example, tokenizer, label2id, max_seq_len=
512, is_infer=False):
    """

    Convert the input samples into input features suitable for the BERT model


    Input:

        - example: a single input sample

        - tokenizer: the tokenizer of the BERT model

        - label2id: the dictionary used to change labels into id

        - max_seq_len: the largest sequence-length for models to process texts

        - is_infer: whether it is a model prediction or not, for predicting data, the labeled data will
not be processed
    """


    # use tokenizer to chanage data into feature forms
    text = example["text"].strip().split(" ")
    encoded_inputs = tokenizer(text, max_seq_len=max_seq_len, is_split_in
to_words="token", return_length=True)


    # process data with labels
    if not is_infer:
        label = [label2id[item] for item in example["label"].split(" ")]
[:max_seq_len-2]
        encoded_inputs["label"] = [label2id["O"]] + label + [label2id["O
"]]
        assert len(encoded_inputs["label"]) == len(encoded_inputs["input_
ids"])

    return encoded_inputs


# initialize tokenizer
model_name = "bert-base-chinese"
label2id = {"O":0, "B":1, "M":2, "E":3, "S":4}
tokenizer = AutoTokenizer.from_pretrained(model_name)
```

The following shows a sample use of convert_example_to_feature:

```
# show a sample
example = {"text":"钱 其 琛 访 问 德 国 ", "label":"S B E B E B E"}
features = convert_example_to_feature(example, tokenizer, label2id,
max_seq_len=512, is_infer=False)

# input feature data
print(features)
```

See Footnote. [8]

The output result is:

{'input_ids': [101, 7178, 1071, 4422, 6393, 7309, 2548, 4408, 102], 'token_type_ids': [0, 0, 0, 0, 0, 0, 0, 0, 0], 'length': 9, 'seq_len': 9, 'label': [0, 4, 1, 3, 1, 3, 1, 3, 0]}.

Next, based on the convert_example_to_feature function, the loaded train set, validation set, and test set will be converted into corresponding feature forms in turn. Here we need to use the partial function to set the parameters in the convert_example_to_feature function, and then convert them based on the map function. The map function provided by datasets supports converting data one by one and in batches. Since the convert_example_to_feature function processes data one by one, we set the parameter batched to False in the map function. Then codes can be implemented as follows:

```
from functools import partial

max_seq_len = 512
trans_fn = partial(convert_example_to_feature, tokenizer=tokenizer, label2
id=label2id, max_seq_len=max_seq_len)


# Convert the input train set, validation set, and test set data into feature forms
columns = ["text", "label"]
train_dataset = dataset["train"].map(trans_fn, batched=False, remove_colum
ns=columns)
dev_dataset = dataset["dev"].map(trans_fn, batched=False, remove_columns=c
olumns)
test_dataset = dataset["test"].map(trans_fn, batched=False, remove_columns
=columns)


# output the samples of the train set, validation set, and test set
print("train_dataset:", len(train_dataset))
print("dev_dataset:", len(dev_dataset))
print("test_dataset:", len(test_dataset))
```

The output result is:

```
train_dataset: 18054
dev_dataset: 1000
test_dataset: 1944.
```

2. **Building DataLoader**

Due to the limitation of computer memory, it is necessary to train data in batches during model training, that is, to select one minibatch of data for training at a time. The DataLoader is used for this purpose. However, there is still a problem here, as the length of data in the same batch may not be the same, which may affect the

[8] Qian Qichen visits Germany.

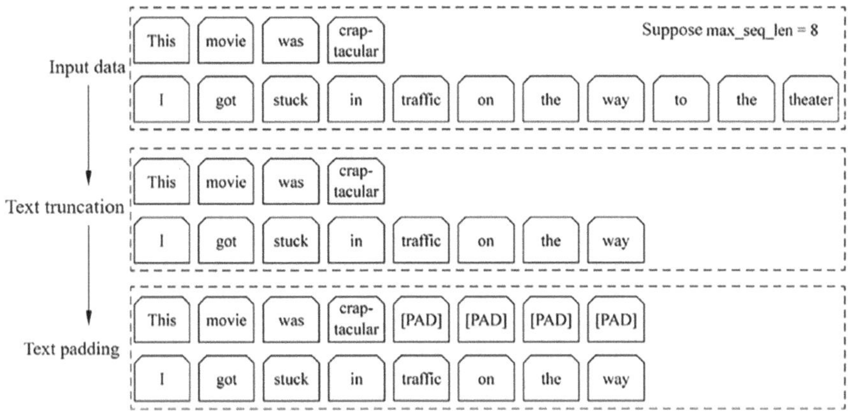

Fig. 7.3 Text truncation and text padding

training effect of the model. Therefore, it is necessary to change the data in the same batch into the same length. There are two common methods: text truncation and text padding.

(1) Text truncation: set a maximum sequence length, max_seq_len, during training to truncate excessively long texts and prevent the overall model training performance from being affected by excessively long data.
(2) Text padding: calculate the maximum length max_len of the text sequence of the minibatch. When the length of the text sequence is less than max_len, use [PAD] to pad the text sequence to the length of max_len.

Figure 7.3 shows examples of text truncation and text padding. By applying these two techniques, the length of each batch of data of DataLoader's upgrading can be guaranteed to be fixed.

The convert_example_to_feature function has already truncated the excessively long texts. Next, we need to process the length of the data in the same minibatch using text padding. There are two methods: manual processing and automatic processing.

(1) Method 1: changing the data in the same batch into the same length

First, let's define the collate_fn function for text truncation and padding, which can be used as a callback function and then it is passed to DataLoader. DataLoader will use collate_fn to process data before returning each batch of data and return the processed text data and corresponding labels. Then codes can be implemented as follows:

```
import paddle

def collate_fn(batch_data, pad_token_id=0, pad_token_type_id=0, pad_label
_id=0):
    """

    Batch data processing function


    Input:

        - batch_data: data in batches to be processed currently

        - pad_token_id: the padding placeholder for the token_id

        - pad_token_type_id: the padding placeholder for the token_type_id

        - pad_label_id: the padding placeholder for the label sequence
    """

    input_ids_list, token_type_ids_list, label_list = [], [], []
    max_len = 0
    for example in batch_data:
        input_ids, token_type_ids, label = example["input_ids"], example
["token_type_ids"], example["label"]


        # carry out content padding for each data
        input_ids_list.append(input_ids)
        token_type_ids_list.append(token_type_ids)
        label_list.append(label)


        # retain the maximum length of sequences
        max_len = max(max_len, len(input_ids))


    # pad data sequences into the largest length
    for i in range(len(input_ids_list)):
        cur_len = len(input_ids_list[i])
        input_ids_list[i] = input_ids_list[i] + [pad_token_id] * (max_len
 - cur_len)
        token_type_ids_list[i] = token_type_ids_list[i] + [pad_token_type
_id] * (max_len - cur_len)
        label_list[i] = label_list[i] + [pad_label_id] * (max_len - cur_l
en)

    return paddle.to_tensor(input_ids_list),  paddle.to_tensor(token_type
_ids_list), paddle.to_tensor(label_list)
```

Next, we need to build a DataLoader for the upgrading of data in batches during model training, and specify the batch size. When building a DataLoader, readers can specify whether to randomize the samples by using the shuffle parameter. If shuffle is set to False, DataLoader will upgrade a batch of data sequentially; if it is set to True, DataLoader will randomly upgrade a batch of data. Then codes can be implemented as follows:

```
from paddle.io import BatchSampler, DataLoader

batch_size= 16
train_sampler = BatchSampler(train_dataset, batch_size=batch_size, shuf
fle=True)
dev_sampler = BatchSampler(dev_dataset, batch_size=batch_size, shuffle=
False)


# build DataLoader, and upgrade the train set, validation set, and test set data in batches based on
the batch size
train_loader = DataLoader(dataset=train_dataset, batch_sampler=train_sa
mpler, collate_fn=collate_fn)
dev_loader = DataLoader(dataset=dev_dataset, batch_sampler=dev_sampler,
 collate_fn=collate_fn)
test_loader = DataLoader(dataset=test_dataset, batch_sampler=dev_sample
r, collate_fn=collate_fn)

# Print the data of the first minibatch in the train set
print(next(iter(train_loader)))
```

The output results are:

```
[Tensor(shape=[16, 317], dtype=int64, place=Place(gpu:0), stop_gradient=Tr
ue,
       [[101 , 1920, 2397, ..., 0    , 0    , 0    ],
        [101 , 4385, 3300, ..., 0    , 0    , 0    ],
        [101 , 794 , 691 , ..., 0    , 0    , 0    ],
        ...,
        [101 , 3315, 2845, ..., 0    , 0    , 0    ],
        [101 , 1343, 2399, ..., 0    , 0    , 0    ],
        [101 , 100 , 100 , ..., 0    , 0    , 0    ]]), Tensor(shape=[16, 31
7], dtype=int64, place=Place(gpu:0), stop_gradient=True,
       [[0, 0, 0, ..., 0, 0, 0],
        [0, 0, 0, ..., 0, 0, 0],
        [0, 0, 0, ..., 0, 0, 0],
        ...,
        [0, 0, 0, ..., 0, 0, 0],
        [0, 0, 0, ..., 0, 0, 0],
        [0, 0, 0, ..., 0, 0, 0]]), Tensor(shape=[16, 317], dtype=int64, pl
ace=Place(gpu:0), stop_gradient=True,
       [[0, 1, 3, ..., 0, 0, 0],
        [0, 1, 3, ..., 0, 0, 0],
        [0, 4, 1, ..., 0, 0, 0],
        ...,
        [0, 1, 3, ..., 0, 0, 0],
        [0, 1, 3, ..., 0, 0, 0],
        [0, 1, 2, ..., 0, 0, 0]])]
```

From the output results we can see that the data length of the same batch is the same.

(2) Method 2: changing the data in the same batch into the same length automatically

In Method 1 described above, the collate_fn function is used to implement the content padding of a batch of data manually. In PaddleNLP, a DataCollator class API which has similar functions is provided to automatically change the data in the same batch into the same length. Because this practice implements Chinese word segmentation based on sequence labeling, which is required to classify each unit in the input sequences, DataCollatorForTokenClassification can be used to automatically change the data in the same batch into the same length. DataCollatorForTokenClassification will receive tokenizer and label_pad_token_id as parameters, in which label_pad_token_id is used to complete the id of the padded label. Then codes can be implemented as follows:

```
from paddle.io import BatchSampler, DataLoader
from paddlenlp.data import DataCollatorForTokenClassification

batch_size= 16
train_sampler = BatchSampler(train_dataset, batch_size=batch_size, shuffl
e=True)
dev_sampler = BatchSampler(dev_dataset, batch_size=batch_size, shuffle=Fa
lse)


# use DataCollator, which is a class API of PaddleNLP
data_collator = DataCollatorForTokenClassification(tokenizer, label_pad_t
oken_id=label2id["O"])
train_loader = DataLoader(dataset=train_dataset, batch_sampler=train_samp
ler, collate_fn=data_collator)
dev_loader = DataLoader(dataset=dev_dataset, batch_sampler=dev_sampler, c
ollate_fn=data_collator)
test_loader = DataLoader(dataset=test_dataset, batch_sampler=dev_sampler,
 collate_fn=data_collator)


# Print the data of the first minibatch in the train set
print(next(iter(train_loader)))
```

The output results are:

```
{'label': Tensor(shape=[16, 256], dtype=int64, place=Place(gpu:0), stop_gr
adient=True,
       [[0, 4, 4, ..., 0, 0, 0],
        [0, 1, 3, ..., 0, 0, 0],
        [0, 1, 2, ..., 0, 0, 0],
        ...,
        [0, 1, 3, ..., 0, 0, 0],
        [0, 4, 1, ..., 0, 0, 0],
        [0, 1, 3, ..., 0, 0, 0]]), 'input_ids': Tensor(shape=[16, 256], dt
ype=int64, place=Place(gpu:0), stop_gradient=True,
       [[101 , 782 , 1762, ..., 0    , 0    , 0    ],
        [101 , 2356, 1767, ..., 0    , 0    , 0    ],
        [101 , 1059, 2157, ..., 0    , 0    , 0    ],
        ...,
        [101 , 7471, 2270, ..., 0    , 0    , 0    ],
        [101 , 3736, 3813, ..., 0    , 0    , 0    ],
        [101 , 704 , 1744, ..., 0    , 0    , 0    ]]), 'token_type_ids': Ten
sor(shape=[16, 256], dtype=int64, place=Place(gpu:0), stop_gradient=True,
       [[0, 0, 0, ..., 0, 0, 0],
        [0, 0, 0, ..., 0, 0, 0],
```

```
        [0, 0, 0, ..., 0, 0, 0],
        ...,
        [0, 0, 0, ..., 0, 0, 0],
        [0, 0, 0, ..., 0, 0, 0],
        [0, 0, 0, ..., 0, 0, 0]]), 'length': Tensor(shape=[16], dtype=int6
4, place=Place(gpu:0), stop_gradient=True,
       [47 , 164, 5  , 153, 67 , 52 , 76 , 90 , 92 , 176, 164, 141, 256, 1
3 ,
        59 , 230]), 'seq_len': Tensor(shape=[16], dtype=int64, place=Place
(gpu:0), stop_gradient=True,
       [47 , 164, 5  , 153, 67 , 52 , 76 , 90 , 92 , 176, 164, 141, 256, 1
3 ,
        59 , 230])}
```

From the output results it is clear that the data length of the same batch is the same.

7.1.4 Model Construction

The BERT model processes each input minibatch data sequentially. First, the feature data of processed texts is passed to the BERT model; then the BERT model encodes it and outputs a corresponding vector sequence. These vector sequences are then passed to a linear layer to obtain the word position tags (BEMSO) corresponding to each word in the input text. Then we can decode and get the final word segmentation result, as shown in Fig. 7.4.

Based on the modeling idea shown in Fig. 7.4, codes can be implemented as follows:

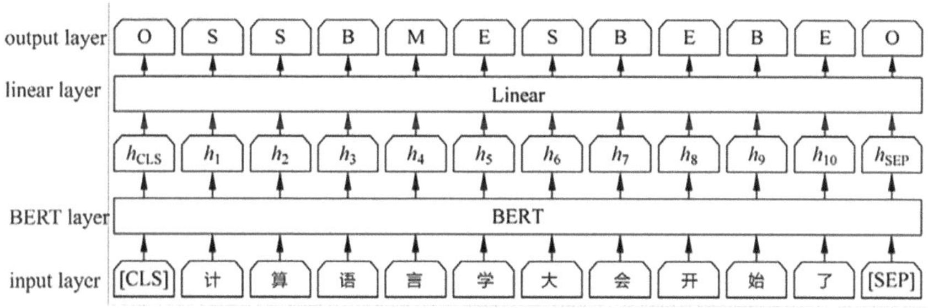

Fig. 7.4 Chinese word segmentation based on BERT by sequence labeling

```
import paddle
import paddle.nn as nn


# Define BertForTokenClassification and inherit the nn.Layer class
class BertForTokenClassification(nn.Layer):
    """

    The BERT model has a linear layer on top of it to classify the tokens in the input sequences,
    such as the NER task

    Input:

        - bert: examples of the BERT model
```

```
- num_classes: the number of categories for classification, with 2 by default

        - dropout: for the dropout probability of the BERT output vectors, if it is "None", Hidde
n_dropout_prob set inside BERT will be used
    """
# Build the function init and the forward calculation function forward
    def __init__(self, bert, num_classes=2, dropout=None):
        super(BertForTokenClassification, self).__init__()
        self.num_classes = num_classes
        self.bert = bert
        self.dropout = nn.Dropout(dropout if dropout is not None else s
elf.bert.config["hidden_dropout_prob"])
        self.classifier = nn.Linear(self.bert.config["hidden_size"], nu
m_classes)

    def forward(self,
                input_ids,
                token_type_ids=None,
                position_ids=None,
                attention_mask=None):


        # Transfer the feature data of the input text sequence into the BERT model for encoding
        outputs = self.bert(input_ids,
                            token_type_ids=token_type_ids,
                            position_ids=position_ids,
                            attention_mask=attention_mask)


        # Obtain the vector representation corresponding to the input sequence
        sequence_output = outputs[0]
        sequence_output = self.dropout(sequence_output)


        # The probability of mapping vectors to word position labels through linear layers
        logits = self.classifier(sequence_output)

        return logits
```

To facilitate usage, PaddleNLP provides some commonly used classes, such as BertForTokenClassification. When instantiating the model, we can simply use these classes to complete the model construction. For BERT models, the commonly used classes supported by PaddleNLP are as follows:

- BertModel: BERT model class.
- BertForMaskedLM: BERT-based masked language model, used for predicting Masked Tokens.
- BertForSequenceClassification: classify a string of text sequences, and its typical task is text classification.
- BertForTokenClassification: it classifies each token in the text sequence, and its typical task is named entity recognition.
- BertForQuestionAnswering: it is applied to question answering tasks, or extracting answers from a piece of text.
- BertForMultipleChoice: it applies to multiple-choice tasks and selects the correct answer from the given multiple answers.

After that, we will call PaddleNLP to use the BertForTokenClassification class to build a Chinese word segmentation model. Codes can be implemented as follows:

```
from paddlenlp.transformers import BertForTokenClassification


# num_classes is set to 5, which is the same as the number of categories of word position labels,
corresponding to five categories: O, B, M, E, and S
model = BertForTokenClassification.from_pretrained(model_name,
num_classes=5)
```

In addition, PaddleNLP also supports automatic model construction. In this practice, readers can use AutoModelForTokenClassification to automatically load BertForTokenClassification by specifying the model. Codes can be implemented as follows:

```
from paddlenlp.transformers import AutoModelForTokenClassification

# Define model_name: bert-base-chinese
model = AutoModelForTokenClassification.from_pretrained(model_name,
num_classes=5)
print(model.__class__)
```

The output results are:

```
<class 'paddlenlp.transformers.bert.modeling.
BertForTokenClassification'>
```

From the output results we can see that AutoModelForTokenClassification automatically finds the BertForTokenClassification class through the specified model_name.

Note

In the subsequent chapters of this book, the Auto module of PaddleNLP is used by default to use related classes.

7.1.5 *Training Configuration*

First, let's define the hyper-parameters, optimizer, loss function, and evaluation metrics during model training. The specific configurations are as follows:

(1) Optimizer: AdamW optimizer.
(2) Loss function: use Cross entropy loss.
(3) Evaluation Metrics: utilize the Precision, Recall, and F1 scores.

PaddleNLP provides an evaluation method named ChunkEvaluator for sequence-labeling tasks, which can evaluate the performance of Chinese word segmentation at the level of word granularity and return the Precision, Recall, and F1 values.

Next, we will illustrate with an example to explain the evaluation method in this practice. Assuming that there is an original sample of "派出所民警在重点乡镇企业进行流动人口登记[9]", and the correct word segmentation results and the word segmentation results predicted by systems are shown as follows:

The correct word segmentation results (marked as set $\mathcal{A}$): 派出所 | 民警 | 在 | 重点 | 乡镇 | 企业 | 进行 | 流动 | 人口 | 登记.

The predicted word segmentation results (marked as set $\mathcal{B}$): 派出所 | 民警 | 在重点 | 乡镇企业 | 进行流动 | 人口 | 登记Then,

$$\mathcal{A} = \{\text{派出所，民警，在，重点，乡镇，企业，进行，流动，人口，登记}\}$$
$$\mathcal{B} = \{\text{派出所，民警，在重点，乡镇企业，进行流动，人口，登记}\}$$
$$\mathcal{A} \cap \mathcal{B} = \{\text{派出所，民警，人口，登记}\}$$

Then $|\mathcal{A}| = 10$, $|\mathcal{B}| = 7$, $|\mathcal{A} \cap \mathcal{B}| = 4$, the calculation methods for the Precision, Recall, and F1 values are separately:

$$Precision = \frac{|\mathcal{A} \cap \mathcal{B}|}{|\mathcal{B}|} = \frac{4}{7} \approx 0.571$$

$$Recall = \frac{|\mathcal{A} \cap \mathcal{B}|}{|\mathcal{A}|} = \frac{4}{10} \approx 0.4$$

$$F1 = \frac{2 \times Precision \times Recall}{Precision + Recall} \approx 0.471$$

[9] Police officers at the local police station are registering the floating population in key township enterprises.

The codes for training configuration are implemented as follows:

```
from paddlenlp.metrics import ChunkEvaluator


# Set the number of training rounds
num_epochs = 3

# Set the learning rate
learning_rate = 3e-5

# Set intervals for model evaluation on the validation set every certain number of steps
eval_steps = 100

# Set intervals for printing the log every certain number of steps
log_steps = 10

# Set the model's saving path and automatically save the model with the best effects during the tra
ining process
save_dir = "./checkpoints"
```

```
# Set the weight decay coefficient during training
weight_decay = 0.01

# Set the proportion of warm-up training during the training process
warmup_proportion = 0.1

# Set the total number of training steps required
num_training_steps = len(train_loader) * num_epochs

# Except for the parameters of bias and LayerNorm, other parameters perform decay operations
during training
decay_params = [
        p.name for n, p in model.named_parameters()
        if not any(nd in n for nd in ["bias", "norm"])
    ]

# Define the optimizer
optimizer = paddle.optimizer.AdamW(
        learning_rate=learning_rate,
        parameters=model.parameters(),
        weight_decay=weight_decay,
        apply_decay_param_fun=lambda x: x in decay_params)

# Define the loss function
loss_fn = nn.CrossEntropyLoss()


# Define the computing method of evaluating indicators
metric = ChunkEvaluator(label_list=label2id.keys())
```

7.1.6 Model Training

During model training, a training log is printed every certain number of training steps (called log_steps). A model evaluation is performed on the validation set every certain number of training steps (called eval_steps). And the model with the best evaluation results is saved during the training process.

The codes for evaluation on the validation set are shown as follows:

```
def evaluate(model, data_loader, metric):
    """

    Model evaluation functions


    Input:

        - model: model examples to be processed

          - data_loader: datasets to be processed

        - metric: class examples used for statistical evaluation indicators
    """
    model.eval()
    metric.reset()
```

```
    precision, recall, f1_score = 0, 0, 0

    # Read the data of dataloader
    for batch_data in data_loader:
        input_ids, token_type_ids, labels, seq_lens = batch_data["input_i
ds"], batch_data["token_type_ids"], batch_data["label"], batch_data["seq_
len"]

        # Forward computation of models
        logits = model(input_ids, token_type_ids)
        predictions = logits.argmax(axis=-1)

        # Model evaluation
        num_infer_chunks, num_label_chunks, num_correct_chunks = metric.c
ompute(seq_lens, predictions, labels)
        metric.update(num_infer_chunks.numpy(), num_label_chunks.numpy(),
 num_correct_chunks.numpy())
    precision, recall, f1_score = metric.accumulate()

    return precision, recall, f1_score
```

During the training process, train_loss_record is used to save the changes in the loss function, and train_score_record is used to save the evaluation scores on the validation set. The codes for model training are implemented as follows:

```
import os

def train(model):
    """

    Model training function


    Input:

        - model: model examples to be trained
    """


    # Start the model's training mode
    model.train()
    global_step = 0
    best_score = 0.

    # Record the loss during training and the model evaluation scores on the validation set
    train_loss_record = []
    train_score_record = []


    # Perform num_epochs rounds of training
    for epoch in range(num_epochs):
        for step, batch_data in enumerate(train_loader):
            inputs, token_type_ids, labels = batch_data["input_ids"], bat
```

```
ch_data["token_type_ids"], batch_data["label"]

            # Obtain the model's predictions
             logits = model(input_ids=inputs, token_type_ids=token_type_id
s)
             loss = loss_fn(logits, labels) # seek mean by default
             train_loss_record.append((global_step, loss.item()))

            # Backpropagation
             loss.backward()
             optimizer.step()
             optimizer.clear_grad()

             if global_step % log_steps == 0:
                 print(f"[Train] epoch: {epoch}/{num_epochs}, step: {globa
l_step}/{num_training_steps}, loss: {loss.item():.5f}")

             if global_step != 0 and (global_step % eval_steps == 0 or glo
bal_step == (num_training_steps-1)):
                 precision, recall, F1 = evaluate(model, dev_loader, metri
c)
                 train_score_record.append((global_step, F1))

                 model.train()


                 # If current indicators are the best, retain this model
                 if F1 > best_score:
                     print(f"[Evaluate] best accuracy performence has been
 updated: {best_score:.5f} --> {F1:.5f}")
                     best_score = F1
                     save_path = os.path.join(save_dir, "best.pdparams")
                     paddle.save(model.state_dict(), save_path)
                 print(f"[Evaluate]  precision: {precision: .5f}, recall:
{recall: .5f}, dev score: {F1:.5f}")

             global_step += 1

    save_path = os.path.join(save_dir, "final.pdparams")
    paddle.save(model.state_dict(), save_path)
    print("[Train] Training done!")

    return train_loss_record, train_score_record

train_loss_record, train_score_record = train(model)
```

The output results are:

```
......
[Evaluate] precision: 0.97553, recall: 0.97483, dev score: 0.97518
[Train] epoch: 2/3, step: 3210/3387, loss: 0.00630
[Train] epoch: 2/3, step: 3220/3387, loss: 0.00668
[Train] epoch: 2/3, step: 3230/3387, loss: 0.00734
[Train] epoch: 2/3, step: 3240/3387, loss: 0.00483
[Train] epoch: 2/3, step: 3250/3387, loss: 0.00531
```

```
[Train] epoch: 2/3, step: 3260/3387, loss: 0.01206
[Train] epoch: 2/3, step: 3270/3387, loss: 0.00988
[Train] epoch: 2/3, step: 3280/3387, loss: 0.00269
[Train] epoch: 2/3, step: 3290/3387, loss: 0.01939
[Train] epoch: 2/3, step: 3300/3387, loss: 0.00770
[Evaluate] precision: 0.97763, recall: 0.97179, dev score: 0.97470
[Train] epoch: 2/3, step: 3310/3387, loss: 0.00763
[Train] epoch: 2/3, step: 3320/3387, loss: 0.01715
[Train] epoch: 2/3, step: 3330/3387, loss: 0.00506
[Train] epoch: 2/3, step: 3340/3387, loss: 0.00459
[Train] epoch: 2/3, step: 3350/3387, loss: 0.01350
[Train] epoch: 2/3, step: 3360/3387, loss: 0.00821
[Train] epoch: 2/3, step: 3370/3387, loss: 0.00393
[Train] epoch: 2/3, step: 3380/3387, loss: 0.01368
[Evaluate] best accuracy performence has been updated: 0.97518 -- >
0.97525
[Evaluate] precision: 0.97649, recall: 0.97402, dev score: 0.97525.
```

And then, the changes in the loss function and F1 score during the training process will be visualized. The codes are implemented as follows:

```
import matplotlib.pyplot as plt

def plot_training_loss(train_loss_record, fig_name, fig_size=(8, 6),
sample_step=10, loss_legend_loc="lower left"):
    """
    Plot the change of the loss function

    Input:
        - train_loss_record: loss records during training
        - fig_name: image saving path
        - fig_size: set the size of images
        - sample_step: A loss point is sampled every sample_step to draw the image
        - loss_legend_loc: The location of legends' descriptions
    """
    plt.figure(figsize=fig_size)

    train_steps = [x[0] for x in train_loss_record][::sample_step]
    train_losses = [x[1] for x in train_loss_record][::sample_step]
```

```
    plt.plot(train_steps, train_losses, color='#e4007f', label="Train
Loss")

    # Draw coordinate axes and legends
    plt.ylabel("Loss", fontsize='large')
    plt.xlabel("Step", fontsize='large')
    plt.legend(loc=loss_legend_loc, fontsize='x-large')

    plt.savefig(fig_name)
    plt.show()

def plot_training_acc(train_score_record, fig_name, fig_size=(8, 6),
sample_step=10, acc_legend_loc="lower left"):
    """

    Create a graph illustrating the changes in assessment scores


    Input:

        - train_score_record: records of evaluation scores during training

        - fig_name: image saving path

        - fig_size: set the size of images

        - sample_step: a loss point is sampled every sample_step to draw the image

        - acc_legend_loc: the location of legends' descriptions
    """
    plt.figure(figsize=fig_size)

    train_steps=[x[0] for x in train_score_record]
    train_losses = [x[1] for x in train_score_record]

    plt.plot(train_steps, train_losses, color='#e4007f', label="Dev
Accuracy")

    # Draw coordinate axes and legends
    plt.ylabel("Accuracy", fontsize='large')
    plt.xlabel("Step", fontsize='large')
    plt.legend(loc=acc_legend_loc, fontsize='x-large')

    plt.savefig(fig_name)
    plt.show()
```

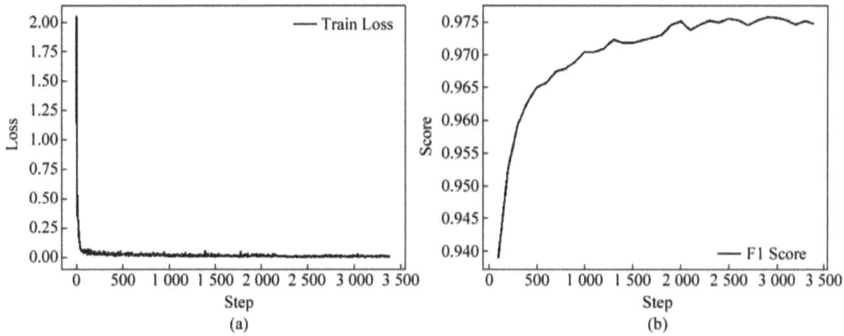

Fig. 7.5 Loss change of the train set and F1 score change of the validation set. **a** Loss change of the train set; **b** F1 score change of the validation set

The output results are shown in Fig. 7.5.

In Fig. 7.5a, the horizontal axis represents the number of training steps, and the vertical axis represents the value of the loss function, which is the cross-entropy between the predicted and true results. In Fig. 7.5b, the horizontal axis and the vertical axis represent the number of training steps and the predicted F1 score, respectively. From Fig. 7.5, it can be seen that as training progresses, the loss function on the training set continues to decrease and gradually converges towards 0. In contrast, the F1 score on the validation set starts to rise continuously and gradually stabilizes after the model converges, with the final F1 score reaching around 97.5%.

Note: The plot_training_loss function and plot_training_acc function are stored in the tools.py file and will not be displayed next. We can simply use the corresponding codes in the tools.py file.

7.1.7 *Model Evaluation*

Next, we will use the test set to evaluate the model with the best performance during training to verify the model's training effects. The codes implemented are as follows:

```
# Load the trained model for prediction, re-instantiate a model, and then load the trained models' parameters into the new model

saved_state = paddle.load("./checkpoints/best.pdparams")
model   =   AutoModelForTokenClassification.from_pretrained(model_name, num_classes=5)
model.load_dict(saved_state)


# Model evaluation
precision, recall, F1 = evaluate(model, test_loader, metric)

# Output models to evaluate results
print(f"[Evaluate]  precision: {precision: .5f}, recall: {recall: .5f}, dev score: {F1:.5f}")
```

The output results are:

```
[Evaluate] precision: 0.97168, recall: 0.96239, dev score: 0.96701.
```

7.1.8 Model Prediction

Given a text to be segmented, such as "Police Station police in key township enterprise conduct mobile population registration," the segmentation results through model prediction will be obtained. Codes are implemented as follows:

```
from seqeval.metrics.sequence_labeling import get_entities

def parsing_label_sequence(tokens, label_sequence):
    """
Obtain the corresponding word segmentation sequences according to the predicted tag sequences

    Input:
        - tokens: use the token sequences after being tokenized
        - label_sequence: predicted word position labeling sequences
    """
    prev = 0
    words = []
    items = get_entities(label_sequence, suffix=False)
    for name, start, end in items:
        if prev != start:
            words.extend(tokens[prev:start])
        words.append("".join(tokens[start:end+1]))
        prev = end + 1

    return words

def infer(model, text, tokenizer, id2label):
    """
    Model prediction function

    Input:
        - model: model examples predicted by Chinese word segmentation
        - text: input texts to be segmented
        - tokenizer: the tokenizer example of BERT
        id2label: the dictionary used to change word position labels into "label"
    """

    model.eval()

    # Data processing
    encoded_inputs = tokenizer(text, max_seq_len=max_seq_len)
```

```
    # Construct data of input models
    input_ids = paddle.to_tensor(encoded_inputs["input_ids"], dtype="int
64").unsqueeze(0)
    token_type_ids = paddle.to_tensor(encoded_inputs["token_type_ids"],
dtype="int64").unsqueeze(0)


    # Compute emission scores
    logits = model(input_ids=input_ids, token_type_ids=token_type_ids)
    predictions = logits.argmax(axis=-1).tolist()[0]
    label_sequence = [id2label[label_id] for label_id in predictions[1:-
1]]


    # Analyze the label with the highest score
    tokens = tokenizer.convert_ids_to_tokens(input_ids[0].tolist()[1:-
1])
    words = parsing_label_sequence(tokens, label_sequence)

    print("The tokenized sequence:", " | ".join(words))

text = "派出所民警在重点乡镇企业进行流动人口登记"
id2label = {v:k for k, v in label2id.items()}
infer(model, text, tokenizer, id2label)
```

The output result is:

```
The tokenized sequence: 派出所 | 民警 | 在 | 重点 | 乡镇企
业 | 进行 | 流动人口 | 登记.
```

7.2 The Implementation of Sub-word Segmentation Based on BPE Algorithm

7.2.1 Task Objectives

As mentioned at the beginning of this Chapter, the sub-word is the granularity between word and character, and it is the fragment of word. BPE is a character compression algorithm that calculates the number of occurrences of two characters next to each other and merges two characters next to each other with the highest number of occurrences to solve the problem of out-of-vocabulary.

Since sub-words can effectively reduce the size of the word list, and their frequency of occurrence is significantly higher than that of words. So in a sense, sub-words can improve the accuracy of semantic representation. From the perspective of form, sub-words are not in line with human language expression habits, and it is only a segmentation result obtained based on the statistical data in the corpus. Figure 7.6 shows the process of sub-word segmentation based on the BPE algorithm.

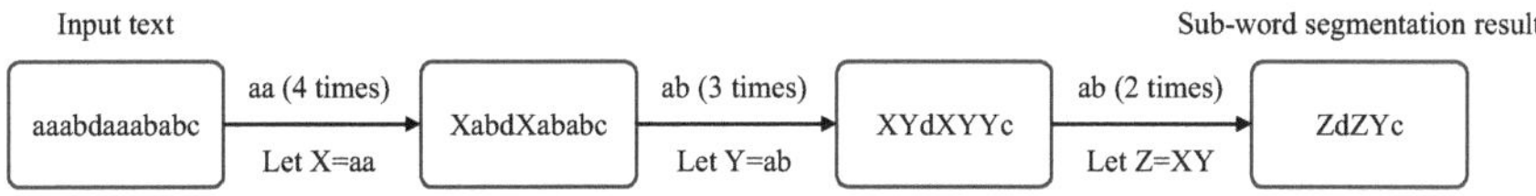

Fig. 7.6 Schematic diagram of BPE algorithm

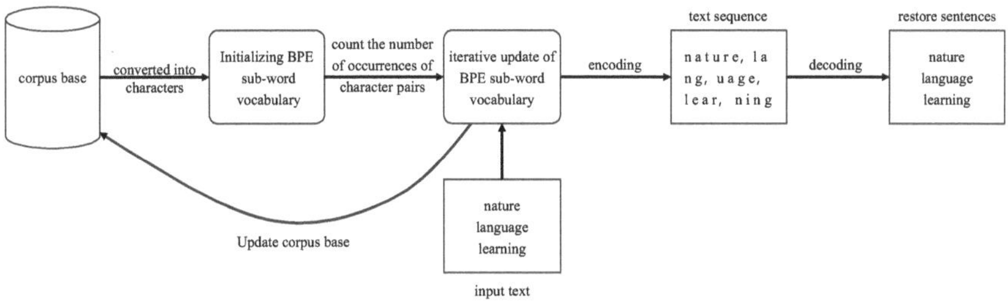

Fig. 7.7 Text segmentation process based on the BPE algorithm

As shown in Fig. 7.6, the BPE sub-word segmentation process is quite different from the Chinese word segmentation based on BERT introduced in Sect. 7.1.

7.2.2 The Implementation Ideas and Processes

The process of segmenting text using the BPE algorithm is shown in Fig. 7.7, which consists of the following two steps:

(1) Building BPE sub-word vocabulary to update the corpus, converting the original corpus into the character form, and initializing BPE sub-word vocabulary. Meanwhile, continuously counting and merging high-frequency character pairs next to each other, and updating BPE sub-word vocabulary and the corpus.
(2) Complementation of the text segmentation based on the BPE sub-word vocabulary. The given text sequence is segmented according to the BPE sub-word vocabulary output in step (1), and then the segmented text sequence is obtained. Then, the text sequence is decoded to restore the original sentence.

7.2.3 Build BPE Vocabulary

Sub-word vocabulary construction is the core of the BPE algorithm. The following examples will introduce its implementation principles and corresponding codes.

1. **Building and Initializing BPE Sub-word Vocabulary**

Figure 7.8a shows a simple corpus example consisting of 5 times of "d e e p", 7 times of "l e a r n i n g", 6 times of "n a t u r a l", 3 times of "l a n g u a g e" and 7 times of "p r o c e s s i n g". The symbol '</w>' is the ending symbol of a word. Then the words in the corpus are converted into characters to obtain the initialized BPE sub-word vocabulary as shown in Fig. 7.8b.

2. **Update the BPE Sub-word vocabulary and Corpus**

The updating processes of the BPE sub-word vocabulary and corpus are as follows:

(1) The first time of updating: count the number of occurrences of consecutive character pairs, merge the character pairs with the highest number of occurrences into new sub-words, and update the vocabulary and corpus. The updating process and results are shown in Fig. 7.9. In the first time of updating, the consecutive character pairs 'n' and 'g' appear 17 times (7 + 3 + 7 = 17), which is the highest number of occurrences, so they are merged into a new word 'ng'.

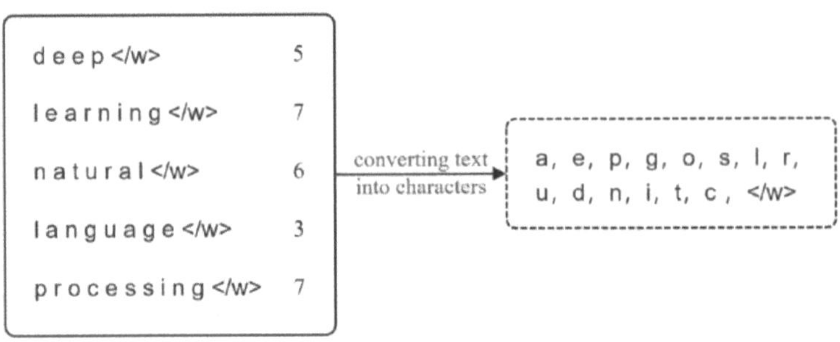

Fig. 7.8 Initialize the BPE sub-word vocabulary

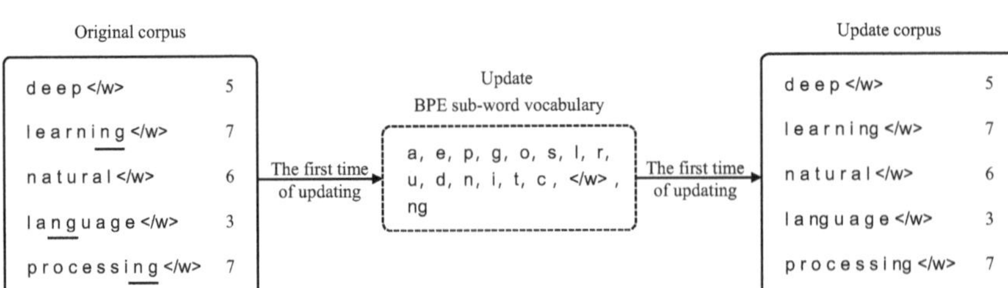

Fig. 7.9 The first updating process and results

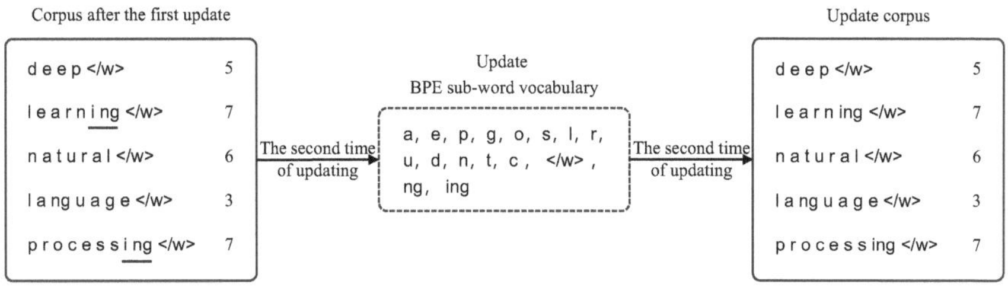

Fig. 7.10 The second updating process and results

(2) The second time of updating: The consecutive characters 'i' and 'ng' appear 14 times (7 + 7 = 14), the most frequently, and they are merged into a new word 'ing'. The updating process and results are shown in Fig. 7.10.
(3) The third time of updating: The consecutive characters 'ing' and '</w>' appear 14 times (7 + 7 = 14), which is the highest number of occurrences. They are merged into 'ing </w>'. At the same time, since 'ing' no longer exists in the corpus, it is deleted from the vocabulary. The updating process and results are shown in Fig. 7.11.
(4) Repeat steps (1)–(3) above until the scale of BPE sub-word vocabulary reaches the preset value, or reaches the set number of mergers, or the number of occurrences of the most frequent character pairs next to each other in the corpus is less than the specified threshold, such as 2.

From the above explanation, it can be seen that word merging is performed within the word itself, and does not merge cross-word borders.

The following describes the code implementation process of the above examples.

(1) Define the function get_subwords to convert words in the corpus into the form of character and store them in the BPE sub-word vocabulary. In the codes, the initialized sub-word vocabulary bpe_vocab is obtained. The codes are implemented as follows:

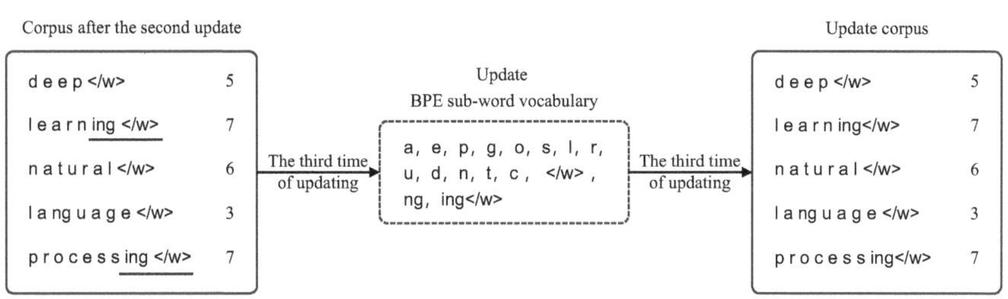

Fig. 7.11 The third updating process and results

```
import collections

def get_subwords(data):


    # Count the sub-words and their corresponding word frequencies
    subwords = collections.defaultdict(int)
    for word, freq in data.items():
        for subword in word.split():
            subwords[subword] += freq

    return subwords

train_data = {'d e e p </w>': 5, 'l e a r n i n g </w>': 7, 'n a t u r a l
 </w>': 6, 'l a n g u a g e </w>': 3,'p r o c e s s i n g </w>':7}
subwords = get_subwords(train_data)


# Get the initialized sub-word word list
bpe_vocab = set(subwords.keys())
print("word list: ", bpe_vocab)
```

The output result is:

```
Vocabulary: {'n', '</w>', 'g', 'c', 'd', 'u', 's', 'i', 'p', 'a',
'o', 't', 'l', 'e', 'r'}.
```

(2) When finishing the BPE sub-word vocabulary, it is necessary to count the occurrences of character pairs next to each other at the same time in order to obtain the highest frequency word pairs. Codes are implemented as follows:

```
def get_pair_with_frequency(data):


    # Count the character pairs and their occurrence times in the corpus
    pairs = collections.defaultdict(int)
    for word, freq in data.items():
        sub_words = word.split()
        for i in range(len(sub_words)-1):
            pair = (sub_words[i],sub_words[i+1])
            pairs[pair] += freq
    return pairs

pairs = get_pair_with_frequency(train_data)
print("character pairs: ", pairs)
best_pair = max(pairs, key=pairs.get)
print("Most frequent character pair: ", best_pair)
```

The output result is:

```
character pairs: defaultdict(<class 'int'>, {('d', 'e'): 5, ('e',
'e'): 5, ('e', 'p'): 5, ('p', '</w>'): 5, ('l', 'e'): 7, ('e', 'a'):
7, ('a', 'r'): 7, ('r', 'n'): 7, ('n', 'i'): 7, ('i', 'n'): 14, ('n',
```

```
'g'): 17, ('g', ' < /w > '): 14, ('n', 'a'): 6, ('a', 't'): 6, ('t',
'u'): 6, ('u', 'r'): 6, ('r', 'a'): 6, ('a', 'l'): 6, ('l', '</w>'):
6, ('l', 'a'): 3, ('a', 'n'): 3, ('g', 'u'): 3, ('u', 'a'): 3, ('a',
'g'): 3, ('g', 'e'): 3, ('e', '</w>'): 3, ('p', 'r'): 7, ('r', 'o'):
7, ('o', 'c'): 7, ('c', 'e'): 7, ('e', 's'): 7, ('s', 's'): 7, ('s',
'i'): 7}).
```

```
Most frequent character pair: ('n', 'g').
```

(3) Merge the character pairs with the highest frequency of occurrence into new sub-words, and update the corpus. Codes are implemented as follows:

```
def merge_data_with_pair(pair, data):
    """

    Merge the most frequent character pairs in the corpus

    Input:

        - pair: the most frequent character pairs

        - data: in the form of dictionary , the corpus updated in the last updating
    """
    result = {}
    bigram = re.escape(' '.join(pair))
    p = re.compile(r'(?<!\S)' + bigram + r'(?!\S)')
    for word in data:
        merged_word = p.sub(''.join(pair), word)
        result[merged_word] = data[word]
    return result

train_data = merge_data_with_pair(best_pair, train_data)
print("Corpus: ", train_data)
```

The output result is:

Corpus: {'d e e p </w>': 5, 'l e a r n i ng </w>': 7, 'n a t u r a l </w>': 6, 'l a ng u a g e </w>': 3, 'p r o c e s s i n g </w>': 7}.

(4) Define the function build_vocab to build the vocabulary, and update steps (1)–(3). Codes are implemented as follows:

```
def build_vocab(train_data, num_merges):
    """

    Build a Vocabulary based on the training corpus

    Input:

        - train_data: in the form of dictionary , the corpus updated in the last updating

        - num_merges: the time of updating
    """


    # Initialize BPE Vocabulary
    subwords = get_subwords(train_data)
    bpe_vocab = set(subwords.keys())
    print(bpe_vocab, len(bpe_vocab))
    i = 1

    # Update BPE Vocabulary
    for _ in range(num_merges):

        # Count the number of occurrences of character pairs next to each other in the corpus
        pairs = get_pair_with_frequency(train_data)

        #   Select the character pairs with the highest frequency of occurrence. If pairs are empty
or the frequency of occurrence of the character pairs is 1, stop
        if not pairs:
            break
        best_pair = max(pairs, key=pairs.get)
        if pairs[best_pair] == 1:
            break

        #   Update the corpus
        train_data = merge_data_with_pair(best_pair, train_data)

        #   Add new sub-words into the Vocabulary
        merged_word = "".join(best_pair)
        bpe_vocab.add(merged_word)

        #   Delete words that do not exist in the corpus from the word list
        subwords = get_subwords(train_data)
        if best_pair[0] not in subwords:
            bpe_vocab.remove(best_pair[0])
        if best_pair[1] not in subwords:
            bpe_vocab.remove(best_pair[1])

        print("Iter - {}, the most frequently occurring character pair: {}".format
(i, best_pair))
        print("training data: ", train_data)
        print("Vocabulary: {}, {}\n".format(len(bpe_vocab), bpe_vocab))
        i += 1
    return bpe_vocab

# Set the number of merging character pairs
num_merges = 14
```

```
train_data = {'d e e p </w>': 5, 'l e a r n i n g </w>': 7, 'n a t u r a
l </w>': 6, 'l a n g u a g e </w>': 3,'p r o c e s s i n g </w>':7}

bpe_vocab = build_vocab(train_data, num_merges)
print("Vocabulary: ", bpe_vocab)
```

The output results are:

```
{'n', '</w>', 'g', 'c', 'd', 'u', 's', 'i', 'p', 'a', 'o', 't', 'l', 'e',
'r'} 15
Iter - 1, the most frequently occurring character pair: ('n', 'g')
Training data: {'d e e p </w>': 5, 'l e a r n i ng </w>': 7, 'n a t u r a l
 </w>': 6, 'l a ng u a g e </w>': 3, 'p r o c e s s i ng </w>': 7}
Vocabulary: 16, {'n', '</w>', 'g', 'c', 'd', 'u', 's', 'i', 'p', 'a', 'o',
't', 'l', 'ng', 'e', 'r'}

Iter - 2, the most frequently occurring character pair: ('i', 'ng')
Training data: {'d e e p </w>': 5, 'l e a r n ing </w>': 7, 'n a t u r a l
</w>': 6, 'l a ng u a g e </w>': 3, 'p r o c e s s ing </w>': 7}
Vocabulary: 16, {'n', '</w>', 'g', 'c', 'ing', 'd', 'u', 's', 'p', 'a', 'o
', 't', 'l', 'ng', 'e', 'r'}

Iter - 3, the most frequently occurring character pair: ('ing', '</w>')
Training data: {'d e e p </w>': 5, 'l e a r n ing</w>': 7, 'n a t u r a l <
/w>': 6, 'l a ng u a g e </w>': 3, 'p r o c e s s ing</w>': 7}
Vocabulary: 16, {'n', '</w>', 'g', 'c', 'd', 'u', 's', 'p', 'a', 'o', 't',
'l', 'ng', 'ing</w>', 'e', 'r'}

Iter - 4, the most frequently occurring character pair: ('l', 'e')
Training data: {'d e e p </w>': 5, 'le a r n ing</w>': 7, 'n a t u r a l </
w>': 6, 'l a ng u a g e </w>': 3, 'p r o c e s s ing</w>': 7}
Vocabulary: 17, {'</w>', 'c', 's', 'ng', 'ing</w>', 'o', 'g', 'd', 'u', 'a
', 't', 'le', 'n', 'l', 'e', 'r', 'p'}

Iter - 5, the most frequently occurring character pair: ('le', 'a')
Training data: {'d e e p </w>': 5, 'lea r n ing</w>': 7, 'n a t u r a l </
w>': 6, 'l a ng u a g e </w>': 3, 'p r o c e s s ing</w>': 7}
Vocabulary: 17, {'</w>', 'c', 's', 'ng', 'ing</w>', 'o', 'g', 'd', 'u', 'a
', 't', 'lea', 'n', 'l', 'e', 'r', 'p'}

Iter - 6, the most frequently occurring character pair: ('lea', 'r')
Training data: {'d e e p </w>': 5, 'lear n ing</w>': 7, 'n a t u r a l </w>
': 6, 'l a ng u a g e </w>': 3, 'p r o c e s s ing</w>': 7}
Vocabulary: 17, {'</w>', 'c', 's', 'ng', 'ing</w>', 'o', 'g', 'd', 'u', 'a
', 't', 'n', 'l', 'e', 'r', 'p', 'lear'}

Iter - 7, the most frequently occurring character pair: ('lear', 'n')
Training data: {'d e e p </w>': 5, 'learn ing</w>': 7, 'n a t u r a l </w>
': 6, 'l a ng u a g e </w>': 3, 'p r o c e s s ing</w>': 7}
Vocabulary: 17, {'</w>', 'c', 's', 'ng', 'ing</w>', 'o', 'g', 'd', 'u', 'a
', 't', 'learn', 'n', 'l', 'e', 'r', 'p'}

Iter - 8, the most frequently occurring character pair: ('learn', 'ing</w>')
```

```
Training data: {'d e e p </w>': 5, 'learning</w>': 7, 'n a t u r a l </w>':
 6, 'l a ng u a g e </w>': 3, 'p r o c e s s ing</w>': 7}
Vocabulary: 17, {'</w>', 'c', 's', 'ng', 'ing</w>', 'o', 'g', 'learning</w>
', 'd', 'u', 'a', 't', 'n', 'l', 'e', 'r', 'p'}

Iter - 9, the most frequently occurring character pair: ('p', 'r')
Training data: {'d e e p </w>': 5, 'learning</w>': 7, 'n a t u r a l </w>':
 6, 'l a ng u a g e </w>': 3, 'pr o c e s s ing</w>': 7}
Vocabulary: 18, {'</w>', 'c', 's', 'ng', 'ing</w>', 'o', 'g', 'learning</w>
', 'd', 'u', 'a', 't', 'n', 'l', 'pr', 'e', 'r', 'p'}

Iter - 10, the most frequently occurring character pair: ('pr', 'o')
Training data: {'d e e p </w>': 5, 'learning</w>': 7, 'n a t u r a l </w>':
 6, 'l a ng u a g e </w>': 3, 'pro c e s s ing</w>': 7}
Vocabulary: 17, {'</w>', 'c', 's', 'ng', 'ing</w>', 'g', 'learning</w>', 'd
', 'u', 'a', 't', 'n', 'pro', 'l', 'e', 'r', 'p'}

Iter - 11, the most frequently occurring character pair: ('pro', 'c')
Training data: {'d e e p </w>': 5, 'learning</w>': 7, 'n a t u r a l </w>':
 6, 'l a ng u a g e </w>': 3, 'proc e s s ing</w>': 7}
Vocabulary: 16, {'</w>', 's', 'ng', 'ing</w>', 'g', 'learning</w>', 'd', 'u
', 'a', 't', 'n', 'proc', 'l', 'e', 'r', 'p'}

Iter - 12, the most frequently occurring character pair: ('proc', 'e')
Training data: {'d e e p </w>': 5, 'learning</w>': 7, 'n a t u r a l </w>':
 6, 'l a ng u a g e </w>': 3, 'proce s s ing</w>': 7}
Vocabulary: 16, {'</w>', 's', 'ng', 'ing</w>', 'g', 'learning</w>', 'd', 'u
', 'a', 't', 'n', 'l', 'proce', 'e', 'r', 'p'}

Iter - 13, the most frequently occurring character pair: ('proce', 's')
Training data: {'d e e p </w>': 5, 'learning</w>': 7, 'n a t u r a l </w>':
 6, 'l a ng u a g e </w>': 3, 'proces s ing</w>': 7}
Vocabulary: 16, {'</w>', 's', 'ng', 'ing</w>', 'g', 'learning</w>', 'd', 'u
', 'a', 't', 'n', 'l', 'e', 'r', 'proces', 'p'}

Iter - 14, the most frequently occurring character pair: ('proces', 's')
Training data: {'d e e p </w>': 5, 'learning</w>': 7, 'n a t u r a l </w>':
 6, 'l a ng u a g e </w>': 3, 'process ing</w>': 7}
Vocabulary: 15, {'</w>', 'ng', 'ing</w>', 'g', 'learning</w>', 'process', '
d', 'u', 'a', 't', 'n', 'l', 'e', 'r', 'p'}

Vocabulary: {'</w>', 'ng', 'ing</w>', 'g', 'learning</w>', 'process', 'd',
 'u', 'a', 't', 'n', 'l', 'e', 'r', 'p'}
```

7.2.4 Text Sub-word Segmentation

The purpose of sub-word segmentation is to segment any input text sequences based on the BPE vocabulary constructed in Sect. 7.2.3. Here, the greedy algorithm is used. First, the BPE vocabulary is sorted by length from the longest to the shortest. Then traverse the sub-words in the vocabulary in the order from front to back for the input text sequences. The entire process can be seen as a maximum matching segmentation method. Its implementation steps are as follows:

(1) The word is segmented based on the sub-word position, resulting in three parts: the sub-word, the part before it, and the part after it. For the part before the sub-word and the part after it, the segmenter continues to traverse using the same idea until all the word sub-strings are found.
(2) If there are still some characters that have not been segmented after the traversal of the sub-word vocabulary is complete, they are replaced by the special token [UNK].

Assuming the BPE sub-word vocabulary is ['</w>', 'ng', 'ing </w>', 'g', 'learning </w>', 'process', 'd', 'u', 'a', 't', 'n', 'l', 'e', 'r', 'p'], the text sequence to be segmented is "natural language processing." The word "processing" is segmented according to the BPE sub-wordvocabulary. The sequence in it will be first divided into 'process', and then continue to traverse in this way. The latter part of sub-words is segmented into 'ing </w>'. Now the word 'processing' is segmented into 'process' and 'ing </w>'.

The following describes the implementation process of segmentation. First, let's define the function tokenize_word to segment words. Codes are implemented as follows:

```
import re

def tokenize_word(word, sorted_vocab, unknown_token='<unk>'):
    """

    Segment a word using the BPE algorithm

    Input:

        - word: words to be segmented

        - sorted_vocab: sub-word dictionary after sequencing

        - unknown_token: substitute characters of sub-words that cannot be replaced
    """

    # If the input word is empty, return an empty list
    if word == "":
        return []

    # If the vocabulary is empty, use [UNK] to replace the input word
    if sorted_vocab == []:
        return [unknown_token] + len(string)

    word_tokens = []

    # Traverse the word list to segment words
    for i in range(len(sorted_vocab)):
        token = sorted_vocab[i]

        # Define the regular expression based on the token, and change the full stop included in
        the token to [.]
        token_reg = re.escape(token.replace('.', '[.]'))

        # Traverse the current word list to find the start and end positions of the matching token
        matched_positions = [(m.start(0), m.end(0)) for m in re.finditer
(token_reg, word)]
```

```
            # If the current token does not match the corresponding sub-words, skip
            if  len(matched_positions) == 0:
                continue


            # Obtain the starting position of the matched sub-words
            end_positions = [matched_position[0] for matched_position in matc
hed_positions]
            start_position = 0

            for end_position in end_positions:
                subword = word[start_position: end_position]
                word_tokens += tokenize_word(subword, sorted_vocab[i+1:], unk
nown_token)
                word_tokens += [token]
                start_position = end_position + len(token)

            # Match the remaining sub-words
            word_tokens += tokenize_word(word[start_position:], sorted_vocab
[i+1:], unknown_token)
            break
    else:

            # If the sub-word is not matched, it is mapped to [UNK]
            word_tokens = [unknown_token] * len(word)

    return word_tokens
```

And then we define the function tokenize to segment texts.

```
def tokenize(text, bpe_vocab):



    # Use BPE to segment the input sentences, and sort the sub-word vocabulary according to the
length of the sub-words
    sorted_vocab = sorted(bpe_vocab, key=lambda subword: len(subword),
reverse=True)
    print("sentences to be segmented: ", text)
    tokens = []
    for word in text.split():
        word = word + "</w>"
        word_tokens = tokenize_word(word, sorted_vocab,
unknown_token='[UNK]')
        tokens.extend(word_tokens)

    return tokens

text = "natural language processing"
tokens = tokenize(text, bpe_vocab)
print("vocabulary: ", bpe_vocab)
print("segmentation results: ", tokens)
```

The output results are:

Sentences to be segmented: natural language processing.

vocabulary: {'</w>', 'ng', 'ing </w>', 'g', 'learning </w>', 'process', 'd', 'u', 'a', 't', 'n', 'l', 'e', 'r', 'p'}.

Segmentation results: ['n', 'a', 't', 'u', 'r', 'a', 'l', '</w>', 'l', 'a', 'ng', 'u', 'a', 'g', 'e', '</w>', 'process', 'ing </w>'].

7.2.5 *Corpus Restoration*

How to restore the segmented texts? The '\ <w>' at the end of the word is crucial. The basic idea is to merge the sub-words until '</w>' appears and a complete word is decoded. Suppose we use BPE to encode the sub-word vocabulary, the segmented text sequence is ['the </w>', 'deep </w>', 'lear', 'ning </w>'], and the corresponding text restoration result is ['the', 'deep', 'learning']. Codes are implemented as follows:

```
def restore(tokens):


    # Restore the result after word segmentation to the original sentence
    text = []
    word = []
    for token in tokens:
        if token[-4:] == "</w>":
            if token != "</w>":
                word.append(token[:-4])
            text.append("".join(word))
            word.clear()
        else:
            word.append(token)
    return text

tokens = ["the</w>", "deep</w>", "lear", "ning</w>"]
text = restore(tokens)
print("restoration result: ", text)
```

The output result is:

Restore the result: ['the', 'deep', 'learning'].

During the segmentation process using the BPE sub-word vocabulary, it is necessary to traverse the entire sub-word vocabulary, which is a time-consuming process. In practical applications, readers can set up a cache to record the segmentation results of commonly used words, thereby speeding up the segmentation process.

7.3 Exercises

(1) In the experiments introduced in this chapter, we used the pre-training model BERT to implement the task of Chinese word segmentation. Please try to replace the model structure, such as using Bi-LSTM + CRF to complete the Chinese word segmentation task. For implementing the codes of Bi-LSTM + CRF, please refer to Chap. 9.

In the experiments described in this Chapter, we usually perform sub-word segmentation on Chinese texts based on the word segmentation results: first it uses a word segmentation tool to segment the given texts into words, and then it applies the BPE algorithm to the segmentation results. Please refer to Sect. 7.2 for more information on Chinese sub-word segmentation or rely on existing word segmentation tools to implement the BPE segmentation algorithm for Chinese texts.

Chapter 8
Text Sentiment Classification

Abstract This chapter introduces text sentiment classification, an essential task in natural language processing that aims to identify emotional polarity in text data. It first outlines the foundations of text classification and distinguishes between rule-based, statistical, and deep learning methods, focusing particularly on neural network approaches. Using the IMDB movie review dataset, the chapter details the implementation of sentiment analysis with both LSTM and BERT models, covering data preprocessing, model construction, training, evaluation, and prediction. Experimental results demonstrate that while LSTM achieves strong baseline performance, BERT significantly improves accuracy, highlighting the effectiveness of pre-trained language models. The chapter further extends to aspect-based sentiment analysis with BERT, enabling fine-grained extraction of aspect–opinion–sentiment triples for more nuanced applications such as product reviews and service evaluation.

Text classification is an important application task of natural language processing, which usually includes text topic classification, sentiment or emotion classification, and spam identification. The text here can be a sentence, a paragraph, or an article. The purpose of text topic classification is to automatically identify the topic of a given text. The purpose of sentiment classification and emotion classification is to automatically identify the sentiment or emotional polarity of a given text. Spam identification is to automatically determine whether a given email belongs to spam information.

Whether it is text topic classification, emotion or sentiment classification, or spam classification, they all belong to the task of pattern classification. The basic methods can be roughly divided into two types: one is knowledge engineering method (prior knowledge based on templates or rules, etc.), statistical learning methods, and other is deep learning method. At present, the text classification method based on deep learning is most widely used. Therefore, this Chapter introduces the sentiment classification method based on LSTM and based on the pre-trained language model BERT respectively.

C. Zong et al., *Natural Language Processing and Large Language Models*,
https://doi.org/10.1007/978-981-92-0682-7_8

It should be noted that text classification tasks can be further subdivided into sentence-level, discourse-level, or multi-document-level, and mono-lingual or cross-lingual and other possible types, depending on the granularity and languages of the input texts. From the perspective of neural network methods, although these tasks are diverse, the neural network structures, objective functions, and training methods used for different text classification tasks are highly similar. Therefore, this Chapter only uses sentence-level sentiment classification as an example for method introduction. Other classification tasks only require replacing the input texts and corresponding labels with the labeled data required for the task.

For the detailed introduction to the basic methods of text classification, please refer to Zong Chengqing et al. (2022).

8.1 Sentiment Analysis Method Based on the LSTM Model

8.1.1 Task Objectives

In natural language, a dialogue or a sentence often contains rich emotions, such as happiness, joy, love, hate, sadness, etc. For enterprises, if they can automatically analyze the emotional tendencies of users based on their comments on their products, it will help them understand consumers' feelings about their products and provide a basis for improving the user experience of their products. It also helps them understand the attitudes of their business partners in a timely manner so as to make better business decisions.

In the sentiment analysis task, emotions are usually divided into three polarities: positive, neutral, and negative. Figure 8.1 shows a simple example of these three emotions.

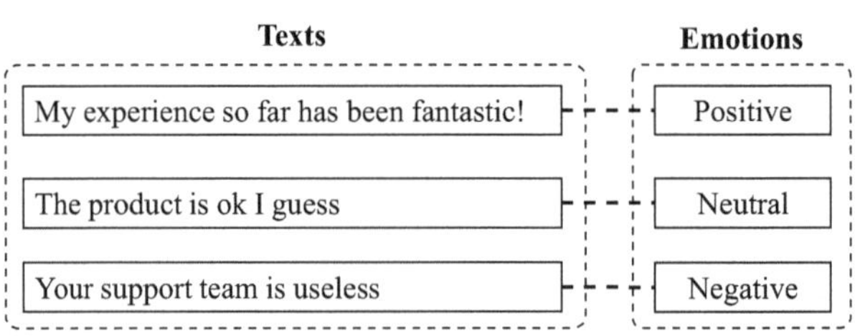

Fig. 8.1 Examples of emotions' categories

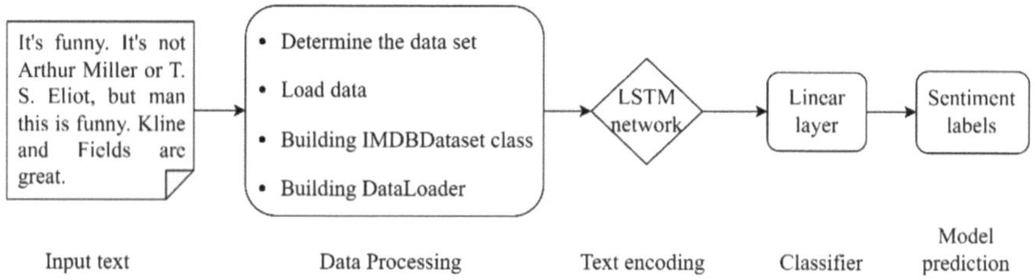

Fig. 8.2 The plan of movie review sentiment analysis

This section takes IMDB movie reviews as an example to introduce how to implement sentiment analysis based on the LSTM model on the PaddlePaddle platform.[1]

8.1.2 *Implementation Ideas and Processes*

The implementation idea of the sentiment analysis method based on the LSTM model is shown in Fig. 8.2. The input of the model is text data, and the output is the sentiment label of the text. During the modeling process, firstly, the input movie review texts need to be processed, and then the LSTM is used to encode the texts to obtain the semantic vector representation of the texts. Finally, the linear layer is used to obtain the score of the text sentiment category label.

According to the above ideas, the main processes of implementing sentiment analysis of review texts based on the PaddlePaddle framework includes the following six steps:

(1) Data processing: process the input text data, including word segmentation, text truncation and padding, word2id, etc., and then assemble the data into a minibatch format and pass it into the LSTM network.
(2) Model construction: according to the agreement, we use the LSTM network to obtain the vector representation of text sequences.
(3) Training configuration: Instantiate the model, select the computing resources (CPU or GPU) for the model, and specify the optimization algorithm for model upgrading.
(4) Model training: train the parameters of the model to achieve optimal results.
(5) Model evaluation: evaluate the trained model and observe the changes in accuracy and loss function.
(6) Model prediction: given a piece of review texts, predict its sentiment label through the model.

The following describes the specific implementation processes of each step.

[1] The IMDB dataset is open-sourced by Kaggle, and the dataset can be obtained at https://www.kaggle.com/c/word2vec-nlp-tutorial/data. Readers can also obtain the dataset on the online course in the AI Studio.

8.1.3 *Data Processing*

Data processing includes the steps of determining the dataset, loading the data, constructing the IMDB Dataset class, and constructing the DataLoader. Finally, the same batch of data is processed into equal-length feature sequences, which are then sent to the LSTM model in batches using the DataLoader.

1. **Determining Datasets**

Before the system is implemented, it is necessary to first clarify the task area, data sources, and the division principle of emotional tendencies in the dataset. According to the previous description, let's use the IMDB dataset in the movie review field as an example for system construction.

When moviegoers comment on a movie, they usually rate it on a scale from 1 to 10. Among them, 1 is the worst rating given by the audience, indicating that they are very dissatisfied with the movie; 10 is the full score, indicating that they are very satisfied with the movie. The IMDB dataset categorizes consumer reviews of movies into positive and negative reviews based on their rating. If the rating is at or above 7, it is considered to be a positive review; if the rating is at or below 4, it is considered to be a negative review. Otherwise, the rating is between 4 and 7, it is considered as neutral. Therefore, IMDB is a good classification dataset for sentiment analysis. Figure 8.1 shows three categories of sentiment analysis, namely positive, neutral, and negative. Here, we take the binary classification of sentiment (positive and negative) as an example for introduction.

The original IMDB dataset has 25,000 training and testing samples, each containing a user's real evaluation of a movie and their emotional orientation towards the movie. This practice generates a vocabulary based on all the IMDB dataset, and divides the testing set into two halves of 50% each: the validation and testing sets. Each contains 12,500 comments. Below is a real positive comment from the training set:

> " The only thing serious about this movie is the humor. Well worth the rental price. I'll bet you watch it twice. It's obvious that Sutherland enjoyed his role."

Data Loading

The following codes read the IMDB data and the constructed vocabulary into memory.

```
# Define data and load functions
def load_dataset(path):
    examples = []
    with open(path, "r", encoding="utf-8") as fr:
        for line in fr:
            sentence_label, sentence = line.strip().lower().split("
\t", maxsplit=1)
            examples.append((sentence, sentence_label))
    return examples

train_data = load_dataset("./dataset/train.txt")
dev_data = load_dataset("./dataset/dev.txt")
test_data = load_dataset("./dataset/test.txt")


# Print the first piece of data and view the data format: (sentence, label)
print(train_data[0])
```

The output result is:

```
('it does seem like this film is polarizing us. you either love it or
hate it. i loved it. I agree with the comment(s) that said, you just
gotta "feel" this one. ··· but if you do like what i described, then
you will surely enjoy it.', '1')
```

From the output result, it can be seen that each sample contains two parts: a text sequence and a corresponding sentiment label ("1" indicates positive sentiment, and "0" indicates negative sentiment).

The following codes load the vocabulary into the dictionary word2id_dict.

```
import os


# Loadvocabulary
def load_dict(path):
    assert os.path.exists(path)
    words = []
    with open(path, "r", encoding="utf-8") as f:
        words = f.readlines()
        words = [word.strip() for word in words if word.strip()]
    word2id = dict(zip(words, range(len(words))))
    return word2id

word2id_dict= load_dict("./dataset/vocab.txt")

for idx, (word, word_id) in enumerate(word2id_dict.items()):
    print("word: {}, word_id: {}".format(word, word_id))
    if idx > 7:
        break
```

The output result is:

```
word: [PAD], word_id: 0
word: [UNK], word_id: 1
word: the, word_id: 2
```

```
word: a, word_id: 3
word: and, word_id: 4
word: of, word_id: 5
word: to, word_id: 6
word: is, word_id: 7
word: in, word_id: 8.
```

There are two special characters in the vocabulary "[PAD]" and "[UNK]." The role of [PAD] is minibatch word padding. The current neural network generally uses the minibatch method, that is, the model trains multiple sentences at a time. Considering that the length of sentences in the same batch is often different, special characters [PAD] are padded behind the shorter sentences to force the length of sentences in the same batch to be the same. For specific implementation methods, please refer to part 4 "Constructing DataLoader" later in this section.

Considering the computing complexity and implementation efficiency of neural network transformation, when constructing the vocabulary, not all words in the corpus are retained, but only the top N words with the highest frequency are selected, and the low-frequency words are replaced with [UNK].

2. **Construction of the IMDBDataset Class**

The purpose of constructing the IMDBDataset class is for data management. It inherits from the paddle.io.DataSet class. The IMDBDataset class includes the following two operations:

(1) Word segmentation: the model cannot directly process text data. The conventional approach is to first segment the text data into words, and then map each word to an index (ID number) in the dictionary to facilitate the model's subsequent search for the vector representation corresponding to the word based on this ID number. Since the IMDB dataset is in English, this practice directly uses spaces as word separators.
(2) word2id: define the convert_tokens_to_ids method in the IMDBDataset class to convert words into index numbers (IDs) in the word table. Use the word2id_dict to map each word in the sequence to a corresponding index number (ID) for further conversion into a word vector. If some words in the text are not included in the vocabulary, they are replaced by [UNK] by default.

The codes to construct the IMDBDataset class are as follows:

```
from paddle.io import Dataset

class IMDBDataset(Dataset):
    def __init__(self, examples, word2id_dict):
        super(IMDBDataset, self).__init__()

        # Construct a dictionary to convert words into numbers for dictionary indexes.
        self.word2id_dict =  word2id_dict
        self.examples = self.convert_tokens_to_ids(examples)

    def convert_tokens_to_ids(self, examples):
        tmp_examples = []
        for idx, example in enumerate(examples):
            seq, label = example

        # Map words to the ID of dictionary indexes, and replace words that do not appear
in the vocabulary with ID corresponding to [UNK].
            seq = [self.word2id_dict.get(word,
self.word2id_dict['[UNK]']) for word in seq.split(" ")]
            label = int(label)
            tmp_examples.append([seq, label])
        return tmp_examples

    def __getitem__(self, idx):
        seq, label = self.examples[idx]
        return seq, label

    def __len__(self):
        return len(self.examples)


# Instantiate the Dataset
train_set = IMDBDataset(train_data, word2id_dict)
dev_set = IMDBDataset(dev_data, word2id_dict)
test_set = IMDBDataset(test_data, word2id_dict)

print('Number of samples of the training set: ', len(train_set))
print('Number of samples of the validation set: ', len(dev_set))
print('Number of samples of the test set: ', len(test_set))
print('Samples: ', train_set[4])
```

The output result is:
Number of samples of the training set: 25000
Number of samples of the validation set: 12500
Number of samples of the test set: 12500
Samples: ([2, 976, 5, 32, 6860, 618, 7673, 8, 2, 13073, 2525, 724, 14, 22837, 18, 164, 416, 8, 10, 24, 701, 611, 1743, 7673, 7, 3, 56391, 21652, 36, 271, 3495, 5, 2, 11373, 4, 13244, 8, 2, 2157, 350, 4, 328, 4118,

```
12, 48810, 52, 7, 60, 860, 43, 2, 56, 4393, 5, 2, 89,
4152, 182, 5, 2, 461, 7, 11, 7321, 7730, 86, 7931, 107,
72, 2, 2830, 1165, 5, 10, 151, 4, 2, 272, 1003, 6, 91,
2, 10491, 912, 826, 2, 1750, 889, 43, 6723, 4, 647, 7,
2535, 38, 39222, 2, 357, 398, 1505, 5, 12, 107, 179, 2,
20, 4279, 83, 1163, 692, 10, 7, 3, 889, 24, 11, 141,
118, 50, 6, 28642, 8, 2, 490, 1469, 2, 1039, 98975,
24541, 344, 32, 2074, 11852, 1683, 4, 29, 286, 478,
22, 823, 6, 5222, 2, 1490, 6893, 883, 41, 71, 3254,
38, 100, 1021, 44, 3, 1700, 6, 8768, 12, 8, 3, 108,
11, 146, 12, 1761, 4, 92295, 8, 2641, 5, 83, 49, 3866,
5352], 0)
```

From the output results, it can be seen that each word in the text has been converted into an ID number in the dictionary after being segmented.

3. **Construction of DataLoader**

When training a model, it is common to pass data into the model in batches for separate training. Each batch of data is passed into the model as a minibatch for processing. Each batch of data contains two parts: text data and corresponding sentiment labels. Readers can use the paddle.io.DataLoader in the PaddlePaddle framework to implement batch loading and update data.

By observing the characteristics of each batch of data, it can be found that a batch of data usually contains several pieces of texts, each of which has a different length. Text truncation and text padding techniques are needed to change the text in each batch of data to a fixed length.

(1) Text truncation: set a maximum sequence length max_seq_len during training to truncate excessively long texts and avoid affecting the overall model training effect due to excessively long data.
(2) Text padding: calculate the maximum sequence length max_seq_len of the mini-batch data. When the sequence length is less than max_seq_len, [PAD] is used to pad the sequence to the length of max_seq_len.

Figure 8.3 shows examples of text truncation and text padding when the batch size is 2. The above two techniques can ensure that the length of each batch of data in the updating of DataLoader is fixed.

We define the collate_fn function for text truncation and padding, which can be passed into DataLoader as the callback function. DataLoader uses collate_fn for data processing before returning each batch of data and returns the processed text data and corresponding labels. Codes to implement the collate_fn function are as follows:

Fig. 8.3 Text truncation and padding

```
import paddle
from functools import partial

# The maximum sequence length max_seq_len is set to 256
def collate_fn(batch_data, pad_val=0, max_seq_len=256):
    seqs, seq_lens, labels = [], [], []
    max_len = 0
    for example in batch_data:
        seq, label = example

        # Truncate the text sequences
        seq = seq[:max_seq_len]

        # Data is truncated and saved in seqs
        seqs.append(seq)
        seq_lens.append(len(seq))
        labels.append(label)

        # Maximum length of saved sequences
        max_len = max(max_len, len(seq))

    # Pad the data sequences to the maximum length
    for i in range(len(seqs)):
        seqs[i] = seqs[i] + [pad_val] * (max_len - len(seqs[i]))

    return paddle.to_tensor(seqs), paddle.to_tensor(labels), paddl
e.to_tensor(seq_lens)
```

The collate_fn function contains two key parameters: pad_val and max_seq_len, which can be fixed using the partial function, and then collate_fn is passed as a callback function to the DataLoader. In this practice, the maximum sequence length (max_seq_len) is set to 256, and the batch size is set to 128. Readers can also try other parameter combinations.

When updating data in batches using DataLoader, the number of samples in the last batch may not meet the batch size limit. Readers can use the drop_last parameter to determine whether to discard the last batch of data. Codes are implemented as follows:

```
max_seq_len = 256
batch_size = 128

collate_fn = partial(collate_fn, pad_val=word2id_dict["[PAD]"], max_
seq_len=max_seq_len)

train_loader = paddle.io.DataLoader(train_set, batch_size=batch_siz
e, shuffle=True, drop_last=False, collate_fn=collate_fn)
dev_loader = paddle.io.DataLoader(dev_set, batch_size=batch_size, sh
uffle=False, drop_last=False, collate_fn=collate_fn)
test_loader = paddle.io.DataLoader(test_set, batch_size=batch_size,
shuffle=False, drop_last=False, collate_fn=collate_fn)
```

8.1.4 Model Construction

The LSTM model processes each input batched data in turn according to the sequential relationship. First, all the words in the text sequences are passed into the LSTM model, and the hidden state output by the LSTM model can be seen as a state vector that integrates all previous words. Therefore, this state variant can also be seen as a semantic vector representation of the text sequence. Then, the semantic vector is passed into the linear layer to obtain scores for each sentiment category (positive or negative) of the input text, as shown in Fig. 8.4.

For a more detailed introduction to the construction of an emotion classification model, please refer to Chap. 7 of Zong Chengqing et al. (2022).

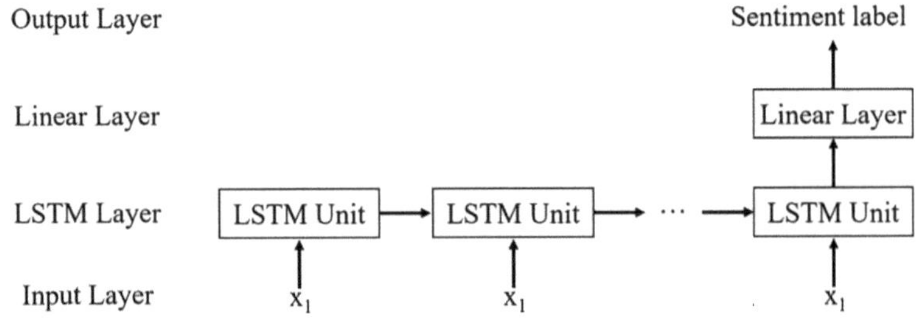

Fig. 8.4 The construction of the LSTM network

In practice, readers can use the paddle.nn.LSTM API to call the LSTM model, which is more concise and efficient. Codes are implemented as follows:

```
import paddle
import paddle.nn.functional as F


# Define a network for sentiment classification, SentimentClassifier
class SentimentClassifier(paddle.nn.Layer):

    def __init__(self, input_size, hidden_size, num_embeddings, num_
classes=2, num_layers=1, init_scale=0.1, dropout_rate=None):
        super(SentimentClassifier, self).__init__()

        # Indicate the dimensions of the input word vectors
        self.input_size = input_size
        # Indicate the dimensions of the state vectors of LSTM
        self.hidden_size = hidden_size
        # Indicate the number of word vectors, corresponding to the size of the vocabulary
        self.num_embeddings = num_embeddings
        # The number of emotional types, this practice is a binary classification task
        self.num_classes = num_classes
        # Indicate the layers of the network
        self.num_layers = num_layers
        # Indicate the proportion of neurons that are inactivated during the use of dropout
        self.dropout_rate = dropout_rate
        # Indicate the initialization range of LSTM network parameters
        self.init_scale = init_scale


        # Define the embedding layer to convert each word in the sentence into a vector
        self.embedding = paddle.nn.Embedding(num_embeddings=num_embe
ddings, embedding_dim=input_size, padding_idx=0)
        self.dropout_layer = paddle.nn.Dropout(p=self.dropout_rate,
mode='upscale_in_train')
        self.lstm_layer = paddle.nn.LSTM(input_size=input_size, hidd
en_size=hidden_size, num_layers=num_layers)
        self.cls_layer = paddle.nn.Linear(self.hidden_size, self.num
_classes)

    def forward(self, inputs, seq_lens=None):

        # Obtain word vectors
        inputs_emb = self.embedding(inputs)
        if self.dropout_rate is not None and self.dropout_rate > 0.
0:
            inputs_emb = self.dropout_layer(inputs_emb)

        # Use LSTM network to obtain semantic information of texts
        sequence_output, (hidden_states, cell_states) = self.lstm_la
yer(inputs_emb, sequence_length=seq_lens)
        hidden_states = hidden_states.squeeze(axis=0)

        # Output labels of sentiment classifications
        logits = self.cls_layer(hidden_states)

        return logits
```

8.1.5 Training Configuration

The following introduces how to determine the computing resources, models, optimizers, loss functions, and optimization indicators used during model training.

(1) Model: use a single-layer LSTM model with num_layers set to 1. Readers can also try setting num_layers to greater than 1 and using a multi-layer LSTM network to observe the effects of model training.
(2) Optimizer: Adam optimizer.
(3) Loss function: cross-entropy.
(4) Evaluation index: accuracy.

Codes are implemented as follows:

```
# Define training parameters
num_epochs = 10
learning_rate = 0.0001
eval_steps = 50
log_steps = 10
save_dir = "./checkpoints"

# Instantiate models
input_size = 256
hidden_size = 256
num_embeddings = len(word2id_dict)
num_layers = 1
dropout_rate = 0.2
num_classes = 2

model = SentimentClassifier(input_size, hidden_size, num_embeddings,
 num_classes=num_classes, num_layers=num_layers, dropout_rate=dropou
t_rate)


# Specify optimization strategies and update models' parameters
optimizer = paddle.optimizer.Adam(learning_rate=learning_rate, beta1
=0.9, beta2=0.999, parameters= model.parameters())


# Specify the loss function
loss_fn = paddle.nn.CrossEntropyLoss()


# Specify the evaluation indicators
metric = paddle.metric.Accuracy()
```

8.1.6 Model Training

During the training process, a training log is printed every log_steps, and model evaluation is performed on the validation set every eval_steps. The model with the best evaluation performances during the training process is saved.

Codes implemented for evaluation on the validation set are as follows:

```
def evaluate(model, data_loader, metric):

    # Set the model to the evaluation mode
    model.eval()

    # Reset evaluations
    metric.reset()


    # Traverse every batch of the validation set
    for batch_id, data in enumerate(dev_loader):
        inputs, labels, seq_lens = data

        # Compute the output of models
        logits = model(inputs, seq_lens)

        # Accumulate evaluations
        correct = metric.compute(logits, labels)
        metric.update(correct)

    dev_score = metric.accumulate()

    return dev_score
```

Codes for model training are given as follows:

```
def train(model):

    # Start the model's training mode
    model.train()
    global_step = 0
    best_score = 0.

    # Record the loss function value during training and the accuracy of the model on the
    validation set
    train_loss_record = []
    train_score_record = []
    num_training_steps = len(train_loader) * num_epochs

    # Perform num_epochs rounds of training
    for epoch in range(num_epochs):
        for step, data in enumerate(train_loader):
            inputs, labels, seq_lens = data
```

```
            # Obtain the model's predictions
            logits = model(inputs, seq_lens)
            # Calculate the average value of the loss function
            loss = loss_fn(logits, labels)
            train_loss_record.append((global_step, loss.item()))

            # Gradient backpropagation
            loss.backward()
            optimizer.step()
            optimizer.clear_grad()

            if global_step % log_steps == 0:
                print(f"[Train] epoch: {epoch}/{num_epochs}, step: {gl
obal_step}/{num_training_steps}, loss: {loss.item():.5f}")

            if global_step != 0 and (global_step % eval_steps == 0 or
global_step == (num_training_steps-1)):
                dev_score = evaluate(model, dev_loader, metric)
                train_score_record.append((global_step, dev_score))
                print(f"[Evaluate]  dev score: {dev_score:.5f}")
                model.train()


                # If current indicators are the best, retain this model
                if dev_score > best_score:
                    save_path = os.path.join(save_dir, "best.pdparams
")
                    paddle.save(model.state_dict(), save_path)
                    print(f"[Evaluate] best accuracy performence has b
een updated: {best_score:.5f} --> {dev_score:.5f}")
                    best_score = dev_score

            global_step += 1

    save_path = os.path.join(save_dir, "final.pdparams")
    paddle.save(model.state_dict(), save_path)
    print("[Train] Training done!")

    return train_loss_record, train_score_record

train_loss_record, train_score_record = train(model)
```

And then, the loss changes during training (train_loss_record) and the score changes on the validation set (train_score_record) are saved, and the training process is visualized. Codes are implemented as follows:

```
import matplotlib.pyplot as plt

def plot_training_loss(train_loss_record, fig_name, fig_size=(8, 6),
sample_step=10, loss_legend_loc="lower left", acc_legend_loc="lower
left"):
    plt.figure(figsize=fig_size)

    train_steps = [x[0] for x in train_loss_record][::sample_step]
    train_losses = [x[1] for x in train_loss_record][::sample_step]

    plt.plot(train_steps, train_losses, color='#e4007f', label="Train
Loss")


    # Draw coordinate axes and legends
    plt.ylabel("Loss", fontsize='large')
    plt.xlabel("Step", fontsize='large')
    plt.legend(loc=loss_legend_loc, fontsize='x-large')
```

```
    plt.savefig(fig_name)
    plt.show()

def plot_training_acc(train_score_record, fig_name, fig_size=(8, 6),
sample_step=10, loss_legend_loc="lower left", acc_legend_loc="lower
left"):
    plt.figure(figsize=fig_size)

    train_steps=[x[0] for x in train_score_record]
    train_losses = [x[1] for x in train_score_record]

    plt.plot(train_steps, train_losses, color='#e4007f', label="Dev
Accuracy")


    # Draw coordinate axes and legends
    plt.ylabel("Accuracy", fontsize='large')
    plt.xlabel("Step", fontsize='large')
    plt.legend(loc=loss_legend_loc, fontsize='x-large')

    plt.savefig(fig_name)
    plt.show()
```

In Fig. 8.5a, the horizontal axis represents the number of training steps, and the vertical axis represents the value of the loss function, which is the cross-entropy between the predicted and true results. In Fig. 8.5b, the horizontal axis and vertical axis represent the number of training steps and prediction accuracy respectively. From Fig. 8.5, it can be seen that as training progresses, the loss function value on the training set continues to decrease and gradually converges until the value tends to be 0, while the accuracy score on the validation set initially increases.

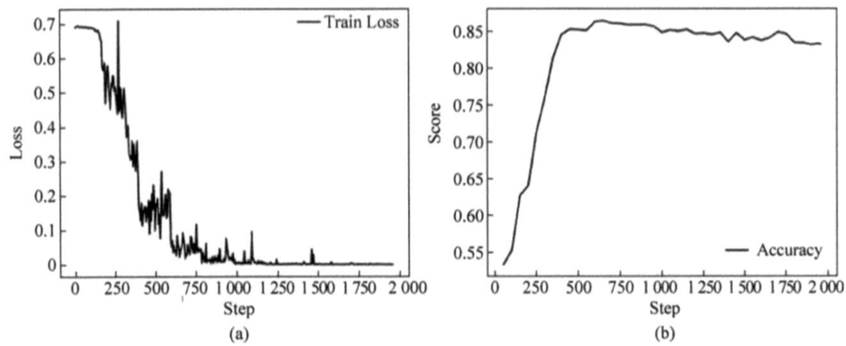

Fig. 8.5 Loss changes in the training set and accuracy changes in the validation set for LSTM models; **a** Loss changes in the training set; **b** Accuracy changes in the validation set

8.1.7 Model Evaluation

The best performing model during training is evaluated by using test set to verify the training effect of the model. Codes are implemented as follows:

```
# Load the trained model for evaluation, re-instantiate a model, and then load the trained models' parameters into the new model
saved_state = paddle.load("./checkpoints/final.pdparams")
model = SentimentClassifier(input_size, hidden_size, num_embeddings, num_classes=num_classes, num_layers=num_layers, dropout_rate=dropout_rate)
model.load_dict(saved_state)


# Evaluate the model
evaluate(model, test_loader, metric)
```

The output result is:`0.83312.`

8.1.8 Model Prediction

If readers input a movie review text, such as "It's funny. It's not Arthur Miller or T.S. Elliot, but man this is funny. Kline and Fields are great," the model will infer the sentiment polarity of the input text. Codes are implemented as follows:

```
def infer(model, text):
    model.eval()

    # Data processing
    words = [word2id_dict.get(word, word2id_dict['[UNK]']) for word in
 text.split(" ")]

    # Construct data of input models
    words = paddle.to_tensor(words, dtype="int64").unsqueeze(0)


    # Calculate the logarithm odds of the output classification of the model
    logits = model(words, paddle.to_tensor([len(words[0])], dtype="int
64"))


    # Analyze the label with the highest score
    id2label={0:"negative emotion", 1:"positive emotion"}
    max_label_id = paddle.argmax(logits, axis=1).numpy()[0]
    pred_label = id2label[max_label_id]

    print("Label: ", pred_label)

text = "It's funny. It's not Arthur Miller or T.S. Elliot, but man this is funny. Kline and Fiel
ds are great.
"
infer(model, text)
```

The output result is:
Label: positive emotion.

8.2 The Implementation of Sentiment Analysis Based on the BERT Model

Section 8.1 introduces the method of implementing sentiment analysis based on long short-term neural networks. This section will implement the sentiment analysis task based on the pre-training model BERT, in which the task objectives, implementation ideas, and data sets are the same as those in Sect. 8.1, and will not be repeated here.

8.2.1 Data Processing

Data processing includes data loading, data indexing (ID), constructing DataLoader, etc. Finally, the same batch of data is processed into feature sequences with equal length, and the fixed-length data is passed into the BERT model in batches using DataLoader.

1. **Data Loading**

The IMDB dataset is loaded by using the datasets tool. The codes are implemented as follows:

```
# Install dataset
%pip install datasets
```

```
import paddlenlp
from datasets import load_dataset


# Use dataset to load the training set, validation set, and test set
train_path = "data/data175000/train.json"
dev_path = "data/data175000/dev.json"
test_path = "data/data175000/test.json"
dataset = load_dataset("json", data_files={"train":train_path, "dev":de
v_path, "test":test_path})


# Print the first data of the training set
print(dataset["train"][:1])
```

The output result is:

```
{'text': ['It does seem like this film is polarizing us. You either
love it or hate it. I loved it.<br /><br />I agree with the comment(s)
that said, you just gotta "feel" this one.<br /><br />Also, early in
the film, Tom Cruise shows his girlfriend a painting done by Monet--
an impressionist painter. Monet\'s style is to paint in little dabs
so up close the painting looks like a mess, but from a distance,
you can tell what the subject is. Cruise mentions that the painting
has a "vanilla sky". I believe this is a hint to the moviegoer.
This movie is like that impressionist painting. It\'s impressionist
filmmaking! And it\'s no coincidence that the title of the movie
refers to that painting.<br /><br />This is not your typical linear
plot. It requires more thought. There is symbolism and there are
scenes that jump around and no, you\'re not always going to be sure
what\'s going on. But at the end, all is explained.<br /><br />You
will need to concentrate on this movie but I think people are making
the mistake of concentrating way too hard on it. After it ends is
when you should think about it. If you try to figure it out as it\'s
unfolding, you will overwhelm yourself. Just let it happen..."go"
with it...keep an open mind. Remember what you see and save the anal-
ysis for later.<br /><br />I found all the performances top notch and
thought it to be tremendously unique, wildly creative, and spell-
binding.<br /><br />But I will not critize the intelligence of those
of you who didn\'t enjoy it. It appeals to a certain taste. If you
like existential, psychedelic, philosophical, thought-provoking,
challenging, spiritual movies, then see it. If you prefer something
```

```
a little lighter, then skip it.<br /><br />But if you DO like what I
described, then you will surely enjoy it.'], 'label': [1]}.
```

From the output results we can see the data structure consists of text data "text" and corresponding sentiment label "label." In the above results, "text" and "label" contain the first data of the training set.

2. **Data Indexing (ID)**

Let's first define the convert_example_to_feature function and convert the input text data into feature form by loading the tokenizer, which facilitates the subsequent reading of the model. As mentioned in the practice of Sect. 7.1, when using the map function of datasets to convert text data into feature form, it can be processed in batches. This section uses the batch processing method, so it is necessary to set up convert_example_to_feature to receive batches of samples and then analyze them. Codes are implemented as follows:

```
from paddlenlp.transformers import AutoTokenizer

def convert_example_to_feature(examples, tokenizer, max_seq_len=512, is
_infer=False):
    encoded_inputs = tokenizer(examples["text"], max_seq_len=max_seq_le
n)

    if not is_infer:
        encoded_inputs["labels"] = [label for label in examples["label
"]]

    return encoded_inputs


# Initialize the tokenizer. Since IMDB is an English dataset, the model_name is set to bert-base-u
ncased. If it is a Chinese dataset, the model_name is set to bert-base-chinese

model_name = "bert-base-uncased"
tokenizer = AutoTokenizer.from_pretrained(model_name)


# Take a piece of data as an example to view the feature data after data conversion
example = {"text":["this movie is so amazing."], "label":[1]}
features = convert_example_to_feature(example, tokenizer)
print(features)
```

The output result is:

```
{'input_ids': [[101, 2023, 3185, 2003, 2061, 6429,
1012, 102]], 'token_type_ids': [[0, 0, 0, 0, 0, 0, 0,
0]], 'labels': [1]}
```

From the output results, the tokenizer converts the text data into three features: input_ids, token_type_ids, and labels.

The following will convert the loaded training set, validation set, and test set into corresponding feature forms based on the convert_example_to_feature function. Here, the partial function is required to set the parameters in the convert_example_to_feature function, and then perform batch data conversion based on the map function. Codes are implemented as follows:

```
from functools import partial
max_seq_len = 512
trans_fn = partial(convert_example_to_feature, tokenizer=tokenizer, max_s
eq_len=max_seq_len)

# Change the input training set, validation set, and test set data into the same feature forms
columns = ["text", "label"]
train_dataset = dataset["train"].map(trans_fn, batched=True, remove_colum
ns=columns)
dev_dataset = dataset["dev"].map(trans_fn, batched=True, remove_columns=c
olumns)
test_dataset = dataset["test"].map(trans_fn, batched=True, remove_columns
=columns)


# Output the samples of every dataset
print("train_dataset:", len(train_dataset))
print("dev_dataset:", len(dev_dataset))
print("test_dataset:", len(test_dataset))
```

The output result is:

```
train_dataset: 25000
dev_dataset: 12500
test_dataset: 12500.
```

3. **Construct DataLoader**

Next, we construct a DataLoader for the upgrading of data in batches. The DataCollatorWithPadding function of PaddleNLP is used to change the data into the same length, the shuffle is set to True, and the data are randomly shuffled. Codes are implemented as follows:

```
from paddle.io import BatchSampler, DataLoader
from paddlenlp.data import DataCollatorWithPadding


batch_size= 16
train_sampler = BatchSampler(train_dataset, batch_size=batch_size, shuffl
e=True)
dev_sampler = BatchSampler(dev_dataset, batch_size=batch_size, shuffle=Fa
lse)


# Use the DataCollatorWithPadding function of PaddleNLP to process the loaded training set,
validation set, and test set data to the same length
data_collator = DataCollatorWithPadding(tokenizer)
# Construct DataLoader, and update the training set, validation set, and test set data in batches base
d
on the batch size
train_loader = DataLoader(dataset=train_dataset, batch_sampler=train_samp
ler, collate_fn=data_collator)
dev_loader = DataLoader(dataset=dev_dataset, batch_sampler=dev_sampler, c
ollate_fn=data_collator)
test_loader = DataLoader(dataset=test_dataset, batch_sampler=dev_sampler,
 collate_fn=data_collator)

for step, batch in enumerate(train_loader):
    if step > 2:
        break
    print(batch)
```

The output result is:

```
{'input_ids': Tensor(shape=[16, 512], dtype=int64, place=Place(gpu:0), sto
p_gradient=True,
       [[101  , 1045 , 2428 , ..., 0    , 0    , 0    ],
        [101  , 2508 , 5980 , ..., 0    , 0    , 0    ],
        [101  , 1045 , 2134 , ..., 0    , 0    , 0    ],
        ...,
        [101  , 2023 , 2003 , ..., 0    , 0    , 0    ],
        [101  , 2054 , 2057 , ..., 14315, 1996 , 102  ],
        [101  , 2023 , 2003 , ..., 0    , 0    , 0    ]]), 'token_type_ids
': Tensor(shape=[16, 512], dtype=int64, place=Place(gpu:0), stop_gradient=
True,
       [[0, 0, 0, ..., 0, 0, 0],
        [0, 0, 0, ..., 0, 0, 0],
        [0, 0, 0, ..., 0, 0, 0],
        ...,
        [0, 0, 0, ..., 0, 0, 0],
        [0, 0, 0, ..., 0, 0, 0],
        [0, 0, 0, ..., 0, 0, 0]]), 'labels': Tensor(shape=[16], dtype=int6
4, place=Place(gpu:0), stop_gradient=True,
       [1, 1, 0, 0, 1, 0, 1, 1, 0, 1, 0, 0, 0, 1, 1, 0])}
```

```
{'input_ids': Tensor(shape=[16, 512], dtype=int64, place=Place(gpu:0), sto
p_gradient=True,
       [[101  , 2043 , 1045 , ..., 0    , 0    , 0    ],
        [101  , 2123 , 1005 , ..., 0    , 0    , 0    ],
        [101  , 1037 , 26352, ..., 0    , 0    , 0    ],
        ...,
        [101  , 10768, 3217 , ..., 12338, 6431 , 102  ],
        [101  , 2292 , 2033 , ..., 0    , 0    , 0    ],
        [101  , 8670 , 8670 , ..., 0    , 0    , 0    ]]), 'token_type_ids
': Tensor(shape=[16, 512], dtype=int64, place=Place(gpu:0), stop_gradient=
True,
       [[0, 0, 0, ..., 0, 0, 0],
        [0, 0, 0, ..., 0, 0, 0],
        [0, 0, 0, ..., 0, 0, 0],
        ...,
        [0, 0, 0, ..., 0, 0, 0],
        [0, 0, 0, ..., 0, 0, 0],
        [0, 0, 0, ..., 0, 0, 0]]), 'labels': Tensor(shape=[16], dtype=int6
4, place=Place(gpu:0), stop_gradient=True,
       [0, 0, 0, 1, 1, 1, 0, 1, 0, 0, 1, 1, 0, 1, 0, 0])}
{'input_ids': Tensor(shape=[16, 512], dtype=int64, place=Place(gpu:0), sto
p_gradient=True,
       [[101  , 1026 , 7987 , ..., 0    , 0    , 0    ],
        [101  , 1996 , 3185 , ..., 2000 , 2022 , 102  ],
        [101  , 17453, 5875 , ..., 0    , 0    , 0    ],
        ...,
        [101  , 2034 , 1010 , ..., 0    , 0    , 0    ],
        [101  , 2026 , 3129 , ..., 0    , 0    , 0    ],
        [101  , 1996 , 3484 , ..., 1037 , 4438 , 102  ]]), 'token_type_ids
': Tensor(shape=[16, 512], dtype=int64, place=Place(gpu:0), stop_gradient=
True,
       [[0, 0, 0, ..., 0, 0, 0],
        [0, 0, 0, ..., 0, 0, 0],
        [0, 0, 0, ..., 0, 0, 0],
        ...,
        [0, 0, 0, ..., 0, 0, 0],
        [0, 0, 0, ..., 0, 0, 0],
        [0, 0, 0, ..., 0, 0, 0]]), 'labels': Tensor(shape=[16,512], dtype=
int64, place=Place(gpu:0), stop_gradient=True,
       [0, 0, 0, 0, 0, 0, 0, 1, 1, 0, 1, 0, 1, 1, 0, 0])}
```

From the output result, the data length of the same batch is the same.

8.2.2 Model Construction

The BERT model processes the input minibatch data sequentially. For the input text sequences, it concatenates [CLS] and [SEP] before sending it into the BERT model. The hidden layer vector at the [CLS] position output by the BERT model can be seen as the semantic vector of this text sequence. This semantic vector is then passed to the linear layer to obtain scores for different sentiment categories (positive or negative) of the text, as shown in Fig. 8.6.

Based on the modeling idea shown in Fig. 8.6, codes are implemented as follows:

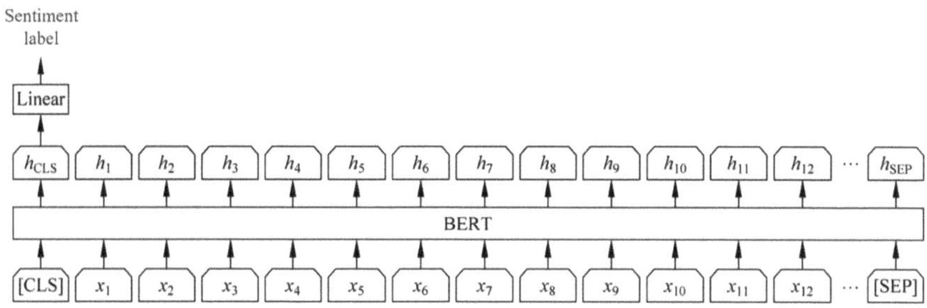

Fig. 8.6 The movie review sentiment analysis

```
import paddle.nn as nn

# Define BertForTokenClassification and inherit the nn.Layer class
class BertForSequenceClassification(nn.Layer):

    def __init__(self, bert, num_classes=2, dropout=None):
        super(BertForSequenceClassification, self).__init__()
        self.num_classes = num_classes
        self.bert = bert
        self.dropout = nn.Dropout(dropout if dropout is not None else sel
f.bert.
                                  config["hidden_dropout_prob"])
        self.classifier = nn.Linear(self.bert.config["hidden_size"],
                                    num_classes)

    def forward(self,
                input_ids,
                token_type_ids=None,
                position_ids=None,
                attention_mask=None):

         # Transfer the feature data of the input text sequence into the BERT model for encoding
        outputs = self.bert(input_ids,
                            token_type_ids=token_type_ids,
                            position_ids=position_ids,
                            attention_mask=attention_mask)

         # Obtain the vector representation corresponding to the [CLS] location
         pooled_output = outputs[1]

        pooled_output = self.dropout(pooled_output)
         # Obtain emotion categories through linear layers
        logits = self.classifier(pooled_output)

        return logits
```

In Subsect. 7.1.4, some commonly used classes in PaddleNLP are introduced. So, we can directly call these classes to complete model construction. Below, we use the AutoModelForSequenceClassification class to directly call and obtain the text classification model. This class can automatically obtain the text classification function of a specified model. Codes are implemented as follows:

```
from paddlenlp.transformers import AutoModelForSequenceClassification


# This experiment contains both positive and negative data, so num_classes is set to 2
num_classes = 2
model = AutoModelForSequenceClassification.from_pretrained(model_name, num
_classes=num_classes)
```

8.2.3 Training Configuration

We first define the hyperparameters, optimizer, loss function, and evaluation indicators during model training.

(1) Model: The BERT model.
(2) Optimizer: AdamW optimizer.
(3) Loss function: cross-entropy.
(4) Evaluation Metrics: Accuracy.

Codes are implemented as follows:

```
import paddle

from paddlenlp.transformers import LinearDecayWithWarmup

# Set the number of training rounds
num_epochs = 2
# Set the learning rate
learning_rate = 1e-5
# Set intervals for model evaluation on the validation set every certain number of steps
eval_steps = 100
# Set intervals for printing the log every certain number of steps
log_steps = 10
# Set the model's saving path and automatically save the model with the best effects during the
training process
save_dir = "./checkpoints"
# Set the weight decay coefficient during training
weight_decay = 0.01
# Set the proportion of warm-up training during the training process
warmup_proportion = 0.1
# Set the total number of training steps required
num_training_steps = len(train_loader) * num_epochs
```

```
# Except for the parameters of bias and LayerNorm, other parameters perform decay operations during training
decay_params = [p.name for n, p in model.named_parameters() if not any(nd
 in n for nd in ["bias", "norm"])]

# Define the optimizer
optimizer = paddle.optimizer.AdamW(learning_rate=learning_rate,
parameters=model.parameters(), weight_decay=weight_decay,
apply_decay_param_fun=lambda x: x in decay_params)

# Define the loss function
loss_fn = nn.CrossEntropyLoss()


# Define the computing method of evaluating indicators
metric = paddle.metric.Accuracy()
```

8.2.4 Model Training

During the training process, a training log is printed every log_steps, and a model evaluation is performed on the validation set every eval_steps. The model with the best evaluation result during the training process is saved.

The codes for evaluation on the validation set are as follows:

```
from tqdm import tqdm

def evaluate(model, data_loader, metric):

    # Set models into the evaluation mode
    model.eval()

    # Reset the evaluation
    metric.reset()


    #  Traverse every batch of the invalidation set
    for data in tqdm(data_loader, desc="[Evaluation Progression]"):
        inputs, token_type_ids, labels = data["input_ids"], data["token_ty
pe_ids"], data["labels"]
        logits = model(input_ids=inputs, token_type_ids=token_type_ids)

        # Accumulate evaluation
        correct = metric.compute(logits, labels)
        metric.update(correct)

    dev_score = metric.accumulate()

    return dev_score
```

The codes for model training are shown as follows:

```
import os

def train(model):

    # Start the model's training mode
    model.train()
    global_step = 0
    best_score = 0.

    # Record the loss function during training and the accuracy of the model on the validation set
    train_loss_record = []
    train_score_record = []
    num_training_steps = len(train_loader) * num_epochs

    # Perform num_epochs rounds of training
    for epoch in range(num_epochs):
        for step, data in enumerate(train_loader):
            inputs, token_type_ids, labels = data["input_ids"], data["toke
n_type_ids"], data["labels"]

            # Obtain the model's predictions
            logits = model(input_ids=inputs, token_type_ids=token_type_id
s)

            loss = loss_fn(logits, labels)
            train_loss_record.append((global_step, loss.item()))


            # Gradient backpropagation
            loss.backward()
            optimizer.step()
            optimizer.clear_grad()

            if global_step % log_steps == 0:
                print(f"[Train] epoch: {epoch}/{num_epochs}, step: {global
_step}/{num_training_steps}, loss: {loss.item():.5f}")

            if global_step != 0 and (global_step % eval_steps == 0 or glob
al_step == (num_training_steps-1)):
                dev_score = evaluate(model, dev_loader, metric)
                train_score_record.append((global_step, dev_score))

                model.train()

                # If current indicators are the best, retain this model
                if dev_score > best_score:
                    print(f"[Evaluate] best accuracy performence has been
updated: {best_score:.5f} --> {dev_score:.5f}")
                    best_score = dev_score
                    save_path = os.path.join(save_dir, "best.pdparams")
                    paddle.save(model.state_dict(), save_path)
                print(f"[Evaluate]  dev score: {dev_score:.5f}")
```

```
            global_step += 1

    save_path = os.path.join(save_dir, "final.pdparams")
    paddle.save(model.state_dict(), save_path)
    print("[Train] Training done!")

    return train_loss_record, train_score_record

train_loss_record, train_score_record = train(model)
```

The output results are:

```
......
[Train] epoch: 0/2, step: 110/3126, loss: 0.47537
[Train] epoch: 0/2, step: 120/3126, loss: 0.31041
[Train] epoch: 0/2, step: 130/3126, loss: 0.35236
[Train] epoch: 0/2, step: 140/3126, loss: 0.29467
[Train] epoch: 0/2, step: 150/3126, loss: 0.37072
[Train] epoch: 0/2, step: 160/3126, loss: 0.24420
[Train] epoch: 0/2, step: 170/3126, loss: 0.25741
[Train] epoch: 0/2, step: 180/3126, loss: 0.20653
[Train] epoch: 0/2, step: 190/3126, loss: 0.42518
[Train] epoch: 0/2, step: 200/3126, loss: 0.34494
[Evaluate] best accuracy performence has been updated: 0.87240 -->
0.90312
[Evaluate] dev score: 0.90312
......
```

Then, we use the plot_training_loss and plot_training_acc defined in the tools.py file to visualize the training process. Codes are implemented as follows:

```
from tools import plot_training_loss, plot_training_acc

fig_path = "./images/chapter7_bert_loss.pdf"
plot_training_loss(train_loss_record, fig_path, loss_legend_loc="upper
right", sample_step=60)

fig_path = "./images/chapter7_bert_acc.pdf"
plot_training_acc(train_score_record, fig_path, sample_step=1,
loss_legend_loc="lower right")
```

The output result is shown in Fig. 8.7.

In Fig. 8.7a, the horizontal axis represents the number of training steps, and the vertical axis represents the value of the loss function, which is the cross-entropy between the predicted result and the true result. In Fig. 8.7b, the horizontal axis and the vertical axis represent the number of training steps and the accuracy of prediction respectively. From Fig. 8.7, it can be seen that as training progresses, the curve of the loss function on the training set generally shows a gradual downward trend, while the accuracy on the validation set also gradually improves.

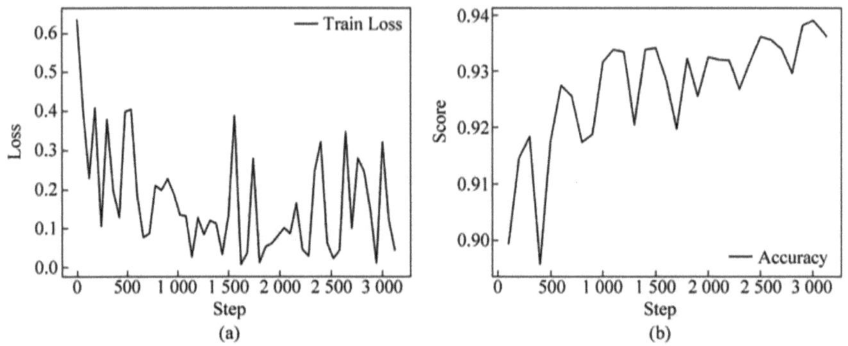

Fig. 8.7 Loss changes in the training set and accuracy changes in the validation set for BERT models; **a** Loss changes in the training set; **b** Accuracy changes in the validation set

8.2.5 Model Evaluation

The best performing model during training is evaluated by using the test set to verify the model's training effect. Codes are implemented as follows:

```
# Load the trained model for prediction, re-instantiate a model, and then load the trained model's parameters into the new model
saved_state = paddle.load("./checkpoints/best.pdparams")
num_classes = 2
model = AutoModelForSequenceClassification.from_pretrained(model_name, num_classes=num_classes)
model.load_dict(saved_state)

# Model evaluation
test_score = evaluate(model, test_loader, metric)
print(f"[Evaluate]  test score: {test_score:.5f}")
```

The output result is:

```
[Evaluate] test score: 0.93960
```

From the output result, the accuracy of the sentiment analysis task based on the LSTM network in Sect. 8.1 on the IMDB dataset is about 86.6% on the test set. In contrast, the accuracy based on the BERT model can reach 94%, which is sufficient to prove that the BERT model has stronger modeling capabilities.

8.2.6 Model Prediction

Readers can input any text related to movie reviews, such as "this movie is so good that I watch it several times," and verify the training effect of the model through model prediction. Codes are implemented as follows:

```
def infer(model, text):
    model.eval()

    # Data processing
    encoded_inputs = tokenizer(text, max_seq_len=max_seq_len)

    # Construct data of input models
    input_ids = paddle.to_tensor(encoded_inputs["input_ids"], dtype="int6
4").unsqueeze(0))
    token_type_ids = paddle.to_tensor(encoded_inputs["token_type_ids"], d
type="int64").unsqueeze(0))


    # Compute scores
    logits = model(input_ids=input_ids, token_type_ids=token_type_ids)


    # Analyze the label with the highest score
    id2label={0:"negative emotion", 1:"positive emotion"}
    max_label_id = paddle.argmax(logits, axis=1).numpy()[0]
    pred_label = id2label[max_label_id]

    print("Label: ", pred_label)

text = "this movie is so good that I watch it several times. "
infer(model, text)
```

The output result is:

Label: positive emotion.

8.3 Aspect-Based Sentiment Analysis Based on BERT

8.3.1 Task Objectives

Sections 8.1 and 8.2 use the LSTM model and the BERT model to complete the task of sentiment classification at the sentence level. This section introduces the implementation method of sentiment classification at the aspect level, which involves extracting aspects, opinion words, and comment opinions from a given text. Sentiment classification at the aspect level also includes multiple sub-tasks, such as aspect term extraction, opinion term extraction, and aspect based sentiment classification. If a given statement to be evaluated is: "The cake has a sweet taste, a beautiful appearance, and the shop's service is also very friendly," then the results of sentiment classification at the sentence level and the aspect level are shown in Fig. 8.8. The aspect term extraction module extracts all aspects in the sentence. The aspects in this example include three items: taste, appearance, and service. The opinion term extraction module extracts opinion words corresponding to each aspect, such as the opinion word "sweet" for the taste aspect, the opinion word "beautiful" for the appearance aspect, and the opinion word "friendly" for the service aspect. The aspect-based sentiment classification module determines the sentiment polarity of each aspect. In this example, the sentiment polarities of all three aspects are positive.

With the rapid development and applications of Internet, sentiment classification tasks have gradually become a hot research topic. Through the analysis, induction and reasoning of users' real evaluations, sentiment classification can provide data support such as corporate consumption and decisions, public opinion analysis, and personalized recommendation. Compared with the sentence-level sentiment analysis tasks, aspect-level sentiment analysis can analyze the emotional polarity contained in text in a more fine-grained manner, and therefore can be applied into more areas. This section introduces how to implement the aspect-level sentiment analysis methods based on the BERT model.

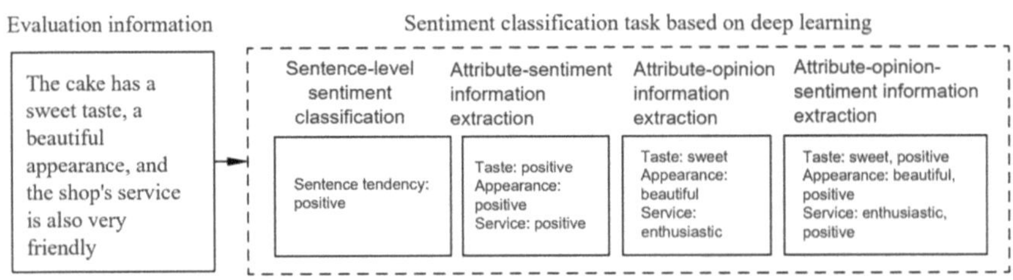

Fig. 8.8 The emotion classification task

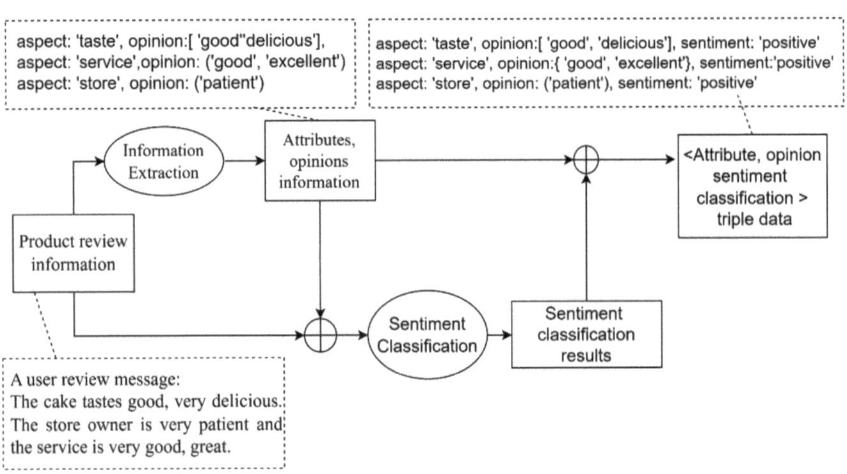

Fig. 8.9 The flow chart of aspect-based sentiment classification

8.3.2 *Implementation Ideas and Processes*

The implementation processes of aspect-level sentiment analysis are shown in Fig. 8.9. The input of the model is text data, and the output is the triples of <aspect, opinion, sentiment label> extracted from the text data. During the modeling process, it consists of the following two sub-tasks:

(1) Aspect and opinion term extraction: for the input text comment, the BERT model adopts a sequence labeling-based method to extract the aspects and corresponding opinion words in the comment.
(2) Sentiment classification: the aspect-opinion words of comments output by task (1) and text comment are concatenated as the input data of task (2). BERT encodes the text, obtain the semantic vector representation of the text, and then passes the hidden layer vector at the [CLS] position into a linear layer to obtain the final sentiment classification result.

Finally, the information obtained from tasks (1) and (2) is summarized to obtain the triples of <attribute, opinion, sentiment classification>.

The following describes the specific implementation processes of each step.

8.3.3 *Aspect Term and Opinion Term Extraction*

For the input text comment, the sequence labeling method is used to extract the aspects and corresponding opinions. In practice, BIO tags are used to model sequence labeling: B-Aspect, I-Aspect, B-Opinion, I-Opinion, and O. The first two are used to label the comment aspects, the second two are used to label the corresponding opinions, and the final O is used to label other words.

The implementation processes of opinion term extraction are the same as that of Sects. 8.1 and 8.2. The specific implementation methods are described below.

1. **Data Processing**

Data processing includes determining the dataset, data loading, data indexing (ID), constructing DataLoader, etc. The same batch of data is finally processed into feature sequences of equal length, and DataLoader is used to pass the data into the BERT model in batches.

(1) Determining the Dataset

To facilitate readers' learning and use, we have processed the data ahead and converted the original Chinese word segmentation data into a sequence labeling format. Below are three sample data from the training set, with one line of data representing one sample. Each sample contains text data and corresponding word position label sequence, and uses spaces for character segmentation.

The dataset used in this practice is self-built by PaddleNLP of PaddlePaddle, and contains 800 data in the training set, 100 data in the validation set, and 100 data in the test set. Each piece of data consists of text data and corresponding word position label sequence, and is separated by spaces. An example of sample data is as follows:

> {'text': 'The product quality is good, and the service is quite thoughtful.', 'label': 'B-Aspect I-Aspect I-Aspect I-Aspect B-Opinion I-Opinion O B-Aspect I-Aspect O B-Opinion I-Opinion O'}.

From the tag information, we can see that this example contains two pieces of "attribute-opinion," namely "product quality-good" and "service-thoughtful."

(2) Data Loading

The data in the training set, validation set, and test set are loaded first. Codes are implemented as follows:

```
 import paddlenlp
from datasets import load_dataset
# Load the data in the training set, validation set, and test set
train_path = "./data/data178684/train_ext.json"
dev_path = "./data/data178684/dev_ext.json"
test_path = "./data/data178684/test_ext.json"
dataset = load_dataset("json", data_files={"train":train_path, "dev":de
v_path, "test":test_path})
# Print the first three data in the training set
print(dataset["train"][:3])
```

The output result is:

```
{'text': [' The exchange is very convenient.', 'I
went into the water to feel the screws. Overall, it
felt good. The air was fresh. It's a good place for a
weekend vacation.', ' A good place to ride a bike'],
'label': ['B-Aspect I-Aspect O B-Opinion I-Opinion O',
'O O O O O O O O B-Aspect I-Aspect I-Aspect I-Aspect O
B-Opinion I-Opinion O B-Aspect I-Aspect O O O B-Opinion
I-Opinion O O O O O O O O B-Opinion B-Aspect I-Aspect',
'O O B-Opinion I-Opinion O B-Aspect I-Aspect']}.
```

From the output result, each sample includes two parts: "text" and corresponding aspect-opinion "label," in which "label" is labeled using the BIO tags.

(3) Data Indexing (ID)

Then, let's define the convert_example_to_feature function, convert the input text data into feature data by loading the Tokenizer, and construct the dictionary label2id to convert the labeled data into a numerical sequence. The mapping rules for BIO word position labels are as follows: "B-Aspect" is mapped to the digit 1, "I-Aspect" is mapped to the digit 2, "B-Opinion" is mapped to 3, and "I-Opinion" is mapped to 4. "O" is mainly used to map non-aspect and opinion characters or some special tokens to 0, such as [CLS], [SEP], [PAD], etc. Codes are implemented as follows:

```
from functools import partial
from paddlenlp.transformers import AutoTokenizer
def convert_example_to_feature(example, tokenizer, label2id, max_seq_len=
512, is_infer=False):
    """
    Convert the input sample into input features suitable for the BERT model

    Input:

        - example: single input sample

        - tokenizer: a tokenizer example of the Bert model

        - label2id: the dictionary used to change labels into id

        - max_seq_len: Maximum sequence length of text processed by the model

        - is_infer: Whether it is a model prediction or not, for predicting data, the labeled
        data will not be processed
    """

    # Use the tokenizer to convert the input data into feature form
    text = example["text"].split()
    label = example["label"].split()
    assert len(text) == len(label), "The length of text is not equal to l
abel's: {} != {}".format(len(text), len(label))
    encoded_inputs = tokenizer(text, is_split_into_words=True, max_seq_le
n=max_seq_len, return_length=True)

    if not is_infer:
        encoded_inputs["label"] = [label2id["O"]] + [label2id[item] for i
tem in label[:max_seq_len - 2]] + [label2id["O"]]

    return encoded_inputs

# The largest sequence length
max_seq_len = 512

# Define the name of models
model_name = "bert-base-chinese"

# Initialize AutoTokenizer
tokenizer = AutoTokenizer.from_pretrained(model_name)

# Define the dictionary label2id and convert the labeled data in the BIO structure into a numerical
sequence
ext_label2id = {"O":0, "B-Aspect":1, "I-Aspect":2, "B-Opinion":3, "I-Opin
ion":4}
ext_id2label = {v: k for k, v in ext_label2id.items()}

# Convert the training set, validation set, and test set data into feature form. Since the amount of
data in this practice is not large, set batched to False and process the data one by one
trans_fn = partial(convert_example_to_feature, tokenizer=tokenizer, label
2id=ext_label2id, max_seq_len=max_seq_len)
columns = ["text", "label"]
train_dataset = dataset["train"].map(trans_fn, batched=False, remove_colu
mns=columns)
```

```
dev_dataset = dataset["dev"].map(trans_fn, batched=False, remove_columns=
columns)
test_dataset = dataset["test"].map(trans_fn, batched=False, remove_column
s=columns)


# Print the number of samples in each dataset
print("train_dataset:", len(train_dataset))
print("dev_dataset:", len(dev_dataset))
print("test_dataset:", len(test_dataset))


# Print the first data in the training set
print("train example:", train_dataset[:1])
```

The output result is:

```
train_dataset: 800
dev_dataset: 100
test_dataset: 100
train example: {'label': [[0, 1, 2, 0, 3, 4, 0, 0]], 'input_ids':
[[101, 1050, 2940, 2523, 3175, 912, 8024, 102]], 'token_type_ids':
[[0, 0, 0, 0, 0, 0, 0, 0]], 'length': [8], 'seq_len': [8]}.
```

From the output result, the text data and labels have been converted into corresponding feature data.

(4) Construction of DataLoader

The purpose to construct DataLoader is to divide the loaded data into batches and pass them to the BERT model. This practice uses the pre-defined DataCollatorForTokenClassification function in PaddleNLP to change a batch of data into the same length. Codes are implemented as follows:

```
from paddle.io import BatchSampler, DataLoader
from paddlenlp.data import DataCollatorForTokenClassification

batch_size= 2
train_sampler = BatchSampler(train_dataset, batch_size=batch_size, shuffl
e=True)
dev_sampler = BatchSampler(dev_dataset, batch_size=batch_size, shuffle=Fa
lse)

# Use the preset DataCollator
data_collator = DataCollatorForTokenClassification(tokenizer, label_pad_t
oken_id=ext_label2id["O"])
# Construct DataLoader, and update the training set, validation set, and test set data in batches base
d
on the batch size
train_loader = DataLoader(dataset=train_dataset, batch_sampler=train_samp
ler, collate_fn=data_collator)
dev_loader = DataLoader(dataset=dev_dataset, batch_sampler=dev_sampler, c
ollate_fn=data_collator)
test_loader = DataLoader(dataset=test_dataset, batch_sampler=dev_sampler,
 collate_fn=data_collator)

# Print the first batch data in the training set
print(next(iter(train_loader)))
```

The output result is:

```
{'label': Tensor(shape=[2, 30], dtype=int64, place=CUDAPlace(0), stop_grad
ient=True,

          [[0, 1, 2, 0, 3, 4, 0, 0, 0, 0, 0, 0, 1, 0, 0, 3, 4, 0, 0, 0, 0
, 0, 0, 0,

            0, 0, 0, 0, 0, 0],

           [0, 0, 0, 0, 0, 0, 0, 0, 0, 0, 0, 0, 0, 0, 0, 0, 1, 2, 3, 4, 0
, 0, 0, 0,

            0, 0, 0, 0, 0, 0]]), 'input_ids': Tensor(shape=[2, 30], dtype
=int64, place=CUDAPlace(0), stop_gradient=True,

          [[101 , 817 , 3419, 2923, 6844, 704 , 4638, 671 , 702 , 1765, 3
175, 8024,

            6983, 738 , 2523, 679 , 7231, 102 , 0   , 0   , 0   , 0   , 0
   , 0   ,

            0   , 0   , 0   , 0   , 0   , 0   ],

           [101 , 1377, 5543, 7479, 6818, 691 , 3175, 3209, 4403, 857 , 2
162, 6963,

            6586, 8043, 7305, 7481, 4692, 4708, 679 , 7231, 852 , 3221, 6
392, 3177,

            1348, 4788, 1348, 3191, 8024, 102 ]]), 'token_type_ids': Tens
or(shape=[2, 30], dtype=int64, place=CUDAPlace(0), stop_gradient=True,

          [[0, 0, 0, 0, 0, 0, 0, 0, 0, 0, 0, 0, 0, 0, 0, 0, 0, 0, 0, 0, 0
, 0, 0, 0,

            0, 0, 0, 0, 0, 0],

           [0, 0, 0, 0, 0, 0, 0, 0, 0, 0, 0, 0, 0, 0, 0, 0, 0, 0, 0, 0, 0
, 0, 0, 0,

            0, 0, 0, 0, 0, 0]]), 'length': Tensor(shape=[2], dtype=int64,
 place=CUDAPlace(0), stop_gradient=True,

          [18, 30]), 'seq_len': Tensor(shape=[2], dtype=int64, place=CUDA
Place(0), stop_gradient=True,

     [18, 30])}
```

2. Model Construction

First, the processed feature data is sent into the BERT model for encoding, and the corresponding hidden layer vector sequence is output. Then, these vector sequences are sent into the linear layer (vector sequence multiplied by weight, plus bias), and after softmax processing, the word position label sequence corresponding to the input is obtained. When outputting the corresponding label for each token position, the greedy algorithm is used by default, which selects the label with the highest score at the current position as the output for that position. As shown in Fig. 8.10.

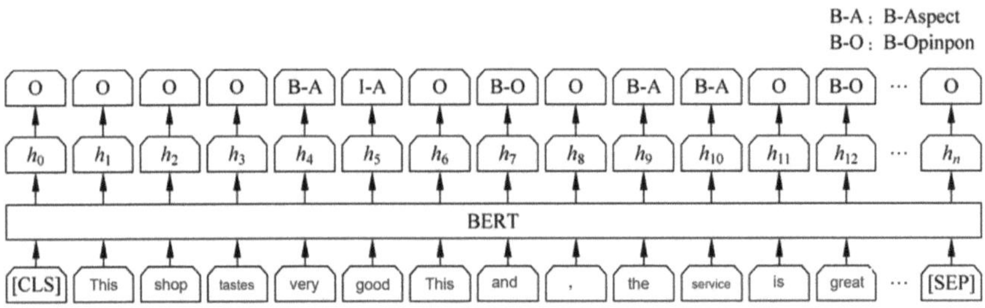

Fig. 8.10 The construction of opinion term extraction model

Based on the modeling idea shown in Fig. 8.10, the AutoModelForTokenClassification class is used to automatically load BertForTokenClassification, and codes are implemented as follows:

```
from paddlenlp.transformers import AutoModelForTokenClassification


# num_classes is set to 5, which is the same as the number of label categories of BIO word
positions
ext_model = AutoModelForTokenClassification.from_pretrained(model_name,
num_classes=5)
print(ext_model.__class__)
```

The output result is:

```
<class 'paddlenlp.transformers.bert.modeling.BertForToken
Classification'>
```

From the output result, AutoModelForTokenClassification can automatically find the BertForTokenClassification class through the specified model_name.

3. **Training Configuration**

Now we define and configure the training environment for the aspect and opinion term extraction model, including: configuring training parameters, configuring model parameters, setting the optimizer during training, defining loss functions and evaluation metrics.

(1) Model: the BERT model.
(2) Optimizer: AdamW optimizer.
(3) Loss function: cross-entropy.
(4) Evaluation indicators: precision, recall, and F1 score.

The paddlenlp.metrics.ChunkEvaluator class is used to perform indicator statistics on the validation and test sets. And then the corresponding aspects and opinion words are located from true label sequences and predicted label sequences, and the corresponding precision, recall, and F1 scores are calculated.

Suppose that for a piece of comment data, its true and predicted label sequences are:

True label sequence	B-Aspect	I-Aspect	O	O	B-Opinion	O	B-Aspect	I-Aspect	O	O	B-Opinion
Predicted label sequence	B-Aspect	I-Aspect	O	O	O	B-Aspect	I-Aspect	O	O	O	B-Opinion

The corresponding aspects and opinion words are located as follows: true label sequence: ('Aspect', 0, 1), ('Opinion', 4, 4), ('Aspect', 6, 7), ('Opinion', 10, 10); predicted label sequence: ('Aspect', 0, 1), ('Aspect', 5, 6), ('Opinion', 10, 10).

From the comparison results of the *true label sequence and predicted label sequence*, it can be seen that there should be 4 words predicted in the true label sequences, but only 3 words are predicted correctly, and 1 word is predicted incorrectly. The corresponding precision, recall, and F1 scores are:

$$\text{Precision} = \frac{2}{3}$$

$$\text{Recall} = \frac{2}{4} = \frac{1}{2}$$

$$\text{F1} = \frac{2 \times \text{Precision} \times \text{Recall}}{\text{Precision} + \text{Recall}} = \frac{4}{7}.$$

The codes for training configuration are as follows:

```
import os
import paddle
import paddle.nn as nn
from paddlenlp.metrics import ChunkEvaluator
from paddlenlp.transformers import LinearDecayWithWarmup


# The number of training rounds
num_epochs = 5

# The learning rate
learning_rate = 3e-5

# Set intervals for model evaluation every certain number of steps
eval_steps = 10

# Set intervals for printing the log every certain number of steps
log_steps = 10

# The model's saving content
save_dir = "./checkpoints"


# Set the weight decay coefficient of AdamW optimizer
weight_decay = 0.01
```

```
# Set the proportion of warm-up training of the learning rate
warmup_proportion = 0.1

# Set the total number of training steps required
num_training_steps = len(train_loader) * num_epochs

# If save_dir doesn't exist, then establish another one

if not os.path.exists(save_dir):
    os.mkdir(save_dir)


# Except for the parameters of bias and LayerNorm, other parameters perform decay operations
during training
decay_params = [
        p.name for n, p in ext_model.named_parameters()
        if not any(nd in n for nd in ["bias", "norm"])
    ]


# Define lr_scheduler, the learning rate increases linearly during the warmup phase, and after the
warmup phase, the learning rate gradually decays
lr_scheduler = LinearDecayWithWarmup(learning_rate=learning_rate, total_
steps=num_training_steps, warmup=warmup_proportion)

# Initialize the optimizer
optimizer = paddle.optimizer.AdamW(
        learning_rate=lr_scheduler,
        parameters=ext_model.parameters(),
        weight_decay=weight_decay,
        apply_decay_param_fun=lambda x: x in decay_params)


# Define the loss function
loss_fn = nn.CrossEntropyLoss()


# Define the computing method of evaluating indicators
metric = ChunkEvaluator(label_list=ext_label2id.keys())
```

4. Model Training

Let's define the train function and evaluate function for model training and evaluation respectively. During the training process, a log is printed every log_steps steps and the model is evaluated every eval_steps steps. During the process, the model that performs best on the validation set is saved. Codes are implemented as follows:

```
import paddle.nn.functional as F

def evaluate(model, data_loader, metric):
    """
```

```
    Model evaluation function


    Input:

        - model: model examples to be evaluated

        - data_loader: datasets to be evaluated

        - metric: class examples used for statistical evaluation indicators
    """
    model.eval()
    metric.reset()
    # Read the data of dataloader
    for idx, batch_data in enumerate(data_loader):
        input_ids, token_type_ids, labels, seq_lens = batch_data["input_
ids"], batch_data["token_type_ids"], batch_data["label"], batch_data["se
q_len"]

        # Define the forward computation
        logits = model(input_ids, token_type_ids=token_type_ids)
        predictions = logits.argmax(axis=2)
        # Model evaluation
        num_infer_chunks, num_label_chunks, num_correct_chunks = metric.
compute(seq_lens, predictions, labels)
        metric.update(num_infer_chunks.numpy(), num_label_chunks.numpy
(), num_correct_chunks.numpy())

    precision, recall, f1 = metric.accumulate()
    return precision, recall, f1

def train(model):

    # Start training models
    model.train()
    global_step, best_f1 = 1, 0.

    # Record the losses during training and the model evaluation scores on the validation set
    train_loss_record = []
    train_score_record = []
    # Perform num_epochs rounds of training
    for epoch in range(1, num_epochs+1):
        for batch_data in train_loader():
            input_ids, token_type_ids, labels = batch_data["input_ids"],
 batch_data["token_type_ids"], batch_data["label"]

            # Model prediction
            logits = model(input_ids, token_type_ids=token_type_ids)

            loss = F.cross_entropy(logits.reshape([-1, len(ext_label2i
d)]), labels.reshape([-1]), ignore_index=-1)
            train_loss_record.append((global_step, loss.item()))
```

```
            # Gradient backpropagation
            loss.backward()
            lr_scheduler.step()
            optimizer.step()
            optimizer.clear_grad()

            if global_step > 0 and global_step % log_steps == 0:
                print(f"epoch: {epoch} - global_step: {global_step}/{num
_training_steps} - loss:{loss.numpy().item():.6f}")
            if (global_step > 0 and global_step % eval_steps == 0) or gl
obal_step == num_training_steps:
                precision, recall, f1  = evaluate(model, dev_loader,  me
tric)
                train_score_record.append((global_step, f1))
                model.train()
            # If the current indicator is the optimal indicator, then save this model
                if f1 > best_f1:
                    print(f"best F1 performence has been updated: {best_
f1:.5f} --> {f1:.5f}")
                    best_f1 = f1
                    paddle.save(model.state_dict(), f"{save_dir}/best_ex
t.pdparams")
                print(f'evalution result: precision: {precision:.5f}, re
call: {recall:.5f},  F1: {f1:.5f}')

            global_step += 1

    paddle.save(model.state_dict(), f"{save_dir}/final_ext.pdparams")

    return train_loss_record, train_score_record

train_loss_record, train_score_record = train(ext_model)
```

The output result is:

```
epoch: 1 - global_step: 10/250 - loss:0.829542
evalution result: precision: 0.00000, recall: 0.00000, F1: 0.00000
......
epoch: 5 - global_step: 240/250 - loss:0.073503
evalution result: precision: 0.73804, recall: 0.84682, F1: 0.78869
epoch: 5 - global_step: 250/250 - loss:0.083309
evalution result: precision: 0.73804, recall: 0.84682, F1: 0.78869.
```

During the training process, the loss train_loss_record and the score train_score_record on the validation set are saved to visualize the changes in the loss function and accuracy. Codes are implemented as follows:

```
from tools import plot_training_acc, plot_training_loss

fig_path = "./images/chapter7_loss.pdf"
plot_training_loss(train_loss_record, fig_path, loss_legend_loc="upper
right", sample_step=5)

fig_path = "./images/chapter7_acc.pdf"
plot_training_acc(train_score_record, fig_path, sample_step=1,
acc_legend_loc="lower right")
```

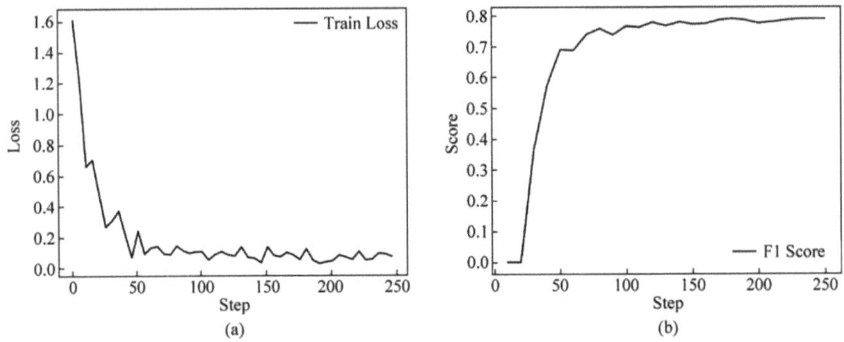

Fig. 8.11 Loss changes of the extraction model in the training set and accuracy changes in the validation set. **a** Loss changes in the training set; **b** Accuracy changes in the F1 scores

The output result is shown in Fig. 8.11.

In Fig. 8.11a, the horizontal axis represents the number of training steps, and the vertical axis represents the value of the loss function, which is the cross entropy between the predicted result and the true result. In Fig. 8.11b, the horizontal axis and vertical axis represent the number of training steps and the predicted F1 scores respectively. From Fig. 8.11, it can be seen that as training progresses, the loss function on the training set continuously decreases and gradually converges towards 0. At the same time, the F1 score on the validation set initially increases continuously, and gradually stabilizes after the model converges, eventually reaching around 77%.

5. **Model Evaluation**

Next, we will load the model that has the best evaluation performance during the training process and test it using the test set. Relevant codes are as follows:

```
# Load the trained model for prediction, re-instantiate a model, and then load the trained model's parameters into the new model
model_path = "./checkpoints/best_ext.pdparams"

loaded_state_dict = paddle.load(model_path)
ext_model = AutoModelForTokenClassification.from_pretrained(model_name, num_classes=5)
ext_model.load_dict(loaded_state_dict)

precision, recall, f1 = evaluate(ext_model, test_loader,  metric)
print(f'evalution result: precision: {precision:.5f}, recall: {recall:.5f},  F1: {f1:.5f}')
```

The output result is:

```
evalution result: precision: 0.72308, recall: 0.84685, F1: 0.78008.
```

6. **Model Prediction**

As an example, we can input a text comment, such as "The taste of the cake is very good, so great, and the service provided by the shop is also very friendly," and then use the model to infer the corresponding prediction label sequences. Codes are implemented as follows:

```
def predict(input_text, ext_model, tokenizer, ext_id2label,
max_seq_len=512):

    encoded_inputs = tokenizer(list(input_text),
is_split_into_words=True, max_seq_len=max_seq_len,)
    input_ids = paddle.to_tensor([encoded_inputs["input_ids"]])
    token_type_ids =
paddle.to_tensor([encoded_inputs["token_type_ids"]])

    logits = ext_model(input_ids, token_type_ids=token_type_ids)
    predictions = logits.argmax(axis=2).numpy()[0]
    tag_seq = [ext_id2label[idx] for idx in predictions][1:-1]

    print("text: ", input_text)
    print("Prediction results: ", tag_seq)

max_seq_len = 512

input_text = "The taste of the cake is very good, so great, and the service provided by the
shop is also very friendly"
predict(input_text, ext_model, tokenizer, ext_id2label,
max_seq_len=max_seq_len)
```

The output result is:

`text:` The taste of the cake is very good, great, and the service provided by the shop is also very friendly Prediction results: `['O', 'O', 'O', 'B-Aspect', 'I-Aspect', 'B-Opinion', 'I-Opinion', 'O', 'O', 'B-Opinion', 'O', 'O', 'O', 'O', 'B-Aspect', 'I-Aspect', 'O', 'O', 'B-Opinion', 'I-Opinion'].`

7. **Matching Aspects and Opinions**

From the example "The taste of the cake is very good, so great, and the service provided by the shop is also very friendly," we can see that text comments often involve multiple aspects and opinions, so it is necessary to match aspects and opinions. For example, in the input results above, the aspect word "taste" is matched with the opinion words "very good" and "great." We use a heuristic method for matching, with the following rules:

(1) If there are aspect words and opinion words in a short sentence, they should be matched first.
(2) If only opinion words appear in a short sentence, they should be matched with the previous aspects in the text sequences.

The function decoding is defined below to implement the above functions. Codes are implemented as follows:

```
from seqeval.metrics.sequence_labeling import get_entities

def decoding(text, tag_seq):
    # Ensure that the lengths of the input sequences and label sequences are the same
    assert len(text) == len(tag_seq), f"text len: {len(text)}, tag_seq
len: {len(tag_seq)}"

    # Find the sentence segmentation positions through punctuation
    puncs = list(",.?;!，。？；！")
    splits = [idx for idx in range(len(text)) if text[idx] in puncs]

    # Split the text and label sequences through the above segmentation positions
    prev = 0
    sub_texts, sub_tag_seqs = [], []
    for i, split in enumerate(splits):
        sub_tag_seqs.append(tag_seq[prev:split])
        sub_texts.append(text[prev:split])
        prev = split
    sub_tag_seqs.append(tag_seq[prev:])
    sub_texts.append((text[prev:]))

    # Analyze the segmented short sentences and label sequences to obtain relevant aspects or
opinion words
    ents_list = []
    for sub_text, sub_tag_seq in zip(sub_texts, sub_tag_seqs):
        ents = get_entities(sub_tag_seq, suffix=False)
        ents_list.append((sub_text, ents))

    # Aspect and opinion word matching strategy
    # Retain analyzed aspects and opinions
    aps = []
    # Save the opinion words that are not matched with the current aspect words
    no_a_words = []
    for sub_text_seq, ent_list in ents_list:
        sub_aps = []
        sub_no_a_words = []
        for ent in ent_list:
            ent_name, start, end = ent
      if ent_name == "Aspect":
                aspect = sub_text_seq[start:end+1]
```

```
                sub_aps.append([aspect])
                if len(sub_no_a_words) > 0:
                    sub_aps[-1].extend(sub_no_a_words)
                    sub_no_a_words.clear()
            else:
                ent_name == "Opinion"
                opinion = sub_text_seq[start:end + 1]
                if len(sub_aps) > 0:
                    sub_aps[-1].append(opinion)
                else:
                    sub_no_a_words.append(opinion)


        # If the current sub_aps is not empty, add it to aps
        if sub_aps:
            aps.extend(sub_aps)
            if len(no_a_words) > 0:
                aps[-1].extend(no_a_words)
                no_a_words.clear()
        elif sub_no_a_words:
            if len(aps) > 0:
                aps[-1].extend(sub_no_a_words)
            else:
                no_a_words.extend(sub_no_a_words)

    # Process the opinion words that are not matched with the attribute words
    if no_a_words:
        no_a_words.insert(0, "None")
        aps.append(no_a_words)

    return aps

text = "The taste of the cake is very good, so great, and the service provided by the shop is also very friendly"
tag_seq = ['O', 'O', 'O', 'B-Aspect', 'I-Aspect', 'B-Opinion', 'I-Opinion', 'O', 'O', 'B-Opinion', 'O', 'O', 'O', 'O', 'B-Aspect', 'I-Aspect', 'O', 'O', 'B-Opinion', 'I-Opinion']

aps = decoding(text, tag_seq)

for ap in aps:
    aspect, opinion = ap[0], set(ap[1:])
    print(f"aspect: {aspect}, opinion: {opinion}\n")
```

The output results are:

```
aspect: taste, opinion: {'very good', great'}
aspect: service, opinion: {'friendly'}.
```

8.3.4 Aspect-Based Sentiment Classification

After extracting aspects and opinions, the aspect-based sentiment classification task is conducted. In this task, we use BERT to encode the text, obtain the semantic vector representation of the text, and obtain the score of the sentiment category label for each aspect through linear layer processing. The implementation processes of the aspect-based sentiment classification task is the same as processes introduced in Subsect. 8.3.3. The specific implementation methods are described below.

1. **Data Processing**

Data processing includes the steps of determining the dataset, loading the data, converting the data into feature form, constructing a DataLoader, and finally processing the same batch of data into an equal-length feature sequence. The DataLoader is used to pass the data into the BERT model in batches.

(1) Determine datasets

In this practice, we use the aspect-based sentiment classification dataset built by PaddleNLP of PaddlePaddle, which includes 800 data in the training sets, 100 data in the validation sets, and 100 data in the test sets. Each piece of data includes three items: the original text, the aspect text, and the labeled data. The aspect texts consist of aspects and opinion words. Below are some examples of training samples.

{"text": "Not bad, a great park", "aspect_text": "good park, very good park", "label": "positive"}

{"text": "The location is particularly good, and the room is slightly larger than a typical express hotel.", "aspect_text": "The room is ordinary.)", "label": "negative"}

{"text": "The location is particularly good, and the room is slightly larger than a typical express hotel.", "aspect_text": "the location is particularly good", "label": "positive"}.

Description:

In this task, we use the already built aspect-based sentiment analysis dataset. Readers can also concatenate the aspects and opinion information into aspect_text according to the output information in Subsect. 8.3.3 and pass it into the model for model training.

(2) Data Loading

The training set, validation set, and test set data are loaded after the dataset determination.

```
import paddlenlp
from datasets import load_dataset


# Load the data of training set, validation set and test set
train_path = "./data/data178685/train_cls.json"
dev_path = "./data/data178685/dev_cls.json"
test_path = "./data/data178685/test_cls.json"
dataset = load_dataset("json", data_files={"train":train_path,
"dev":dev_path, "test":test_path})


# Print the first data of the training set
print(dataset["train"][:1])
```

The output result is:

{'text': ['The service and environment are both good,......The service attitude is quite good,], 'aspect_text': ['the environment is good', 'Nice shape', 'the service attitude is quite good'], 'label': ['positive', 'positive', 'positive']}.

(3) Conversion from data into the feature form

Nest, we define the convert_example_to_feature function, and convert the input text data into feature form using the loaded AutoTokenizer, and then define a label dictionary cls_label2id, and convert the label sequence into the ID form. Codes are implemented as follows:

```
from functools import partial
from paddlenlp.transformers import AutoTokenizer

def convert_example_to_feature(example, tokenizer, label2id, max_seq_len
=512, is_infer=False):
    encoded_inputs = tokenizer(text=example["aspect_text"], text_pair=ex
ample["text"],  max_seq_len=max_seq_len, return_length=True)

    if not is_infer:
        encoded_inputs["label"] = label2id[example["label"]]

    return encoded_inputs


# The largest sequence length
max_seq_len = 512

# Define the name of models
model_name = "bert-base-chinese"

# Define Bert Tokenizer
tokenizer = AutoTokenizer.from_pretrained(model_name)

# Define the dictionary
cls_label2id = {"negative":0, "positive":1}
cls_id2label = {v: k for k, v in cls_label2id.items()}


# Change data into the ID form
trans_fn = partial(convert_example_to_feature, tokenizer=tokenizer, labe
l2id=cls_label2id, max_seq_len=max_seq_len)
columns = ["text", "aspect_text", "label"]
train_dataset = dataset["train"].map(trans_fn, batched=False, remove_col
umns=columns)
dev_dataset = dataset["dev"].map(trans_fn, batched=False, remove_columns
=columns)
```

```
test_dataset = dataset["test"].map(trans_fn, batched=False, remove_colum
ns=columns)


# Print samples of every dataset
print("train_dataset:", len(train_dataset))
print("dev_dataset:", len(dev_dataset))
print("test_dataset:", len(test_dataset))


# Print the first data in the training set
print("train example:", train_dataset[:1])
```

The output result is:

```
train_dataset: 800
dev_dataset: 100
test_dataset: 100
```

```
train example: {'label': [1], 'input_ids': [[101,
4384, 1862, 679, 7231, 102, 3302, 1218, 1469, 4384,
1862, 6963, 679, 7231, 100, 100, 102]], 'token_type_
ids': [[0, 0, 0, 0, 0, 0, 1, 1, 1, 1, 1, 1, 1, 1, 1, 1,
1]], 'length': [17], 'seq_len': [17]}.
```

(4) Construction of DataLoader

Next, it needs to construct DataLoader and use the predefined DataCollatorWithPadding function in PaddleNLP to change a batch of data into the same length. Codes are implemented as follows:

```
from paddle.io import BatchSampler, DataLoader
from paddlenlp.data import DataCollatorWithPadding

batch_size= 16
train_sampler = BatchSampler(train_dataset, batch_size=batch_size, shuff
le=True)
dev_sampler = BatchSampler(dev_dataset, batch_size=batch_size, shuffle=F
alse)
# Use DataCollatorWithPadding to change data into the same length in batches
data_collator = DataCollatorWithPadding(tokenizer)
train_loader = DataLoader(dataset=train_dataset, batch_sampler=train_sam
pler, collate_fn=data_collator)
dev_loader = DataLoader(dataset=dev_dataset, batch_sampler=dev_sampler,
collate_fn=data_collator)
test_loader = DataLoader(dataset=test_dataset, batch_sampler=dev_sample
r, collate_fn=data_collator)

# Print data of the first batch in the training set
print(next(iter(train_loader)))
```

The output result is:

```
{'input_ids': Tensor(shape=[16, 81], dtype=int64, place=CUDAPlace(0), stop
_gradient=True,
       [[101 , 1962, 1456, ..., 0    , 0    , 0    ],
        [101 , 3302, 1218, ..., 0    , 0    , 0    ],
        [101 , 1914, 3952, ..., 5763, 8024, 102 ],
        ...,
        [101 , 6392, 3177, ..., 0    , 0    , 0    ],
        [101 , 7650, 1501, ..., 0    , 0    , 0    ],
        [101 , 5439, 2360, ..., 0    , 0    , 0    ]]), 'token_type_ids': Ten
sor(shape=[16, 81], dtype=int64, place=CUDAPlace(0), stop_gradient=True,
       [[0, 0, 0, ..., 0, 0, 0],
        [0, 0, 0, ..., 0, 0, 0],
        [0, 0, 0, ..., 1, 1, 1],
        ...,
        [0, 0, 0, ..., 0, 0, 0],
        [0, 0, 0, ..., 0, 0, 0],
        [0, 0, 0, ..., 0, 0, 0]]), 'length': Tensor(shape=[16], dtype=int6
4, place=CUDAPlace(0), stop_gradient=True,
       [73, 18, 81, 21, 47, 22, 33, 15, 16, 22, 13, 35, 72, 17, 72, 18]),
'seq_len': Tensor(shape=[16], dtype=int64, place=CUDAPlace(0), stop_gradie
nt=True,
       [73, 18, 81, 21, 47, 22, 33, 15, 16, 22, 13, 35, 72, 17, 72, 18]),
'labels': Tensor(shape=[16], dtype=int64, place=CUDAPlace(0), stop_gradien
t=True,
       [1, 1, 1, 1, 1, 1, 1, 1, 1, 0, 0, 1, 1, 1, 0, 1])}
```

2. **Model Building**

The modeling method in this section is shown in Fig. 8.12. First, each aspect text (i.e., aspect words and opinion words) is concatenated with the user evaluation text to construct the input sentence of "[CLS] aspect text [SEP] overall evaluation text [SEP]." In this example, the input sentence is: "[CLS] the taste is good) [SEP] The food in this store tastes great. [SEP]," in which "[CLS]" represents the features of the entire sentence and "[SEP]" represents the separator between two sentences. In this way, we can convert the aspect-level sentiment classification task into a classification problem of an input sentence. Then, we can input the concatenated sequence into the BERT model and pass the vector at the position of [CLS] into the linear layer to obtain the final aspect-level sentiment classification result.

To conduct the tasks in this section, we use the AutoModelForSequenceClassification class to implement the above model construction function. At the same time, we need to classify text comments as positive or negative, which is a binary classification process, so it is necessary to set the number of classification categories to 2. The corresponding codes are as follows:

```
from paddlenlp.transformers import AutoModelForSequenceClassification

# There are positive and negative data in this case, so it is necessary to set the number of categories to 2
num_classes = 2
cls_model = AutoModelForSequenceClassification.from_pretrained(model_name, num_classes=num_classes)
```

The output results are:

```
2022-11-2220 35 37 708 [INFO] - We are using <class 'paddlenlp.transformers.bert.modeling.BertForSequence Classification'> to load 'bert-base-chinese'.

2022-11-2220 35 37 710 [INFO] - Already cached /home/aistudio/.paddlenlp/models/bert-base-chinese/bert-base-chinese.pdparams.
```

3. **Training Configuration**

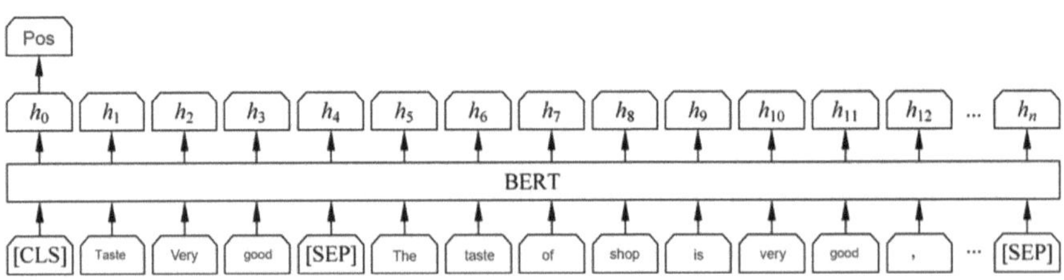

Fig. 8.12 The aspect-based sentiment classification model

Next, we define and configure the training environment for the aspect-based sentiment classification model, including configuring training parameters, configuring model parameters, setting the optimizer during training, defining loss functions, and evaluating metrics.

(1) Model: the BERT model.
(2) Optimizer: AdamW optimizer.
(3) Loss function: cross-entropy.
(4) Evaluation indicator: accuracy.

Codes are implemented as follows:

```
import os
import paddle
import paddle.nn as nn
from paddlenlp.transformers import LinearDecayWithWarmup


# Set the number of training rounds
num_epochs = 5


# The learning rate
learning_rate = 3e-5


# Set intervals for model evaluation every certain number of steps
eval_steps = 10


# Set intervals for printing the log every certain number of steps
log_steps = 10


# The model's saving content
save_dir = "./checkpoints"



# The weight decay coefficient of the AdamW optimizer during training

weight_decay = 0.01


# The proportion of warm-up training during the training process
warmup_proportion = 0.1


# The total number of training steps required
num_training_steps = len(train_loader) * num_epochs



# If save_dir doesn't not exist, then establish another one
if not os.path.exists(save_dir):
    os.mkdir(save_dir)
```

```
# Except for the parameters of bias and LayerNorm, other parameters perform decay operations during training
decay_params = [
        p.name for n, p in cls_model.named_parameters()
        if not any(nd in n for nd in ["bias", "norm"])
    ]

# Define lr_scheduler, the learning rate increases linearly during the warmup phase, and after the warmup phase, the learning rate gradually decays
lr_scheduler = LinearDecayWithWarmup(learning_rate=learning_rate,
total_steps=num_training_steps, warmup=warmup_proportion)

# Initialize the optimizer
optimizer = paddle.optimizer.AdamW(
        learning_rate=lr_scheduler,
        parameters=cls_model.parameters(),
        weight_decay=weight_decay,
        apply_decay_param_fun=lambda x: x in decay_params)

# Define the loss function
loss_fn = nn.CrossEntropyLoss()

# Define the computing method of evaluating indicators
metric = paddle.metric.Accuracy()
```

4. **Model Training**

Next, we define the train function and evaluate function to perform training and evaluate the model during training respectively. The codes are as follows:

```
def evaluate(model, data_loader, metric):

    model.eval()
    metric.reset()
    for batch_data in data_loader:
        input_ids, token_type_ids, labels = batch_data["input_ids"],  ba
tch_data["token_type_ids"], batch_data["labels"]
        logits = model(input_ids, token_type_ids=token_type_ids)

        correct = metric.compute(logits, labels)
        metric.update(correct)

    accuracy = metric.accumulate()

    return accuracy

def train(model):
```

```
    # Start the training mode of models
    model.train()
    global_step, best_acc = 1, 0.

    # Record the loss function during training and the accuracy score on the validation set
    train_loss_record = []
    train_score_record = []
    for epoch in range(1, num_epochs+1):
        for batch_data in train_loader():
            input_ids, token_type_ids, labels = batch_data["input_ids"],
  batch_data["token_type_ids"], batch_data["labels"]
            logits = model(input_ids, token_type_ids=token_type_ids)

            loss = F.cross_entropy(logits, labels)
            train_loss_record.append((global_step, loss.item()))

            loss.backward()
            lr_scheduler.step()
            optimizer.step()
            optimizer.clear_grad()

            if global_step > 0 and global_step % log_steps == 0:
                print(f"epoch: {epoch} - global_step: {global_step}/{num
_training_steps} - loss:{loss.numpy().item():.6f}")
            if (global_step > 0 and global_step % eval_steps == 0) or gl
obal_step == num_training_steps:
                accuracy  = evaluate(model, dev_loader,  metric)
                train_score_record.append((global_step, accuracy))
                model.train()
                if accuracy > best_acc:
                    print(f"best accuracy performence has been updated:
{best_acc:.5f} --> {accuracy:.5f}")
                    best_acc = accuracy
                    paddle.save(model.state_dict(), f"{save_dir}/best_cl
s.pdparams")
                print(f'evalution result: accuracy:{accuracy:.5f}')

            global_step += 1

    paddle.save(model.state_dict(), f"{save_dir}/final_cls.pdparams")

    return train_loss_record, train_score_record

train_loss_record, train_score_record = train(cls_model)
```

The output results are:

```
epoch: 1 - global_step: 10/250 - loss:0.632661
best accuracy performence has been updated: 0.00000 --> 0.79000
evalution result: accuracy:0.79000
epoch: 1 - global_step: 20/250 - loss:0.524497
evalution result: accuracy:0.79000
epoch: 1 - global_step: 30/250 - loss:0.463366
evalution result: accuracy:0.79000
epoch: 1 - global_step: 40/250 - loss:0.394368
best accuracy performence has been updated: 0.79000 --> 0.91000
evalution result: accuracy:0.91000
epoch: 1 - global_step: 50/250 - loss:0.285122
best accuracy performence has been updated: 0.91000 --> 0.92000
evalution result: accuracy:0.92000
```

```
……
epoch: 5 - global_step: 210/250 - loss:0.011691
evalution result: accuracy:0.96000
epoch: 5 - global_step: 220/250 - loss:0.005244
evalution result: accuracy:0.95000
epoch: 5 - global_step: 230/250 - loss:0.005973
evalution result: accuracy:0.95000
epoch: 5 - global_step: 240/250 - loss:0.008304
evalution result: accuracy:0.95000
epoch: 5 - global_step: 250/250 - loss:0.026805
evalution result: accuracy:0.95000.
```

During the training process, the loss train_loss_record and the score train_score_record on the validation set are saved, and the loss function and accuracy change of the training process are visualized. Codes are implemented as follows:

The output result is shown in Fig. 8.13.

In Fig. 8.13a, the horizontal axis represents the number of training steps, and the vertical axis represents the value of the loss function, which is the cross entropy between the predicted result and the true result. In Fig. 8.13b, the horizontal axis and vertical axis represent the number of training steps and the prediction accuracy respectively. From Fig. 8.13, it can be seen that as training progresses, the loss function on the training set continues to decrease, and the overall convergence tends to be 0. At the same time, the accuracy score on the validation set initially increases, and gradually stabilizes after the model converges. The final accuracy reaches around 97.5%.

5. **Model Evaluation**

The model is loaded with the best evaluation effect during training, and test it using the test set. The codes are as follows:

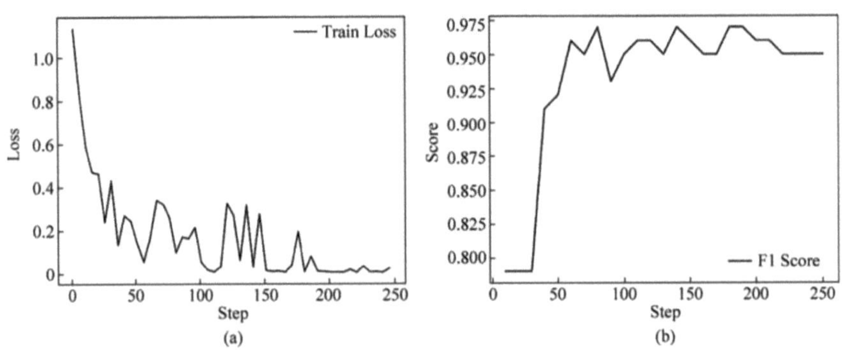

Fig. 8.13 The training of the classification model. **a** Loss changes in the training set; **b** Accuracy changes in the validation set

```
# Load the saved model parameters
model_path = "./checkpoints/best_cls.pdparams"

loaded_state_dict = paddle.load(model_path)
cls_model = AutoModelForSequenceClassification.from_pretrained(model_
name, num_classes=2)
cls_model.load_dict(loaded_state_dict)

accuracy = evaluate(cls_model, test_loader,  metric)
print(f'evalution result: accuracy:{accuracy:.5f}')
```

The output result is:
`evalution result: accuracy: 0.94000.`

6. Model Prediction

We can input any piece of text comment and attribute text, such as “The taste of the cake is good, but the appearance is a bit lacking.” and “Poor appearance,” and then obtain the corresponding sentiment labels for the attributes through model reasoning. Codes are implemented as follows:

```
def predict(input_text, aspect_text, cls_model, tokenizer, cls_id2lab
el, max_seq_len=512):

    # Change the input text information into the ID form
    encoded_inputs = tokenizer(text=aspect_text, text_pair=text, max_
seq_len=max_seq_len)
    input_ids = paddle.to_tensor([encoded_inputs["input_ids"]])
    token_type_ids = paddle.to_tensor([encoded_inputs["token_type_ids
"]])


    # Compute the sentiment classification
    logits = cls_model(input_ids, token_type_ids=token_type_ids)
    label_id = logits.argmax(axis=-1).numpy()[0]
    sentiment = cls_id2label[label_id]

    print("text: ", text)
    print("aspect_text: ", aspect_text)
    print("Prediction result: ", sentiment)

max_seq_len = 512
input_text = "The taste of the cake is good, but the appearance is a bit lacking."
aspect_text = "Poor appearance"

predict(input_text, aspect_text, cls_model, tokenizer, cls_id2label,
 max_seq_len=max_seq_len)
```

The output result is:

text: The taste of the cake is good, but the appearance is a bit lacking
aspect_text: Poor appearance
Prediction result: negative.

8.3.5 *Full-process Model Reasoning*

Subsections 8.3.3 and 8.3.4 respectively complete the training, evaluation, and prediction of the aspect-based sentiment classification model and the aspect-opinion extraction model, and save the model with the best evaluation results during the training process. Next, in this Sub-section, we will implement the full-process reasoning based on the saved model, that is, given any input of a piece of comment text, the model can predict the corresponding aspects, opinions, and sentiment tendencies.

First, the aspect-opinion extraction model, sentiment classification model, and other required data and components are loaded into memory. Codes are implemented as follows:

```
# Load the dictionary data
ext_label2id = {"O":0, "B-Aspect":1, "I-Aspect":2, "B-Opinion":3, "I-
Opinion":4}
ext_id2label = {v: k for k, v in ext_label2id.items()}
cls_label2id = {"negative":0, "positive":1}
cls_id2label = {v: k for k, v in cls_label2id.items()}


# Load Tokenizer
model_name = "bert-base-chinese"
tokenizer = AutoTokenizer.from_pretrained(model_name)
print("tokenizer loaded.")


# Load the aspect&opinion extraction model
ext_model_path = "./checkpoints/best_ext.pdparams"
ext_state_dict = paddle.load(ext_model_path)
ext_model = AutoModelForTokenClassification.from_pretrained(model_nam
e, num_classes=5)
ext_model.load_dict(ext_state_dict)
print("extraction model loaded.")

# Load the aspect-based sentiment classification model
cls_model_path = "./checkpoints/best_cls.pdparams"
cls_state_dict = paddle.load(cls_model_path)
cls_model = AutoModelForSequenceClassification.from_pretrained(model_
name, num_classes=2)
cls_model.load_dict(cls_state_dict)
print("classification model loaded.")
```

Then, we use the loaded dictionary, tokenizer, and model to make predictions. Codes are implemented as follows:

```
def concate_aspect_and_opinion(text, aspect, opinions):
   # Concatenate aspect words and opinion words
  aspect_text = ""
    for opinion in opinions:
        if text.find(aspect) <= text.find(opinion):
      aspect_text += aspect + opinion + ", "
        else:
            aspect_text += opinion + aspect + ", "
    aspect_text = aspect_text[:-1]

    return aspect_text

def predict(input_text, ext_model, cls_model, tokenizer, ext_id2label, c
ls_id2label, max_seq_len=512):

    ext_model.eval()
    cls_model.eval()

    # Process input text data
    encoded_inputs = tokenizer(list(input_text), is_split_into_words=Tru
e, max_seq_len=max_seq_len,)
    input_ids = paddle.to_tensor([encoded_inputs["input_ids"]])
    token_type_ids = paddle.to_tensor([encoded_inputs["token_type_ids
"]])


    # Extract aspects and opinion words
    logits = ext_model(input_ids, token_type_ids=token_type_ids)
    predictions = logits.argmax(axis=2).numpy()[0]
    tag_seq = [ext_id2label[idx] for idx in predictions][1:-1]
    aps = decoding(input_text, tag_seq)


    # Perform sentiment classification according to aspects
    results = []
    for ap in aps:
        aspect = ap[0]
        opinion_words = list(set(ap[1:]))
        aspect_text = concate_aspect_and_opinion(input_text, aspect, opi
nion_words)

        # Concatenate aspects and opinion words
        aspect_text = ""
        for opinion_word in opinion_words:
            if text.find(aspect) <= text.find(opinion_word):
                aspect_text += aspect+opinion_word+", "
            else:
                aspect_text += opinion_word+aspect+", "
        aspect_text = aspect_text[:-1]

        encoded_inputs = tokenizer(aspect_text, text_pair=input_text, ma
```

```
x_seq_len=max_seq_len, return_length=True)
        input_ids = paddle.to_tensor([encoded_inputs["input_ids"]])
        token_type_ids = paddle.to_tensor([encoded_inputs["token_type_id
s"]])

        logits = cls_model(input_ids, token_type_ids=token_type_ids)
        prediction = logits.argmax(axis=1).numpy()[0]

        result = {"aspect": aspect, "opinions": opinion_words, "sentimen
t": cls_id2label[prediction]}
        results.append(result)

    # Output prediction results
    for result in results:
        aspect, opinions, sentiment = result["aspect"], result["opinions
"], result["sentiment"]
        print(f"aspect: {aspect}, opinions: {opinions}, sentiment: {sent
iment}")

max_seq_len = 512
input_text = "The taste of the cake is very good, so great, and the service provided by the sho
p is also very friendly."
predict(input_text, ext_model, cls_model, tokenizer, ext_id2label, cls_i
d2label,  max_seq_len=max_seq_len)
```

The output result is:

aspect: taste, opinions: ['very good)', 'so great)'], sentiment: positive)
aspect: service, opinions: ['friendly)'], sentiment: positive).

It can be seen that in the above example, the model successfully extracted the opinion words of the aspects "taste" and "service" and correctly predicted the corresponding sentiment polarity.

8.4 Exercises

(1) In the practice of this chapter, we use a single-layer LSTM for sentiment analysis tasks. However, LSTMs can be stacked in multiple layers in practical applications. Readers are encouraged to experiment using multiple layers of LSTMs for sentiment analysis tasks.
(2) In the practice of this chapter, we use the LSTM API provided by PaddlePaddle to build models. Readers are encouraged to implement an LSTM model and conduct sentiment analysis experiments based on the models.
(3) In the experiments in this Chapter, some heuristic methods are used to match aspects and opinions. Please consider whether there are other ways to match aspects and opinions.

Chapter 9
Information Extraction

Abstract This chapter introduces information extraction (IE) as a key technique for transforming unstructured text into structured knowledge, with a focus on named entity recognition (NER) and relation extraction. It first presents a Bi-LSTM-CRF-based approach for NER, detailing data processing, model construction, and evaluation using the CLUENER2020 dataset, where entities are identified through sequence labeling. The chapter then explores relation extraction using ERNIE-UIE, a prompt-based pre-trained model framework that supports low-resource and zero-shot scenarios, demonstrating its ability to extract semantic relations as structured triples. Together, these methods illustrate practical deep learning strategies for advancing automated information extraction tasks.

Information extraction (IE) refers to the process of processing and extracting important information contained in unstructured texts and transforming it into the structured form. The input of this task is text, and the output is information fragments with a fixed structure. Information fragments are extracted from different documents and then integrated together in a unified format. Information extraction technology does not require understanding all the information in the entire document, but only analyzes the relevant information contained in the document, depending on the task goal. It mainly includes sub-tasks such as named entity recognition, relation extraction, and event extraction.

This chapter focuses on the implementation methods of the first two sub-tasks.

Named entity recognition (NER) is also known as named entity recognition and classification (NERC), and its purpose is to identify named entities and their corresponding categories in text. The concept of named entity was first proposed at the Sixth Message Understanding Conferences, MUC-6. It mainly focused on the extraction and recognition of entities such as people's names, place names, and organization names. With the rapid development of named entity recognition technology, the types of entities have been continuously increased and refined. MUC-7 defines three major categories of named entities: named entities (such as people's names, place names, organization names, etc.), time expressions (such as dates, times, duration, etc.),

C. Zong et al., *Natural Language Processing and Large Language Models*,
https://doi.org/10.1007/978-981-92-0682-7_9

and numerical expressions (such as percentages, lengths, areas, volumes, monetary amounts, and other measurement data, as well as cardinal numbers). For example:

> The 24th Winter Olympic Games was opened in Beijing on February 4, 2022.

Among them, "Winter Olympic Games" is the name of the event, "Beijing" is the name of the place, and "February 4, 2022" is the date. The current named entity recognition task mainly identifies the entity information contained in the text through sequence modeling.

Entity relation extraction refers to the identification of entities from unstructured or semi-structured texts and the extraction of semantic relations between entities, such as the positional relation between places and the social relation between people's names. This relation is represented by <subject, predicate, object>. For example:

> Zhao Hongyin passed away during the Later Zhou Dynasty, and it was not until his son Zhao Kuangyin established the Song Dynasty.

This sentence can be extracted into two triples: Zhao Hongyin, child, Zhao Kuangyin、 Zhao Kuangyin, father, Zhao Hongyin.

This chapter mainly introduces information extraction methods based on deep learning, focusing on two key approaches. The first is the entity extraction method based on the bi-directional long short-term memory, Bi-LSTM, and CRF, which models texts' context features through neural networks to complete the task of named entity recognition. The second is the entity relation extraction method based on the pre-training language model ERNIE-UIE (Lu et al. 2022).

9.1 The Method of NER Based on Bi-LSTM and CRF

9.1.1 Task Objectives and Implementation Processes

The process of NER can be seen as a sequence labeling problem. The basic idea is to first segment the given text into words or characters. For the English NER, this Chapter uses word as labeling unit, and for the Chinese NER, the characters are treated as labeling units.

Taking Chinese NER as example, we use CRF model to convert the character sequences of a given text into entity labeling sequences. The BISO tagging system is used, as shown in Table 9.1. In the sentence "Yao Ming serves as the President of the Chinese Basketball Association," "Yao Ming" corresponds to the person entity, in which "Yao" is the starting character of the person entity, represented by "B-person," and "Ming" belongs to the person entity, represented by "I-person." "the Chinese Basketball Association"is an organization entity. The tags "B-organization" and "I-organization" are used to indicate the initial character and belonging characters of this entity, the same as the labeling method of person entities. "S" indicates that a single word independently constitutes an entity. Other non-entity types of characters are represented by the tag "O," indicating that the character does not belong to any type of entity. After identifying the tag sequences of the given text, the entity boundaries and types in the text can be obtained through the labeling sequences.

CRF is shown in Fig. 9.1. The input of the model is text, and the output is the label sequence corresponding to the given text. During the modeling process, firstly, the data of input text is processed, and then Bi-LSTM is used to encode the text sequences. At each step of LSTM, a semantic vector representation corresponding to the input is obtained. These vectors are passed to the linear layer for processing, and the emission score of the semantic vector is obtained. Finally, the emission score is passed to the CRF model for decoding, so as to obtain the entity label corresponding to the original input.

Please note that assuming the length of the text sequence is n and the number of labels is k, and there will be k^n different label sequences. The role of CRF is to select the path with the highest score among many candidate paths, which is the optimal solution, as shown in Fig. 9.2.

9.1.2 Data Processing

Data processing includes the steps of determining the dataset, loading the data, constructing the CLUENERDataset class, constructing the DataLoader, etc. The same batch of data is finally processed into an equal-length feature sequence, which is sent into the Bi-LSTM model in batches using DataLoader.

1. **Determine the Dataset**

This practice uses the CLUENER2020[1] dataset to realize the method of NER. The CLUENER2020 is selected for more fine-grained named entity labeling based on the text classification dataset THUCTC open-sourced by Tsinghua University. The dataset contains 12,091 pieces of data, including 9405 training samples, 1343 test samples, and 1343 validation samples. It contains 10 different entity categories: address, book, company, game, government, movie, name, organization, position, and scene. In this experiment, we use BISO tags to label entities. "B" represents the beginning character of a multi-word entity word, "I" represents the middle or end character of a multi-character entity word, "S" represents the character of a single-character entity word, and "O" represents a non-entity word. See Table 9.2 for more details.

[1] The CLUENER2020 dataset can be obtained at https://www.cluebenchmarks.com/introduce.html. Readers can also obtain the dataset from the online course in the AI Studio.

Table 9.1 Corresponding relations between words and labels

Original sentence	姚明担任中国篮协主席 Yao Ming serves as the president of the Chinese Basketball Association)									
Word segmentation	姚 (Yao)	明 (Ming)	担 任 (serves as)		中 国 (China)		(Basketball Association)		主 席 (president)	
Mark	B-person	I-person	O	O	B-organization	I-organization	B-organization	I-organization	O	O

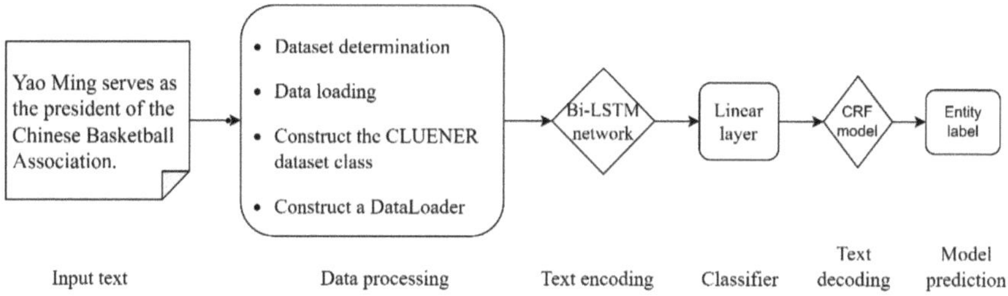

Fig. 9.1 The implementation flowchart of named entity recognition task

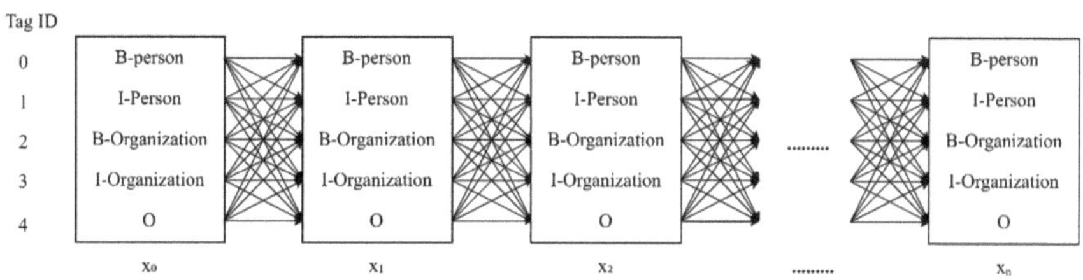

Fig. 9.2 The decoding of the CRF model

Table 9.2 The CLUENER2020 dataset and label set

Starting with B	Starting with I	S single character	Other
B-address	I-address	S-address	O
B-book	I-book	S-book	
B-company	I-company	S-company	
B-government	I-government	S-government	
B-movie	I-movie	S-movie	
B-game	I-game	S-game	
B-name	I-name	S-name	
B-organization	I-organization	S-organization	
B-position	I-position	S-position	
B-scene	I-scene	S-scene	

To facilitate readers' learning and use, we have pre-processed the data and generated a tag dictionary "tag.dict" and a character dictionary "vocab.dict." The relevant data includes: train.json training set, dev.json validation set, test.json test set, tag.dict tag dictionary, and vocab.dict character dictionary. Below is a sample data from the training set, containing text data and corresponding word position label sequences, separated by spaces.

{'text': 'The Housing Security and Real Estate Management Bureau of Shijiazhuang City has provided answers to the hot issues that citizens have been concerned about recently.)', 'tag': 'B-government I-government I-government I-government I-government I-government I-government I-government I-government I-government I-government I-government I-government I-government O O O O O O O O O O O O O O O O O O O O O O O O '}.

In this example, the first 14 characters compose a organization name and others characters don't belong to any named entity.

2. **Data Loading**

The following codes read the CLUENER2020 data and the constructed tag dictionary and character dictionary into memory:

```
import os
import json

# Define data and load functions
def load_dataset(path,mode="train"):
    assert mode in ["train", "dev", "test"]
    data_path = os.path.join(path, mode+".json")
    examples = []
    with open(data_path, "r", encoding="utf-8") as f:
        for idx, line in enumerate(f):
            example = {}
            line = json.loads(line.strip())
            text = line["text"]
            tag_entities = line.get("label", None)
            words = list(text)
            tags = ["O"] * len(words)
            if tag_entities is not None:
                for tag_name, tag_value in tag_entities.items():
                    for entity_name, entity_index in tag_value.items():
                        for start_index, end_index in entity_index:
                            assert "".join(words[start_index:end_index+
1]) == entity_name
                            if start_index == end_index:
                                tags[start_index] = "S-" + tag_name
                            else:
                                tags[start_index] = "B-" + tag_name
                                tags[start_index + 1:end_index + 1] = ["I
-" + tag_name] * (len(entity_name) - 1)
            example["text"] = " ".join(words)
            example["tag"] = " ".join(tags)
            examples.append(example)
    return examples
```

```
# Load the training set, test data and validation data
root_path = "./dataset/"
train_data = load_dataset(root_path,mode="train")
dev_data = load_dataset(root_path,mode="dev")
test_data = load_dataset(root_path,mode="test")

# Print the first piece of data and view the data format: (sentence, label)
print(train_data[0])
```

The output result is:

'text': 'The live broadcast report of the "2009 Urban Building Complex High end Financial Intelligence Summit" by Jinmao Center leads netizens into the "Top of Xicheng" Jinmao Center.)', 'tag': 'B-address I-address I-address I-address O O O O O O O O O O O O O O O O O O O O O O O O O O O O O O O O O O O O O O B-address I-address I-address I-address O'}.

From the output results, it can be seen that each sample contains two parts: text sequence and label, where the character "jin" corresponds to the label B-address, the character "mao" corresponds to the label I-address, and so on.

The following codes load thevocabulary and labels into the dictionaries word2id_dict and tag2id_dict.

```
import os
# Load the dictionaries
def load_dict(path,dict_name):
    assert dict_name in ["tag", "vocab"]
    data_path = os.path.join(path, dict_name+".dict")
    data_dict = {}
    with open(data_path, "r", encoding="utf-8") as f:
        lines = [item.strip().split("\t") for item in f.readlines()]
        data_dict = dict([(item[1], int(item[0])) for item in lines])
    return data_dict
word2id_dict = load_dict(root_path,dict_name="vocab")
tag2id_dict = load_dict(root_path,dict_name="tag")
# View the mapping relationship between the entity labels of the dataset and id
train_example_list = []
for train_tag in train_data[0]['tag'].split():
    if train_tag in tag2id_dict:
        train_example_list.append(str(tag2id_dict[train_tag]))
    else:
        train_example_list.append(str(train_tag))
print('tag2id:', ' '.join(train_example_list))
```

The output result is:

tag2id: 1 11 11 11 0 0 0 0 0 0 0 0 0 0 0 0 0 0 0 0 0 0 0 0 0 0 0 0 0 0 0 0 0 0 0 0 0 0 0 0 0 0 0 1 11 11 11 0.

3. **Construct the CLUENERDataset Class**

The CLUENERDataset class is constructed for data management and inherits from the paddle.io.DataSet class. Codes are implemented as follows:

```
from paddle.io import Dataset

class CLUENERDataset(Dataset):
    def __init__(self, examples, word2id_dict, tag2id_dict):
        super(CLUENERDataset, self).__init__()

        # Construct a dictionary to convert words into numbers for dictionary indexes.
        self.word2id_dict =  word2id_dict
        self.tag2id_dict =  tag2id_dict

        # The dataset after loading
        self.examples = self.convert_tokens_to_ids(examples, self.word2
id_dict, self.tag2id_dict)
        self.examples = self.split_data()

    def split_data(self):

        # Sort the data according to the length of the samples
        self.examples = sorted(self.examples, key=lambda x: x[2], rever
se=True)
        return self.examples

    def convert_tokens_to_ids(self,data, word2id_dict, tag2id_dict):
        examples = []
        for example in data:
            text = example['text']
            tokens = [word2id_dict.get(w, word2id_dict["[UNK]"]) for w
in text.split(" ")]
            text_real_len = len(tokens)
            tag = example['tag']
            tag_ids = [tag2id_dict.get(t, tag2id_dict["O"]) for t in ta
g.split(" ")]
            examples.append((tokens, tag_ids, text_real_len))
        return examples

    def __getitem__(self, idx):
        tokens, tag_ids, text_real_len = self.examples[idx]
        return tokens, tag_ids, text_real_len

    def __len__(self):
        return len(self.examples)

# Instantiate the Dataset
train_set = CLUENERDataset(train_data, word2id_dict, tag2id_dict)
dev_set = CLUENERDataset(dev_data, word2id_dict, tag2id_dict)
test_set = CLUENERDataset(test_data, word2id_dict, tag2id_dict)

print('Number of samples of the training set: ', len(train_set))
print('Number of samples of the validation set: ', len(dev_set))
print('Number of samples of the test set: ', len(test_set))
print('Samples: ', train_set[4])
```

The output result is:

Number of samples of the training set: 9405
Number of samples of the validation set: 1343
Number of samples of the test set: 1343.
Samples: ([2616, 2617, 37, 18, 27, 630, 281, 765, 37, 18, 27, 288, 1481, 37, 18, 27, 24, 387, 173, 1284, 37, 6, 691, 274, 5, 153, 56, 5, 24, 391, 22, 11, 263, 26, 222, 200, 1729, 11, 388, 314, 585, 484, 639, 51, 13, 391, 426, 538, 133, 12], [0, 0, 0, 0, 0, 0, 0, 0, 0, 0, 0, 0, 0, 0, 0, 0, 0, 0, 0, 0, 0, 0, 0, 0, 0, 0, 0, 0, 0, 0, 0, 0, 0, 0, 0, 0, 7, 17, 17, 0, 9, 19, 0, 0, 0, 0, 6, 16, 16, 16, 16, 16], 50).

From the output result we can see that after the text is segmented by characters, each character is converted into an ID number in the dictionary.

4. **Construct DataLoader**

When training models, data is usually passed into the model in batches for separate training. Each batch of data is passed into the model as a minibatch for processing. Each batch of data contains two parts: text data and corresponding labels. The paddle.io.DataLoader in the PaddlePaddle framework can be used to implement functions of loading in batches and upgrading data. In addition, we also need to define the collate_fn function for text truncation and padding, which can be passed into the DataLoader as a callback function. The DataLoader calls collate_fn for data processing before returning each batch of data, and then returns the processed text data and corresponding labels. Codes are implemented as follows:

```
import paddle
from functools import partial
def collate_fn(batch, pad_val, mask=None):
    # By adding [Padding], the samples in a minibatch are padded to the maximum length
    batch = list(zip(*batch))
    tokens_list, tags_list, lens = batch
    max_len = max(lens)
    batch_size = len(lens)
    # Initialize the data needed to be passed into
    batch_token = paddle.full(shape=[batch_size, max_len], fill_value=0,
dtype="int64")
    batch_tag = paddle.full(shape=[batch_size, max_len], fill_value=0, dt
ype="int64")
    batch_mask = paddle.full(shape=[batch_size, max_len], fill_value=0, d
type="int64")
    # Padding the data in each minibatch
    for i in range(batch_size):
        batch_token[i, :lens[i]] = paddle.to_tensor(tokens_list[i], dtype
="int64")
        batch_tag[i, :lens[i]] = paddle.to_tensor(tags_list[i], dtype="in
t64")

        if mask:
            batch_mask[i, :lens[i]] = paddle.to_tensor([1] * lens[i], dty
pe="int64")
    if mask:
        batch_lens = paddle.to_tensor(lens)
        return (batch_token, batch_tag, batch_mask, batch_lens)
    batch_lens = paddle.to_tensor(lens)
    return (batch_token, batch_tag, batch_lens)
```

Next, we need to construct the DataLoader for upgrading data in batches and specify the batch size. When upgrading the data, we should specify whether to shuffle the samples using the parameter shuffle. Codes are implemented as follows:

```
from paddle.io import DataLoader as DataLoader

batch_size = 32
collate_fn = partial(collate_fn, pad_val=0, mask=True)

# Construct the DataLoader, and upgrade the training set data, validation set data, and test set data in batches according to the batch size
train_loader = DataLoader(train_set, batch_size=batch_size, shuffle=True, drop_last=False, collate_fn=collate_fn)
dev_loader = DataLoader(dev_set, batch_size=batch_size, shuffle=False, drop_last=False, collate_fn=collate_fn)
test_loader = DataLoader(test_set, batch_size=batch_size, shuffle=False, drop_last=False, collate_fn=collate_fn)

# Print the data of the first minibatch in the training set
print(next(iter(train_loader)))
```

The output result is:

```
[Tensor(shape=[32, 49], dtype=int64, place=CUDAPlace(0), stop_gradient=True,
       [[540 , 609 , 866 , ..., 0   , 0   , 0   ],
        [146 , 60  , 25  , ..., 0   , 0   , 0   ],
        [17  , 9   , 9   , ..., 0   , 0   , 0   ],
        ...,
        [103 , 78  , 63  , ..., 209 , 1412, 5   ],
        [25  , 75  , 213 , ..., 0   , 0   , 0   ],


        [123 , 243 , 328 , ..., 0   , 0   , 0   ]]), Tensor(shape=[32, 49], dtype=int64, place=CUDAPlace(0), stop_gradient=True,
       [[0 , 1 , 11, ..., 0 , 0 , 0 ],
        [0 , 0 , 0 , ..., 0 , 0 , 0 ],
        [0 , 0 , 0 , ..., 0 , 0 , 0 ],
        ...,
        [0 , 0 , 0 , ..., 0 , 0 , 0 ],
        [8 , 18, 18, ..., 0 , 0 , 0 ],
        [0 , 0 , 0 , ..., 0 , 0 , 0 ]]), Tensor(shape=[32, 49], dtype=int64, place=CUDAPlace(0), stop_gradient=True,
       [[1, 1, 1, ..., 0, 0, 0],
        [1, 1, 1, ..., 0, 0, 0],
        [1, 1, 1, ..., 0, 0, 0],
        ...,
        [1, 1, 1, ..., 1, 1, 1],
        [1, 1, 1, ..., 0, 0, 0],
        [1, 1, 1, ..., 0, 0, 0]]), Tensor(shape=[32], dtype=int64, place=CUDAPlace(0), stop_gradient=True,
       [46, 36, 42, 32, 36, 38, 16, 32, 47, 47, 42, 46, 37, 45, 47, 26, 41, 43, 47, 47, 49, 39, 14, 41, 32, 23, 19, 32, 49, 49, 21, 46])]
```

From the output results, we can see that the data lengths of the same batch are the same.

9.1.3 Model Construction

The NER method based on Bi-LSTM and CRF is shown in Fig. 9.3. The specific processes are as follows: The Bi-LSTM module and the CRF module process the data sent in batches in turn. First, the Bi-LSTM module encodes the processed text and obtains the context vector sequences of the text. Then, the context vector at each position is passed to the linear layer (vector sequences multiplied by weight, plus bias), and the emission score corresponding to each vector is output, representing the label score corresponding to that position. Then the model passes the emission score at each position to the CRF module. During the model training stage, CRF calculates the score of the true label path and the scores of all paths, and outputs the value of the loss function for model upgrading. During the model testing stage, CRF analyzes the path with the highest score from all potential label paths, and outputs the corresponding label sequences.

Based on the modeling idea shown in Fig. 9.3, codes are implemented as follows:

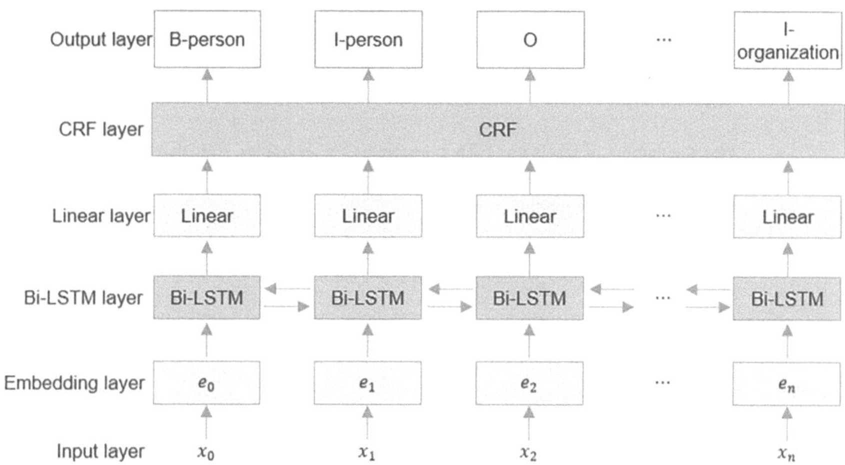

Fig. 9.3 The framework of the named entity recognition method based on BiLSTM and CRF

```
import paddle
import paddle.nn as nn
from paddlenlp.layers import crf
from utils.metric import SeqEntityScore

class NERModel(nn.Layer):
    def __init__(self, vocab_size, embedding_size, hidden_size, label2i
d, n_layers=2, drop_p=0.1):
        super(NERModel, self).__init__()
        self.vocab_size = vocab_size
        self.embedding_size = embedding_size
        self.hidden_size = hidden_size
        self.label2id = label2id

        self.embedding = paddle.nn.Embedding(vocab_size, embedding_siz
e)
        self.bilstm = paddle.nn.LSTM(input_size=embedding_size, hidden_
size=hidden_size, direction="bidirectional", num_layers=n_layers, dropo
ut=drop_p)
        self.layer_norm = paddle.nn.LayerNorm(hidden_size*2)
        self.dropout_emb = paddle.nn.Dropout(p=drop_p)

        # When implementing CRF, two auxiliary tags, <START> and <STOP>, are introduced
        self.classifier = paddle.nn.Linear(hidden_size*2, len(label2id)
+2)

        """

        Pass the number of labels into CRF to generate CRF examples. Note that n_labels does
not include the <START> and <STOP> labels
        num_labels (int):the number of labels


        crf_lr (float, optional):The default learning rate for the CRF layer is 0.1


        with_start_stop_tag (bool, optional):If set to True, include the <START> and <STOP>

tags    """
```

```
        self.crf = crf.LinearChainCrf(len(label2id), crf_lr=0.001, with
_start_stop_tag=True)

        """
        Define the negative log-likelihood loss function for CRF

        crf (LinearChainCrf): LinearChainCf object, whose parameters will be used to calculate
the loss
        """
        self.crf_loss = crf.LinearChainCrfLoss(self.crf)

        """

        ViterbiDecoder can decode the  tag sequence with highest score, which should be used

during testing


        transitions (Tensor): Transition matrix. Its data type is float32


        with_start_stop_tag (bool, optional): If set to True, include <START> and <STOP> tags
        """
        self.viterbi_decoder = crf.ViterbiDecoder(self.crf.transitions)

    def forward(self, input_ids, input_mask):


        # This forward calculation will output the sequences' hidden states of the last layer of

Bi-LSTM
        # input_ids: [batch_size, seq_len]
        # embs: [batch_size, seq_len, embedding_size]
        embs = self.embedding(input_ids)
        embs = self.dropout_emb(embs)
        embs = embs * paddle.to_tensor(input_mask, dtype="float32").uns
queeze(2)
        last_layer_hiddens, _ = self.bilstm(embs)
        last_layer_hiddens = self.layer_norm(last_layer_hiddens)
        features = self.classifier(last_layer_hiddens)

        return features

    def forward_loss(self, input_ids, input_mask, input_lens, input_tag
s=None):

        features = self.forward(input_ids, input_mask)

        if input_tags is not None:
            return features, self.crf_loss(features, input_lens, input_
tags)
        return features
```

9.1.4 Training Configuration

Now we need to determine the computing resources, models, optimizers, loss functions, and optimization indicators used during model training.

(1) Model: Utilize the Bi-LSTM model and the CRF model.
(2) Loss function: Negative log-likelihood loss function.
(3) Evaluation index: F1 value.

Codes are implemented as follows:

```
from utils.utils import set_seed
# Set the rounds of training
num_epochs = 30
vocab_size = len(word2id_dict.keys())
embedding_size = 256
hidden_size = 256
n_layers = 2
dropout_rate = 0.1
# Set the learning rate
learning_rate = 0.001
seed = 0
# Set the interval for printing the log every certain number of steps
log_steps = 100
# Set the interval for model evaluation on the validation set
eval_steps = 500
# Set the model's saving path and automatically save the model with the best effect during the

training process
checkpoint = "./checkpoint"
# Use the fixed random seed method
set_seed(seed)
# Conversely convert the mapping relationship tag2id to obtain the id2tag dictionary
id2tag = dict([items[1], items[0]] for items in tag2id_dict.items())
# Instantiate models
ner_model = NERModel(vocab_size=vocab_size, embedding_size=embedding_si
ze,
                     hidden_size=hidden_size,label2id=tag2id_dict, n_la
yers=n_layers, drop_p=dropout_rate)
# Specify optimization strategies and update models' parameters
optimizer = paddle.optimizer.Adam(learning_rate=learning_rate, beta1=0.
9, beta2=0.99,
                                  parameters=ner_model.parameters())
# Instantiate the evaluation indicator object, which is used to receive data, real labels, and

calculate the accuracy, recall, and F1 score of the evaluation indicator
metric = SeqEntityScore(id2tag)
```

9.1.5 Model Training

During the model training process, a training log is printed every certain log_steps, and model evaluation is conducted on the validation set every certain eval_steps. The model with the best evaluation performance during the training process is saved.

During model training, the accuracy, recall, and F1 score for each entity type are calculated. Assuming that the number of true person entities in the set of samples is T, and the number of predicted person entities is P, and the number of correctly predicted person entities is C, the calculation method is:

$$\text{Precision} = \frac{C}{P} \times 100\%$$

$$\text{Recall} = \frac{C}{T} \times 100\%$$

$$\text{F1} = \frac{2 \times \text{precision} \times \text{recall}}{\text{precision} + \text{recall}}$$

The calculation method for other types of entities is the same. The statistics for all entity types' indicators are calculated by summarizing the data of all types of tags. We will not repeat this here. The codes for evaluation on the validation set are as follows:

```
def evaluate(model,test_loader, metric):
    # Reset the class of evaluation indicators
    metric.reset()
    model.eval()
    with paddle.no_grad():
        for step, batch in enumerate(test_loader):
            # Obtain data
            batch_ids, batch_tags, batch_mask, batch_lens = batch
            features, loss = model.forward_loss(batch_ids, batch_mask,
batch_lens, batch_tags)
            # Decode using CRF based on the emission score
            scores, pred_paths = model.viterbi_decoder(features,
batch_lens)
            pred_paths = pred_paths.tolist()
            # Convert these predicted tag sequences into id2tag, that is, convert them into
corresponding tags
            pred_paths = [[id2tag[tag_id] for tag_id in tag_seq] for
tag_seq in pred_paths]
            batch_tags = batch_tags.numpy().tolist()
            real_paths = [tag_seq[:tag_len] for tag_seq, tag_len in
zip(batch_tags, batch_lens)]
            # Update the data related to statistical indicators
            metric.update(pred_paths=pred_paths, real_paths=real_paths)
        # Calculate the final F1 value based on data of metric
        result = metric.get_result()
    return result
```

The codes for model training are implemented as follows:

```
def train(model, train_loader):
    global_step, best_f1 = 1, 0.

    # Record the loss function value during training and the score of the model on the

    validation set
    train_loss_record = []
    train_score_record = []
    for epoch in range(1, 1 + num_epochs):
        model.train()
        for batch in train_loader:
            batch_ids, batch_tags, batch_mask, batch_lens = batch

            # Perform forward computation of the model and calculate the loss
            features, loss = model.forward_loss(batch_ids, batch_mask,
batch_lens, batch_tags)
            loss = paddle.mean(loss)
            train_loss_record.append((global_step, loss.item()))


            # Gradient calculation and reverse parameter updating
            loss.backward()
            optimizer.step()
            optimizer.clear_gradients()

            # Print information during training
            if global_step % log_steps == 0:
                print(f"epoch: {epoch}, global_step: {global_step}, loss:
```

```
{loss.numpy()[0]}")
            if global_step % eval_steps == 0 or global_step ==
len(train_loader) * num_epochs:
                result = evaluate(model, dev_loader, metric)
                model.train()
                f1 = result["Total"]["F1"]
                train_score_record.append((global_step, f1))
                if f1 > best_f1:
                    print(f"best F1 performence has been updated:
{best_f1:.5f} --> {f1:.5f}")
                    best_f1 = f1
                    paddle.save(model.state_dict(),
f"{checkpoint}/best.pdparams")
                metric.format_print(result)
            global_step += 1

    paddle.save(model.state_dict(), f"{checkpoint}/final.pdparams")
    return train_loss_record, train_score_record

train_loss_record, train_score_record = train(ner_model, train_loader)
```

The output result is:

```
epoch: 1, global_step: 100, loss: 23.395023345947266
epoch: 1, global_step: 200, loss: 16.94906234741211
epoch: 2, global_step: 300, loss: 16.112838745117188
epoch: 2, global_step: 400, loss: 9.73039722442627
epoch: 2, global_step: 500, loss: 12.321440696716309
best F1 performence has been updated: 0.00000 --- > 0.51510

Total: Precision: 0.574 - Recall: 0.4671 - F1: 0.5151
Entity: name - Precision: 0.6543 - Recall: 0.529 - F1: 0.585
Entity: position - Precision: 0.8045 - Recall: 0.4088 - F1: 0.5421
Entity: address - Precision: 0.3651 - Recall: 0.185 - F1: 0.2456
Entity: organization - Precision: 0.5656 - Recall: 0.5286 - F1:
0.5465
Entity: company - Precision: 0.5117 - Recall: 0.463 - F1: 0.4861
Entity: government - Precision: 0.5857 - Recall: 0.498 - F1: 0.5383
Entity: movie - Precision: 0.6643 - Recall: 0.6291 - F1: 0.6463
Entity: book - Precision: 0.7097 - Recall: 0.4286 - F1: 0.5344
Entity: scene - Precision: 0.3654 - Recall: 0.4545 - F1: 0.4051
Entity: game - Precision: 0.6019 - Recall: 0.661 - F1: 0.63
......
Epoch: 30, global_step: 8600, loss: 0.7725200653076172
epoch: 30, global_step: 8700, loss: 0.4402198791503906
epoch: 30, global_step: 8800, loss: 0.17098569869995117
best F1 performence has been updated: 0.65140 -- > 0.65500

Total: Precision: 0.6735 - Recall: 0.6374 - F1: 0.655
Entity: name - Precision: 0.7252 - Recall: 0.6925 - F1: 0.7085
Entity: position - Precision: 0.7411 - Recall: 0.6744 - F1: 0.7062
Entity: address - Precision: 0.4837 - Recall: 0.437 - F1: 0.4592
Entity: organization - Precision: 0.7455 - Recall: 0.6703 - F1:
0.7059
Entity: company - Precision: 0.6925 - Recall: 0.6376 - F1: 0.6639
Entity: government - Precision: 0.637 - Recall: 0.7247 - F1: 0.678
Entity: movie - Precision: 0.7436 - Recall: 0.5762 - F1: 0.6493
Entity: book - Precision: 0.7254 - Recall: 0.6688 - F1: 0.6959
Entity: scene - Precision: 0.4973 - Recall: 0.445 - F1: 0.4697
Entity: game - Precision: 0.7095 - Recall: 0.7864 - F1: 0.746.
```

Then, the loss train_loss_record and the score train_score_record on the validation set during the training process are saved, and the training process is visualized.

```
from utils.utils import plot_training_loss, plot_training_acc

fig_path = "./images/chapter8_loss.pdf"
plot_training_loss(train_loss_record, fig_path, loss_legend_loc="upper ri
ght", sample_step=20)

fig_path = "./images/chapter8_acc.pdf"
plot_training_acc(train_score_record, fig_path, sample_step=20, loss_lege
nd_loc="lower right")
```

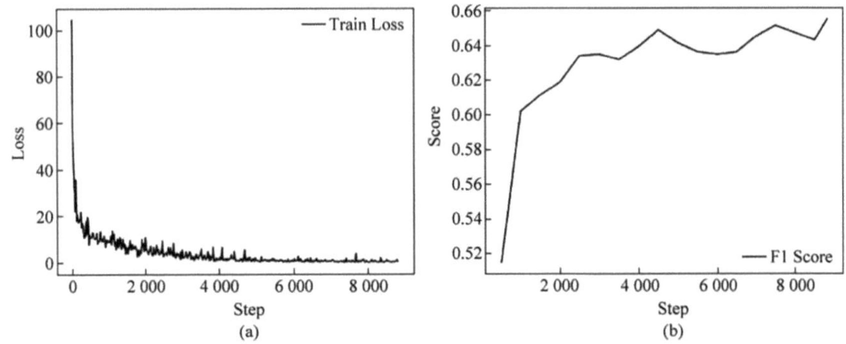

Fig. 9.4 The trend of the loss function of the model on the training set and the trend of the F1 score of the model on the validation set. **a** Loss changes in the training set. **b** F1 score changes in the validation set

The output result is shown in Fig. 9.4.

From the output results, it is clear that as the training progresses, the loss function value on the training set continues to decrease and then converges until the value tends to 0. In contrast, the F1 value on the validation set continues to increase and gradually converges towards a stable value as the number of training steps increases.

9.1.6 Model Evaluation

Then, the model that performs best during training is loaded, and the F1 score of the model on the test set is calculated. Codes are implemented as follows:

```
# Load the trained model for evaluation
model_path = "./checkpoint/best.pdparams"
loaded_state_dict = paddle.load(model_path)
ner_model = NERModel(vocab_size=vocab_size, embedding_size=embedding_siz
e, hidden_size=hidden_size,label2id=tag2id_dict, n_layers=n_layers, drop_
p=dropout_rate)
ner_model.load_dict(loaded_state_dict)

metric = SeqEntityScore(id2tag)
result = evaluate(ner_model, test_loader, metric)
metric.format_print(result)
```

The output result is:

```
Total: Precision: 0.6671 - Recall: 0.6213 - F1: 0.6434
Entity: address - Precision: 0.4611 - Recall: 0.4313 - F1: 0.4457
Entity: government - Precision: 0.6667 - Recall: 0.6726 - F1: 0.6696
Entity: movie - Precision: 0.6641 - Recall: 0.6204 - F1: 0.6415
Entity: name - Precision: 0.7182 - Recall: 0.6717 - F1: 0.6942
```

```
Entity: company - Precision: 0.6618 - Recall: 0.6285 - F1: 0.6447
Entity: position - Precision: 0.7343 - Recall: 0.659 - F1: 0.6946
Entity: book - Precision: 0.685 - Recall: 0.5577 - F1: 0.6148
Entity: organization - Precision: 0.7297 - Recall: 0.6553 - F1:
0.6905
Entity: scene - Precision: 0.4973 - Recall: 0.4667 - F1: 0.4815
Entity: game - Precision: 0.7835 - Recall: 0.7729 - F1: 0.7782.
```

9.1.7 Model Prediction

Readers can enter any text, such as "In January of this year, China Everbright Bank Shenyang Branch called and said that due to his good credit record, the bank could upgrade his regular credit card to a platinum card for free." After reasoning, the model can obtain all the entities and corresponding categories of the input text. Codes are implemented as follows:

```
metric = SeqEntityScore(id2tag)

def infer(model, text):
    model.eval()

    # Data processing
    tokens = [word2id_dict.get(w, word2id_dict["[UNK]"]) for w in list(te
xt)]
    tokens_len = len(tokens)

    # Construct data of input models
    tokens = paddle.to_tensor(tokens, dtype="int64").unsqueeze(0)
    tokens_mask = paddle.to_tensor([1] * tokens_len, dtype="int64").unsqu
eeze(0)
    tokens_len = paddle.to_tensor(tokens_len, dtype="int64")

    # Calculate the emission score
    features = model.forward_loss(tokens, tokens_mask, tokens_len)


    # Decode according to the emission score
    _, pred_paths = model.viterbi_decoder(features, tokens_len)
    pred_paths = pred_paths.tolist()

    # Decode the entities in the path
    entities = metric.get_entities(pred_paths[0])
    for entity in entities:
        entity_type, start, end = entity
        print(f"{text[start:end+1]} : {entity_type}")

text="今年 1 月中国光大银行沈阳分行致电，称由于他的信用记录良好，银行可把普通信
用卡免费升级为白金卡。(In January of this year, China Everbright Bank Shenyang Branch called
and said that due to his good credit record, the bank could upgrade his regular credit card to a platin
um card for free.)"
infer(ner_model, text)
```

The output result is:

China Everbright Bank Shenyang Branch: company.

9.2 The Implementation of Relation Extraction Based on ERNIE-UIE

9.2.1 Task Objectives and Implementation Processes

Entity relation extraction is an important task for building knowledge bases or knowledge graphs, and is also one of the key tasks in the field of NLP. In this practice, we use ERNIE-UIE (universal information extraction, UIE) to implement entity relation extraction. UIE is a unified framework for universal information extraction proposed by Lu et al. in ACL 2022 (Lu et al. 2022). This method is based on the pre-training model for prompt learning, and can support information extraction tasks without restricting the domain and target. It has the ability to extract information with small samples or even zero samples. Drawing on the method of (Lu et al. 2022), PaddleNLP has trained and open-sourced the model ERNIE-UIE for Chinese universal information extraction, which is based on the knowledge-enhanced pre-training model ERNIE 3.0 (Sun et al. 2021). In this chapter, we take ERNIE-UIE integrated in PaddleNLP as an example to introduce the entity relation extraction method based on pre-training models.

The entity relation extraction method based on ERNIE-UIE is shown in Fig. 9.5, using a two-stage prediction approach.

The first stage identifies the entities based on prompt. Usually, the first stage's prompt is all candidate entity categories. In the example in Fig. 9.5, the prompt

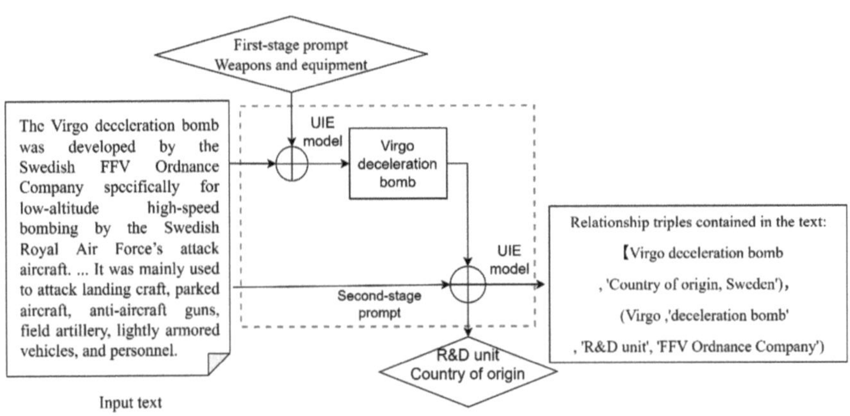

Fig. 9.5 The relation extraction task

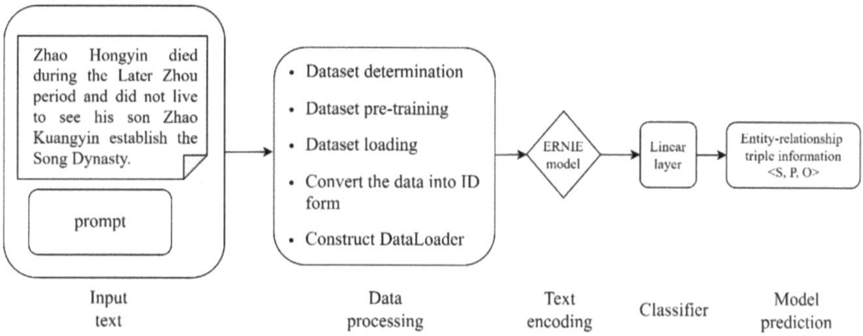

Fig. 9.6 The process of relation extraction based on ERNIE-UIE

is "weapons and equipment," indicating that the UIE model needs to generate all entities with the category "weapons and equipment" (such as the example "Virgo deceleration bomb").

The second stage extracts relations based on prompt. The relation category and the entity output from the first stage are concatenated as a prompt for the second stage, so as to predict the entity relation <head entity, relation, tail entity>, which can also be referred to as <subject, predicate, object> (or <S, P, O>).

The process of relation extraction based on ERINE-UIE model is shown in Fig. 9.6. The input of the model is text data and prompt information, and the output of the model is entity relation triplet. In the modeling process, the input text is first processed, then different prompts are added, and the pre-training model is used to encode the text and prompt sequences to obtain the corresponding semantic vector representation. Finally, through linear layer decoding with two pointers, the probabilities of the starting and ending positions of the subject (entity) in the input text under different prompt are obtained. By analyzing the probabilities of the starting and ending positions, the positions of the subject and object in the text are determined, and finally the triplet information is output.

9.2.2 *Data Processing*

Data processing includes determining the dataset, data pre-processing, converting labeled data into the training data format, constructing negative sample data, data loading, converting data into the ID format, constructing DataLoader, etc. Finally, the same batch of data is processed into equal-length feature sequences, which are passed into the ERNIE-UIE model using DataLoader in batches.

1. **Determine the Dataset**

Compared to the traditional tasks that require thousands or even more data, ERNIE-UIE has excellent fine-tuning capabilities using limited samples, and can usually

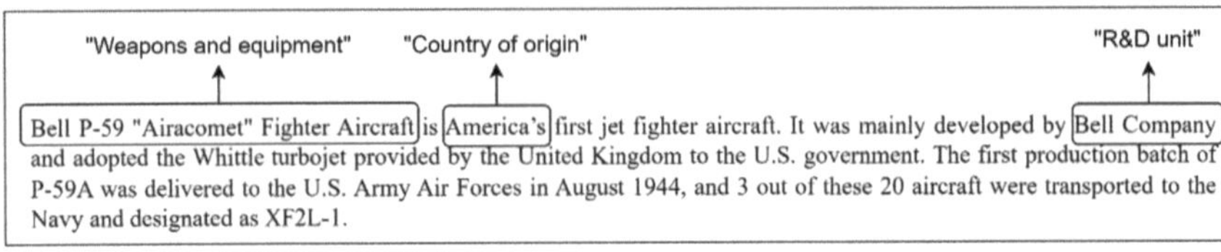

Fig. 9.7 An example of sample data

achieve good results in specific scenarios with only dozens of samples. In this practice, we use a dataset from Baidu, with a total of 52 labeled data. The dataset contains three entity types: weapon equipment, country, and enterprise, as well as two relationship types: country of origin and research and development institution. The training set has 20 samples, the validation set has 16 samples, and the test set has 16 samples. In UIE, it can be considered as extracting the target based on a predefined set of cascading operations, and extracting the corresponding tuple fragments. The schema for this scenario is set as follows:

schema = {weapon equipment": ["country of production", "research and development institution"]}

An example of sample data is shown in Fig. 9.7.

2. **Data Pre-processing**

The purpose of data preprocessing is to convert the labeled data into the format of training data and construct the negative samples at a ratio of 1:5. The negative samples are labeled as null, meaning that the text to be extracted does not contain any entities. In the fine-tuning of the ERNIE-UIE model, the ratio of positive and negative samples is very important. If the dataset only has positive samples, it is easy to cause low accuracy, while if there are too many negative samples, it may cause low recall rate of the model. Codes are implemented as follows:

```
with open("./data/data175124/military_extraction.jsonl", "r", encoding="u
tf-8") as f:
    raw_examples = f.readlines()

def _create_ext_examples(examples,
                         negative_ratio,
                         is_train=True):
```

```
    entities, relations = convert_ext_examples(
        examples, negative_ratio, is_train=is_train)

    # Build training data of entities and relationship types
    examples = entities + relations
    return examples

def _save_examples(save_dir, file_name, examples):
    count = 0
    save_path = os.path.join(save_dir, file_name)
    with open(save_path, "w", encoding="utf-8") as f:
        for example in examples:

            # Save the training data after constructing negative samples
            f.write(json.dumps(example, ensure_ascii=False) + "\n")
            count += 1
    logger.info("Save %d examples to %s." % (count, save_path))

np.random.seed(1000)

# Generate a random index
indexes = np.random.permutation(len(raw_examples))
index_list = indexes.tolist()

# Randomize samples
raw_examples = [raw_examples[i] for i in indexes]

# The proportion of samples in the training set, validation set, and test set
splits = [0.4, 0.3, 0.3]

p1 = int(len(raw_examples) * splits[0])
p2 = int(len(raw_examples) * (splits[0] + splits[1]))

train_ids = index_list[:p1]
dev_ids = index_list[p1:p2]
test_ids = index_list[p2:]


# Build the training set, validation set, and test set
train_examples = _create_ext_examples(raw_examples[:p1], 5)
dev_examples = _create_ext_examples(raw_examples[p1:p2], -1, is_train=Fal
se)
test_examples = _create_ext_examples(raw_examples[p2:], -1, is_train=Fals
e)

_save_examples("./data", "train.txt", train_examples)
_save_examples("./data", "dev.txt", dev_examples)
_save_examples("./data", "test.txt", test_examples)
```

3. **Data Loading**

The data is read into the memory, as shown in the following codes:

```
from paddlenlp.datasets import load_dataset

# Define data and load functions
def load_data(path):
    examples = []
    with open(path, 'r', encoding='utf-8') as f:
        for line in f:
            json_line = json.loads(line)
            content = json_line['content'].strip()
            prompt = json_line['prompt']
            examples.append(json_line)
    return examples

train_dataset = load_dataset(load_data, path='./data/train.txt', lazy=Fal
se)
dev_dataset = load_dataset(load_data, path='./data/dev.txt', lazy=False)
test_dataset = load_dataset(load_data, path='./data/test.txt', lazy=Fals
e)

print('Number of samples of the training set: ', len(train_dataset))
print('Number of samples of the validation set: ', len(dev_dataset))
print('Number of samples of the test set: ', len(test_dataset))
print('Samples: ', train_dataset[0])
```

The output result is:

Number of samples of the training set: 234
Number of samples of the validation set: 80
Number of samples of the test set: 80

Samples: {'content': 'The P-47 Thunderbolt fighter jet was manufactured by Republic Aircraft Company and was one of the main fighter jets of the United States Army Air Force (predecessor of the United States Air Force) in the mid to late stages of World War II. It was also the largest single engine fighter jet at the time, and its later M/N model ranked among the fastest propeller fighters of the Allies during World War II. In addition to performing well in aerial combat, the P-47 is more suitable for carrying out ground attack missions. In addition to the United States Army Air Forces, other allied air forces also use P-47 fighter jets. Due to its significantly larger and more obese body compared to other types of fighter jets, it was nicknamed "Jug" at the time. The production of 15,686 P-47s ranks first in the United States' fighter jet production, with 12,602 D-type aircraft being the world's largest single aircraft production.', 'result_list': [{'text': 'The P-47 Thunderbolt fighter jet', 'start': 0, 'end': 11}], 'prompt': ' weaponry'}.

4. **Convert Data into the ID Form**

Then, the text is segmented by the tokenizer and each word is mapped to an index ID in the dictionary. Codes are implemented as follows:

```
import functools

def map_offset(ori_offset, offset_mapping):

    for index, span in enumerate(offset_mapping):
        if span[0] <= ori_offset < span[1]:
            return index
    return -1

def preprocess_function(example, tokenizer, max_seq_len):

    # Perform tokenization on the text data and prompts
    encoded_inputs = tokenizer(text=[example["prompt"]],
                               text_pair=[example["content"]],
                               truncation=True,
                               max_seq_len=max_seq_len,
                               pad_to_max_seq_len=True,
                               return_attention_mask=True,
                               return_position_ids=True,
                               return_offsets_mapping=True)
    offset_mapping = [list(x) for x in encoded_inputs["offset_mapping"]
[0]]
    bias = 0
    for index in range(1, len(offset_mapping)):
        mapping = offset_mapping[index]
        if mapping[0] == 0 and mapping[1] == 0 and bias == 0:
            bias = offset_mapping[index - 1][1] + 1

        if mapping[0] == 0 and mapping[1] == 0:
            continue
        offset_mapping[index][0] += bias
        offset_mapping[index][1] += bias
    start_ids = [0 for x in range(max_seq_len)]
    end_ids = [0 for x in range(max_seq_len)]
    for item in example["result_list"]:
        start = map_offset(item["start"] + bias, offset_mapping)
        end = map_offset(item["end"] - 1 + bias, offset_mapping)
        start_ids[start] = 1.0
        end_ids[end] = 1.0

    tokenized_output = [
        encoded_inputs["input_ids"][0], encoded_inputs["token_type_ids"]
[0],
        encoded_inputs["position_ids"][0],
        encoded_inputs["attention_mask"][0], start_ids, end_ids
    ]
    tokenized_output = [np.array(x, dtype="int64") for x in tokenized_out
put]
    return tuple(tokenized_output)

convert_example_to_feature = functools.partial(preprocess_function,
    tokenizer=tokenizer, max_seq_len=512)

# Data pre-processing for the training set, validation set and test set
train_dataset = train_dataset.map(convert_example_to_feature)
dev_dataset = dev_dataset.map(convert_example_to_feature)
test_dataset = test_dataset.map(convert_example_to_feature)
```

```
# Print the processed data samples
print('Samples: ', train_dataset[0])
```

The output result is:

```
    Samples: (array([    1,   587,   361,   371,   366,     2,  2524, 12051,
2565,
        23,  1048,   128,    24,   267,   765,    98,    78, 10892,
      9740,  9904,    77,   190,   300,    14,   706,    98,    53,
       230,   108,   294,     4,    10,   188,    20,   891,   261,
       928,   411,   261,    78,   188,    20,   411,   261,   152,
       262,    77,    11,   177,   218,    19,   267,    12,    49,
      ……
```

5. Construction of DataLoader

When training models, data need to be passed into the model in batches for training. Each batch of data is passed into the model as a minibatch for computation and processing. Each minibatch of data contains two parts: text data and corresponding span labels (i.e., the start and end positions of the entity). The paddle.io.DataLoader in the PaddlePaddle framework can be used to implement the functions of updating data in batches. Codes are implemented as follows:

```
from paddle.io import DataLoader, BatchSampler

train_batch_sampler = paddle.io.BatchSampler(train_dataset, batch_size=8,
 shuffle=True)
dev_batch_sampler = paddle.io.BatchSampler(dev_dataset, batch_size=8)
test_batch_sampler = paddle.io.BatchSampler(test_dataset, batch_size=8)


# Use DataLoader to load the training set, validation set, and test set
train_loader = DataLoader(train_dataset, batch_sampler=train_batch_sample
r, return_list=True)
dev_loader = DataLoader(dev_dataset, batch_sampler=dev_batch_sampler, ret
urn_list=True)
test_loader = DataLoader(test_dataset, batch_sampler=test_batch_sampler,
return_list=True)
```

9.2.3 Model Construction

The ERNIE-UIE model processes a small batch of data at a time. The overall model framework is shown in Fig. 9.8. The model first passes the processed text feature data into the ERNIE 3.0 module, which encodes it and outputs the corresponding vector sequences. These vector sequences are then passed into two linear layers (vector sequences multiplied by weights, plus bias), which are used to represent the start and end positions of the extraction target. Finally, the sigmoid function is used to obtain the probabilities of the start and end positions.

Based on the modeling idea shown in Fig. 9.8, codes are implemented as follows:

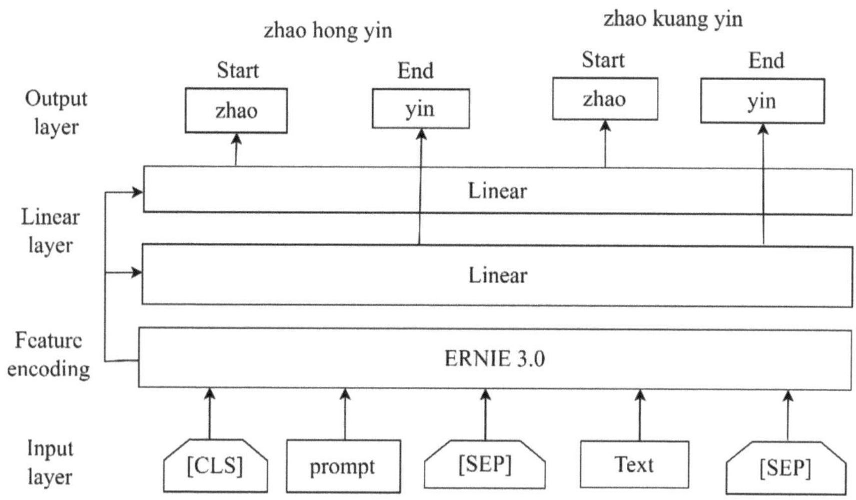

Fig. 9.8 Relation extraction based on the UIE model

```
import paddle
import paddle.nn as nn
from paddlenlp.transformers import ErniePretrainedModel


# ERINE-UIE uses ERNIE 3.0 as the model base and inherits the ErniePretrainedModel class
class UIE(ErniePretrainedModel):

    def __init__(self, encoding_model):
        super(UIE, self).__init__()
        self.encoder = encoding_model
        hidden_size = self.encoder.config["hidden_size"]

# Get the dimension of the hidden vectors
        self.linear_start = paddle.nn.Linear(hidden_size, 1)
        self.linear_end = paddle.nn.Linear(hidden_size, 1)
        self.sigmoid = nn.Sigmoid() # 初始化 sigmoid

    def forward(self, input_ids, token_type_ids, pos_ids, att_mask):
        sequence_output, _ = self.encoder(input_ids=input_ids,
                                          token_type_ids=token_type_ids,
                                          position_ids=pos_ids,
                                          attention_mask=att_mask)
        start_logits = self.linear_start(sequence_output)
        start_logits = paddle.squeeze(start_logits, -1)

        # Obtain the probability value of the starting position
        start_prob = self.sigmoid(start_logits)
        end_logits = self.linear_end(sequence_output)
        end_logits = paddle.squeeze(end_logits, -1)

        # Obtain the probability value of the end position
        end_prob = self.sigmoid(end_logits)
        return start_prob, end_prob
```

Next, the UIE-Base pre-training model, vocabulary, and related configuration files are Downloaded. Codes are implemented as follows:

```
import os
import json
import numpy as np

# Introduce tools for downloading files

from paddle.utils.download import get_path_from_url
from paddlenlp.transformers import AutoTokenizer
from utils import convert_ext_examples


# Download the UIE-Base pre-training model, 词表, and configuration files
pretrained_model_name = "uie-base"
resource_file_urls = {
    "model_state.pdparams":
    "https://bj.bcebos.com/paddlenlp/taskflow/information_extraction/uie_
base_v1.0/model_state.pdparams",
    "model_config.json":
    "https://bj.bcebos.com/paddlenlp/taskflow/information_extraction/uie_
base/model_config.json",
    "vocab_file":
    "https://bj.bcebos.com/paddlenlp/taskflow/information_extraction/uie_
base/vocab.txt",
    "special_tokens_map":
    "https://bj.bcebos.com/paddlenlp/taskflow/information_extraction/uie_
base/special_tokens_map.json",
    "tokenizer_config":
    "https://bj.bcebos.com/paddlenlp/taskflow/information_extraction/uie_
base/tokenizer_config.json"
}


# Download ERNIE-UIE related models, 词表, and configuration files
for key, val in resource_file_urls.items():
    file_path = os.path.join(pretrained_model_name, key)
    if not os.path.exists(file_path):
        get_path_from_url(val, pretrained_model_name)


# Load the pre-training model ERNIE-UIE and Tokenizer
model = UIE.from_pretrained(pretrained_model_name)
tokenizer = AutoTokenizer.from_pretrained(pretrained_model_name)
```

9.2.4 Training Configuration

The purpose of training configuration is to set the computing resources, models, optimizers, loss functions, and optimization indicators during model training.

(1) Model: the ERNIE-UIE model.
(2) Optimizer: AdamW optimizer.
(3) Loss function: BCELoss (Binary Cross Entropy Loss), which is often used in multi-label classification problems and is often used with the sigmoid function.
(4) Evaluation indicators: Precision, Recall, and F1.

Codes are implemented as follows:

```
from paddlenlp.metrics import SpanEvaluator

# Define training parameters
num_epochs = 10
log_steps = 10
eval_steps = 20
learning_rate = 1e-5
save_dir = "./best_model"
if not os.path.exists(save_dir):
    os.makedirs(save_dir)

# Specify optimization strategies
optimizer = paddle.optimizer.AdamW(learning_rate=learning_rate, parameter
s=model.parameters())

# Specify the loss function
loss_fn = paddle.nn.BCELoss()



# Specify evaluation indicators
metric = SpanEvaluator()
```

9.2.5 Model Training

During the training process, a training log is printed every log_steps, and model evaluation is performed on the validation set every eval_steps. The model with the highest F1 score during the training process is saved. The codes for evaluation on the validation set are implemented as follows:

```
def evaluate(model, metric, data_loader):
    # Set the model to the evaluation mode
    model.eval()
    # Reset evaluations
    metric.reset()
    # Traverse every batch of the validation set
    for batch in data_loader:
        input_ids, token_type_ids, position_ids, attention_mask, start_
ids, end_ids = batch
        # Compute the output of models
        start_prob, end_prob = model(input_ids, token_type_ids, positio
n_ids, attention_mask)
        start_ids = paddle.cast(start_ids, 'float32')
        end_ids = paddle.cast(end_ids, 'float32')
        num_correct, num_infer, num_label = metric.compute(
            start_prob, end_prob, start_ids, end_ids)
        # Update indicators
        metric.update(num_correct, num_infer, num_label)
    precision, recall, f1 = metric.accumulate()
    model.train()
    return precision, recall, f1
```

The codes for training the model are as follows:

```
def train(model):
    # Start the training mode of models
    model.train()
    global_step = 0
    best_f1 = 0
    num_training_steps = len(train_loader) * num_epochs
    # Record the loss function value during training and the F1 value on the validation set
    train_loss_record = []
    train_score_record = []
    for epoch in range(num_epochs):
        for step, batch in enumerate(train_loader):
            input_ids, token_type_ids, position_ids, attention_mask, st
art_ids, end_ids = batch
            start_prob, end_prob = model(input_ids, token_type_ids, pos
ition_ids, attention_mask)
            start_ids = paddle.cast(start_ids, 'float32')
            end_ids = paddle.cast(end_ids, 'float32')
            loss_start = loss_fn(start_prob, start_ids)
            loss_end = loss_fn(end_prob, end_ids)
            loss = (loss_start + loss_end) / 2.0
            train_loss_record.append((global_step, loss.item()))
            # Gradient backpropagation
            loss.backward()
            optimizer.step()
            optimizer.clear_grad()
            if global_step > 0 and global_step % log_steps == 0:
                print(f"- epoch: {epoch} - global_step: {global_step}/
{num_training_steps} -loss: {loss.numpy().item():.6f}")
            if global_step > 0 and global_step % eval_steps == 0:
```

```
                precision, recall, f1 = evaluate(model, metric, dev_loa
der)
                train_score_record.append((global_step, f1))
                if f1 > best_f1:
                    print(f"best F1 performance has been updated: {best
_f1 :.5f} --> {f1 :.5f}")
                    best_f1 = f1
                    paddle.save(model.state_dict(), f"{save_dir}/model_
state.pdparams")
                print(f"evaluation result: Precision: {precision:.5f},
Recall: {recall:.5f}, F1: {f1:.5f}")
            global_step += 1
    paddle.save(model.state_dict(), f"{save_dir}/final.pdparams")
    return train_loss_record, train_score_record

train_loss_record, train_score_record = train(model)
```

The output result is:

```
- epoch: 0 - global_step: 10/300 -loss: 0.004818
- epoch: 0 - global_step: 20/300 -loss: 0.000344
best F1 performance has been updated: 0.00000 -- > 0.79710
evaluation result: Precision: 0.72368, Recall: 0.88710, F1: 0.79710
- epoch: 1 - global_step: 30/300 -loss: 0.000573
- epoch: 1 - global_step: 40/300 -loss: 0.000093
best F1 performance has been updated: 0.79710 -- > 0.83333
evaluation result: Precision: 0.78571, Recall: 0.88710, F1: 0.83333
- epoch: 1 - global_step: 50/300 -loss: 0.000359
- epoch: 2 - global_step: 60/300 -loss: 0.000329
evaluation result: Precision: 0.76389, Recall: 0.88710, F1: 0.82090
- epoch: 2 - global_step: 70/300 -loss: 0.000026
- epoch: 2 - global_step: 80/300 -loss: 0.000059
evaluation result: Precision: 0.75342, Recall: 0.88710, F1: 0.81481
- epoch: 3 - global_step: 90/300 -loss: 0.000355
- epoch: 3 - global_step: 100/300 -loss: 0.000003
evaluation result: Precision: 0.78571, Recall: 0.88710, F1: 0.83333
- epoch: 3 - global_step: 110/300 -loss: 0.000022
- epoch: 4 - global_step: 120/300 -loss: 0.000016
evaluation result: Precision: 0.78571, Recall: 0.88710, F1: 0.83333
- epoch: 4 - global_step: 130/300 -loss: 0.000014
- epoch: 4 - global_step: 140/300 -loss: 0.000037
best F1 performance has been updated: 0.83333 -- > 0.83582
evaluation result: Precision: 0.77778, Recall: 0.90323, F1: 0.83582
- epoch: 5 - global_step: 150/300 -loss: 0.000008
- epoch: 5 - global_step: 160/300 -loss: 0.000010
evaluation result: Precision: 0.76712, Recall: 0.90323, F1: 0.82963
- epoch: 5 - global_step: 170/300 -loss: 0.000157
- epoch: 6 - global_step: 180/300 -loss: 0.000006
evaluation result: Precision: 0.77778, Recall: 0.90323, F1: 0.83582
- epoch: 6 - global_step: 190/300 -loss: 0.000001
- epoch: 6 - global_step: 200/300 -loss: 0.000001
evaluation result: Precision: 0.78571, Recall: 0.88710, F1: 0.83333
- epoch: 7 - global_step: 210/300 -loss: 0.000003
```

```
- epoch: 7 - global_step: 220/300 -loss: 0.000000
evaluation result: Precision: 0.78571, Recall: 0.88710, F1: 0.83333
- epoch: 7 - global_step: 230/300 -loss: 0.000002
- epoch: 8 - global_step: 240/300 -loss: 0.000008
evaluation result: Precision: 0.78571, Recall: 0.88710, F1: 0.83333
- epoch: 8 - global_step: 250/300 -loss: 0.000006
- epoch: 8 - global_step: 260/300 -loss: 0.000015
evaluation result: Precision: 0.78571, Recall: 0.88710, F1: 0.83333
- epoch: 9 - global_step: 270/300 -loss: 0.000003
- epoch: 9 - global_step: 280/300 -loss: 0.000002
evaluation result: Precision: 0.78571, Recall: 0.88710, F1: 0.83333
- epoch: 9 - global_step: 290/300 -loss: 0.000002.
```

Then, the loss train_loss_record and the score train_score_record on the validation set during the training process are saved, and the training process is visualized. Codes are as follows:

```
import matplotlib.pyplot as plt

def plot_training_loss(train_loss_record, fig_name, fig_size=(8, 6), samp
le_step=10, loss_legend_loc="lower left", acc_legend_loc="lower left"):
    plt.figure(figsize=fig_size)

    train_steps = [x[0] for x in train_loss_record][::sample_step]
    train_losses = [x[1] for x in train_loss_record][::sample_step]

    plt.plot(train_steps, train_losses, color='#e4007f', label="Train Los
s")

    # Draw coordinate axes and legends
    plt.ylabel("Loss", fontsize='large')
    plt.xlabel("Step", fontsize='large')
    plt.legend(loc=loss_legend_loc, fontsize='x-large')

    plt.savefig(fig_name)
    plt.show()

def plot_training_f1(train_score_record, fig_name, fig_size=(8, 6), sampl
e_step=10, loss_legend_loc="lower left", acc_legend_loc="lower left"):
    plt.figure(figsize=fig_size)

    train_steps=[x[0] for x in train_score_record]
    train_losses = [x[1] for x in train_score_record]

    plt.plot(train_steps, train_losses, color='#e4007f', label="Dev F1")

    # Draw coordinate axes and legends
    plt.ylabel("Macro F1", fontsize='large')
    plt.xlabel("Step", fontsize='large')
    plt.legend(loc=loss_legend_loc, fontsize='x-large')

    plt.savefig(fig_name)
    plt.show()
plot_training_loss(train_loss_record, 'loss.jpg')
plot_training_f1(train_score_record, 'f1.jpg')
```

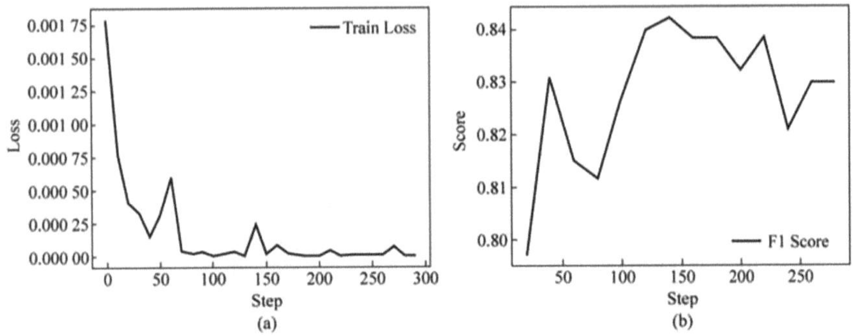

Fig. 9.9 The trend of loss function changes in the training set and the trend of F1 value changes in the validation set. **a** Loss changes in the training set. **b** F1 score changes in the validation set

The output result is shown in Fig. 9.9.

From the output results, it can be seen that as the training progresses, the loss function on the training set continuously decreases and then converges, with the value tending towards 0. At the same time, the F1 score on the validation set initially increases continuously, and gradually stabilizes after the model converges, remaining above 83%.

9.2.6 Model Evaluation

Then, the best performing model during training is evaluated by using the validation set to verify the model's training effect. Codes are implemented as follows:

```
# Load the trained model for evaluation. Re-instantiate a model, and then load the trained model's
parameters into the new model
saved_state = paddle.load("./best_model/model_state.pdparams")
model.load_dict(saved_state)

# Evaluate models
precision, recall, f1 = evaluate(model, metric, test_loader)
print(f"evaluation result: Precision: {precision:.5f}, Recall:
{recall:.5f}, F1: {f1:.5f}")
```

The output result is:
evaluation result: Precision: 0.91549, Recall: 0.98485, F1: 0.94891.

9.2.7 Model Prediction

Readers can input any text to be predictedand verify the model's training effect through model inference. The codes are implemented as follows:

```
def get_bool_ids_greater_than(probs, limit=0.5):
    probs = np.array(probs)
    dim_len = len(probs.shape)
    if dim_len > 1:
        result = []
        for p in probs:
            result.append(get_bool_ids_greater_than(p, limit))
        return result
    else:
        result = []
        for i, p in enumerate(probs):
            if p > limit:
                result.append(i)
        return result

def get_span(start_ids, end_ids):
    start_ids = sorted(start_ids)
    end_ids = sorted(end_ids)

    start_pointer = 0
    end_pointer = 0
    len_start = len(start_ids)
    len_end = len(end_ids)
    couple_dict = {}
    while start_pointer < len_start and end_pointer < len_end:
        if start_ids[start_pointer] == end_ids[end_pointer]:
            couple_dict[end_ids[end_pointer]] = start_ids[start_pointer]
            start_pointer += 1
            end_pointer += 1
            continue
        if start_ids[start_pointer] < end_ids[end_pointer]:
            couple_dict[end_ids[end_pointer]] = start_ids[start_pointer]
            start_pointer += 1
            continue
        if start_ids[start_pointer] > end_ids[end_pointer]:
            end_pointer += 1
            continue
    result = [(couple_dict[end], end) for end in couple_dict]
    result = set(result)
    return result

def get_span_ids(span_set, offset_mapping):
    prompt_end_token_id = offset_mapping[1:].index([0, 0])
    bias = offset_mapping[prompt_end_token_id][1] + 1
    for idx in range(1, prompt_end_token_id + 1):
        offset_mapping[idx][0] -= bias
        offset_mapping[idx][1] -= bias
```

```
    span_ids = []
    for start, end in span_set:
        start_id = offset_mapping[start][0]
        end_id = offset_mapping[end][1]
        span_ids.append((start_id, end_id))
    return span_ids

def infer(model, text, prompt):
    """
    predict inference functions
  Input:
    -model: models
    -text: texts
    """
    model.eval()
    # Data processing
    encoded_inputs = tokenizer(text=[prompt],
                               text_pair=[text],
                               truncation=True,
                               max_seq_len=512,
                               pad_to_max_seq_len=True,
                               return_attention_mask=True,
                               return_position_ids=True,
                               return_offsets_mapping=True)

    input_ids =
paddle.to_tensor(encoded_inputs["input_ids"][0]).unsqueeze(0)
    token_type_ids =
paddle.to_tensor(encoded_inputs["token_type_ids"][0]).unsqueeze(0)
    position_ids =
paddle.to_tensor(encoded_inputs["position_ids"][0]).unsqueeze(0)
    attention_mask =
paddle.to_tensor(encoded_inputs["attention_mask"][0]).unsqueeze(0)
    offset_mapping = [
        list(x) for x in encoded_inputs["offset_mapping"][0]
    ]

    start_prob, end_prob = model(input_ids, token_type_ids, position_ids,
attention_mask)
    start_prob = start_prob.tolist()
    end_prob = end_prob.tolist()

    start_ids_list = get_bool_ids_greater_than(start_prob, limit=0.5)
    end_ids_list = get_bool_ids_greater_than(end_prob, limit=0.5)

    for start_ids, end_ids in zip(start_ids_list, end_ids_list):
        span_set = get_span(start_ids, end_ids)
        span_ids = get_span_ids(span_set, offset_mapping)
        results = [text[start:end] for start, end in span_ids]
    return results
```

```
text = " The Nevada battleship (hull number BB-36) was a battleship
    belonging to the United States Navy and was the lead ship of the
    Nevada class battleships. It is the second US military ship named
    after Nevada and also the first batch of super dreadnoughts in the
    United States. The Nevada was built at the Ho River Shipyard in 1912,
    launched in 1914, and commissioned in 1916. At that time, World War I
    had already broken out, and the Nevada had been staying near the sea
    for training. After the United States entered the war, the Nevada was
    sent to British waters and joined the British mainland fleet, mainly
    responsible for escort missions."
    # Extract entities with the category of weapon equipment from the text, using weapon
    equipment as the prompt here
    prompt_ent = "weapon equipment"

    ent_results = infer(model, text, prompt_ent)

    rel1_name = "Country of origin"
    rel2_name = "Research and development institutions "


    # Define a list for storing SPO triplets
    spo_list = []

    # Extract the country of origin and research and development institutions of weapon
    equipment in the text
    for ent_result in ent_results:
        prompt_rel1 = ent_reuslt + "的(de)" + rel1_name
        prompt_rel2 = ent_reuslt + "的(de)" + rel2_name

        rel1_results = infer(model, text, prompt_rel1)
        rel2_results = infer(model, text, prompt_rel2)

        for rel1_result in rel1_results:
            spo_list.append((ent_result, rel1_name, rel1_result))

        for rel2_result in rel2_results:
            spo_list.append((ent_result, rel2_name, rel2_result))

    print("The text contains a relation triplet (S, P, O) :", spo_list)
```

The output result is:

The corresponding extraction result of weapon equipment is `['Nevada-class battleship ']`

The text contains a relation triplet (S, P, O): `[('Nevada-class battleship', 'Country of origin', 'America'), ('Nevada-class battleship ', 'R&D unit ', 'Huohe Shipyard ')]`.

9.3 Exercises

(1) In the experiments in this chapter, the negative examples are added to the training data. Please consider the role of negative examples and experiment with different negative example ratios to affect the model's performance.
(2) In this chapter, the entity relation extraction is implemented based on ERNIE-UIE. Readers are encouraged to try to design their own prompts and train a UIE model for extracting entities.

Chapter 10
Text Semantic Matching

Abstract This chapter introduces text semantic matching, a fundamental NLP task that determines semantic similarity between texts and underpins applications such as Q&A systems and information retrieval. It first discusses the limitations of early string-matching methods and highlights the advantages of neural networks and pre-trained models in capturing deeper semantic relationships. The chapter presents two approaches: SimNet, a GRU-based neural framework for short text matching using the LCQMC dataset, and RocketQA, a pre-trained model leveraging ERNIE with advanced training strategies to improve generalization. Through detailed implementation of data processing, model construction, training, and evaluation, the results demonstrate that while SimNet achieves moderate accuracy, RocketQA significantly enhances performance, underscoring the effectiveness of pre-training and attention mechanisms in semantic matching.

Text semantic matching is one of the fundamental tasks of NLP. Its target is to determine whether two texts are similar in semantics, commonly referred to as text similarity calculation. Given the following three sentences, judging from the semantics they contain, the semantics of sentence (1) and (2) are more similar.

(1) What mountain is the highest mountain in the world?
(2) Which mountain is the tallest in the world?
(3) Which mountains are the highest in the world?

The task of text semantic matching is not completely independent. It is widely used in areas such as Q&A systems and information retrieval. For example, in the Q&A scenario, the most accurate answers are given by calculating the similarity between the question texts entered by the user and the questions stored in the standard base.

In early times, string-matching techniques are usually used in text semantic matching. They can be used to judge the similarity between two text strings by counting the frequency of characters or words. This approach had a clear drawback: it did not take into account the semantic relationships between words. For instance, in the examples above, if the string-matching method is used, sentence (2) and sentence (3) would be considered more semantically similar, but clearly this is

C. Zong et al., *Natural Language Processing and Large Language Models*,
https://doi.org/10.1007/978-981-92-0682-7_10

not the most accurate result. In recent years, text semantic matching methods based on neural network models have been proposed, in which sequence modeling is used to obtain semantic vectors, thereby more accurately capturing the semantic relationships between two texts and effectively advancing the development of semantic matching technology. However, the text semantic matching method based on RNN suffers from the long-distance dependency challenges. With the rise of the Transformer model and pre-training language models, especially the proposal of the BERT model, using attention mechanism to capture the long-term dependencies of sentences can effectively improve the performances of semantic matching tasks.

This chapter introduces the text semantic matching method based on neural networks and pre-training models, in which the approach based on neural networks adopts the SimNet framework (the backbone network adopts the GRU model), and the pre-training model RocketQA is adopted in the method based on pre-training models. RocketQA is commonly used in information retrieval scenarios. LCQMC is used as dataset in this chapter.

10.1 SimNet-Based Text Semantic Matching

10.1.1 Task Objectives and Implementation Processes

The task of text semantic matching can be broadly classified into three categories: semantic matching between given two short texts, matching between given a short text and a long text, and matching between given two long texts. Matching between short texts is most widely used in the current industry, such as matching queries with titles in information retrieval. In this task, texts are usually labeled with two labels: 1 and 0, in which 1 indicates two texts are very similar and 0 indicates that they are very dissimilar. For the three texts in the examples above, an ideal text semantic matching model can achieve an output of 1 when sentence (1) and sentence (2) are input, and an output of 0 when sentence (1) and sentence (3) are input.

This section introduces a semantic matching method for two short texts based on SimNet (SimilarityNet). SimNet is a framework independently developed by Baidu for calculating the similarity of two short texts. It can use a variety of neural network models, such as convolutional neural network (CNN) or recurrent neural network (RNN) to extract semantic features of texts and achieve good matching results.

The idea of matching text semantics based on SimNet is shown in Fig. 10.1. The input of the model is two sentences (text), and the output is the result of their semantic matching (similar or not similar). In the modeling process, data processing is first carried out, and then SimNet is used to obtain the text semantic vector representation. Finally, the result of semantic matching is obtained through the fully connected layer and regression processing.

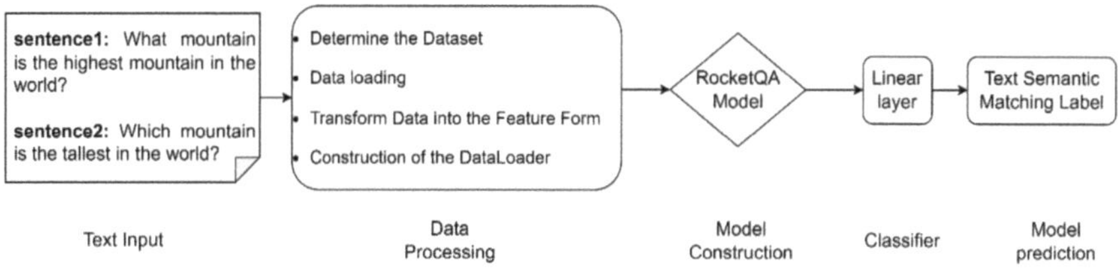

Fig. 10.1 The implementation flowchart of the text semantic matching task

10.1.2 Data Processing

Data processing includes steps such as determining the dataset, data loading, constructing the LCQMCDataset class, and constructing the DataLoader. The same batch of data is finally processed into feature sequences with the equal length, and DataLoader is used to send data into the SimNet framework (the backbone network uses the GRU model) one by one.

1. **Determine the Dataset**

In this practice, we use the LCQMC (a large-scale Chinese question matching corpus, LCQMC) dataset to conduct text semantic matching tasks (Liu et al., 2018). LCQMC is a Chinese question matching dataset, and the corpus is extracted from user questions in different fields on Baidu Zhidao (the world's leading Chinese Q&A interactive platform). The dataset contains a total of 247,568 data items, in which there are 238,766 items in the training set, 4,401 items in the validation set, and 4,401 ones in the test set.

The following is a sample from the training set. Each sample contains two sentences (representing query and title, respectively) and the corresponding text semantic matching label (label). The number "1" represents "similar."

```
{'query': ' What kind of girls do boys who like to play basketball like?',
'title': ' What kind of girls do boys who love playing basketball like?',
'label': 1}
```

2. **Data Loading**

First, let's read the training set, validation set, and test set data into memory. Codes are implemented as follows:

```
# Install datasets

from paddlenlp.datasets import load_dataset

def load_data(data_path):
    with open(data_path,'r') as f:
        for item in f.readlines():
            query, title, label = item.split('\t')
            yield {'query': query, 'title': title, 'label': int(label)}


# Load the training set, test data and validation data
train_dataset = load_dataset(load_data,data_path='lcqmc/train.tsv',lazy=F
alse)
dev_dataset = load_dataset(load_data,data_path='lcqmc/dev.csv',lazy=Fals
e)
test_dataset = load_dataset(load_data,data_path='lcqmc/test.csv',lazy=Fal
se)



# Print the first piece of data in the training set
print('samples: ', train_dataset[0])
```

The output result is:

Samples: {'query': 'What kind of girls do boys who like to play basketball like?', 'title': ' What kind of girls do boys who love playing basketball like?', 'label': 1}

From the output results, it can be seen that each sample contains two parts: two text sequences (indicated by query and title, respectively) and the corresponding semantic matching labels (1 means "very similar" and 0 means "completely dissimilar").

3. **Constructing the LCQMCDataset Class**

The purpose of constructing the LCQMCDataset class is for data management and it inherits from the paddle.io.DataSet class. The LCQMCDataset class includes the following two operations: word segmentation and converting words to IDs in the character table.

Since the LCQMC dataset is in Chinese, and there are no separators between words, in this practice, we use the JiebaTokenizer API in Jieba Toolkit to perform word segmentation and convert the segmented words into the ID format. Here is an example of the codes:

```
# Load JiebaTokenizer, which is a tool class for Jieba word segmentation

from paddlenlp.data import JiebaTokenizer, Vocab
import os


# Load the vocabulary
vocab_path="./simnet_vocab.txt"
vocab = Vocab.load_vocabulary(vocab_path, unk_token='[UNK]', pad_token='
[PAD]')
tokenizer = JiebaTokenizer(vocab)

# Word segmentation of JiebaTokenizer and the sample y of conversion ID
text='What kind of girls do boys who love playing basketball like?'

# Use the cut class to segment text sequences into corresponding words
tokenized_text=tokenizer.cut(text)

# Use the encode class to convert tokens into corresponding IDs
tokenized_ids=tokenizer.encode(text)
print(tokenized_text)
print(tokenized_ids)
```

The output result is:

['What', 'Kind', 'of', 'Girls', 'Do', 'Boys', 'Who', 'Love', 'Playing', 'Basketball', 'like', '?']

[227006, 369,452, 415,919, 423,468, 322,727, 349,331, 58,909, 423,468, 242161]

The vocabulary contains two special characters: [PAD] and [UNK]. [PAD] is used for padding words in minibatch processing. In current neural networks, we use batch processing, in which the model trains multiple sentences at once. Considering that the lengths of sentences in the same batch are often different, [PAD] is padded at the end of shorter sentences to make the lengths of sentences in the same batch the same. For specific implementation methods, please refer to part 4 "Constructing DataLoader" in this sub-section.

Due to the high computing complexity of neural networks, the model does not retain all the words in the corpus when constructing the vocabulary, but only selects the top N most frequently occurring words, and the low-frequency words following them are all replaced with [UNK].

And then we construct the LCQMCDataset class to perform word segmentation and ID conversion on the dataset. Codes are implemented as follows:

```
# Load the dataset

from paddle.io import Dataset

class LCQMCDataset(Dataset):
    def __init__(self, examples, tokenizer):
        super(LCQMCDataset, self).__init__()

        # The dictionary is used to convert words into numbers indexed by the dictionary
        self.tokenizer =  tokenizer

        # Convert the loaded dataset from texts into the ID format
        self.examples = self.convert_tokens_to_ids(examples)

    def convert_tokens_to_ids(self,examples):
        list_examples = []
        for example in examples:
            query, title = example["query"], example["title"]

            # Generate query ID
            query_ids = self.tokenizer.encode(query)
            query_seq_len = len(query_ids)

            # Generate title ID
            title_ids = self.tokenizer.encode(title)
            title_seq_len = len(title_ids)

            label = example["label"]
            list_examples.append([query_ids, query_seq_len, title_ids, ti
tle_seq_len, label])
        return list_examples

    def __getitem__(self, idx):
        query_ids, query_seq_len, title_ids, title_seq_len, label = self.
examples[idx]
        return query_ids, query_seq_len, title_ids, title_seq_len, label

    def __len__(self):
        return len(self.examples)

# Instantiate the Dataset
train_dataset = LCQMCDataset(train_dataset, tokenizer)
dev_dataset = LCQMCDataset(dev_dataset, tokenizer)
test_dataset = LCQMCDataset(test_dataset, tokenizer)

print('Number of samples of the training set: ', len(train_dataset))
print('Number of samples of the validation set: ', len(dev_dataset))
print('Number of samples of the test set: ', len(test_dataset))
print('Samples: ', train_dataset[0])
```

The output result is:

Number of samples of the training set: `238766`

Number of samples of the validation set: `4401`

Number of samples of the test set: `4401`

Samples: ([349331, 369452, 415919, 423468, 322727, 349331, 58909, 423468, 242161], 9, [227006, 369452, 415919, 423468, 322727, 349331, 58909, 423468, 242161], 9, 1)

From the output results, it can be seen that after text segmentation, each word has been converted into an ID number in the dictionary.

4. **Construction of the DataLoader**

When training models, data is sent into the model in batches for separate training. Each batch of data is passed into the model for processing, and each batch contains two parts: text data and corresponding semantic matching labels. Now we use Paddle.io.DataLoader in the PaddlePaddle framework to achieve the function of loading in batches and data upgrading. Text truncation and text padding techniques are used to change the texts in each batch to a fixed length.

We define the collate_fn function for text truncation and padding, which is used as a callback function to be sent into DataLoader. Before returning each batch of data, DataLoader calls the collate_fn function to process data and returns the processed text data and corresponding labels. Below are the codes to implement the collate_fn function:

```
import paddle
import numpy as np


# Set the maximum sequence length max_seq_len to 64
def collate_fn(batch_data, pad_val=0, max_seq_len=64):
    list_query_ids, query_seq_lens, list_title_ids, title_seq_lens, label
s = [], [], [], [], []
    for example in batch_data:
        query_ids, query_seq_len, title_ids, title_seq_len, label = examp
le

        # Truncate data sequences
        query_ids = query_ids[:max_seq_len]
        title_ids = title_ids[:max_seq_len]

        # Save the ID and the actual length of the query
        list_query_ids.append(query_ids)
        query_seq_lens.append(len(query_ids))

        # Save the ID and actual length of the title
        list_title_ids.append(title_ids)
        title_seq_lens.append(len(title_ids))

        # Save labels
        labels.append(label)


    # Pad the data series to the maximum length
    for i in range(len(list_query_ids)):
        list_query_ids[i] = list_query_ids[i] + [pad_val] * (max_seq_len
- len(list_query_ids[i]))
        list_title_ids[i] = list_title_ids[i] + [pad_val] * (max_seq_len
- len(list_title_ids[i]))

    return np.array(list_query_ids), \
           np.array(query_seq_lens), \
           np.array(list_title_ids), \
           np.array(title_seq_lens), \
           np.array(labels)
```

Next, we construct the DataLoader for iterating data in batches and specify the batch size. When iterating the data, we need to specify whether to shuffle the samples by using the parameter "shuffle." If "*shuffle*" *is set to False*, the DataLoader upgrades a batch of data sequentially; if "*shuffle*" *is* set to True, the DataLoader upgrades a batch of data randomly. Codes are implemented as follows:

```
from paddle.io import BatchSampler, DataLoader

batch_size = 64

# Randomize the training set and iterate data in batches
train_sampler = BatchSampler(
        train_dataset, batch_size=batch_size, shuffle=True)
train_loader = DataLoader(dataset=train_dataset,
                                  batch_sampler=train_sampler,
                                  collate_fn=collate_fn)

# The validation set and iterate data in batches
dev_sampler = BatchSampler(
        dev_dataset, batch_size=batch_size, shuffle=False)
dev_loader = DataLoader(dataset=dev_dataset,
                                  batch_sampler=dev_sampler,
                                  collate_fn=collate_fn)

# The test set and iterate data in batches
test_sampler = BatchSampler(
        test_dataset, batch_size=batch_size, shuffle=False)
test_loader = DataLoader(dataset=test_dataset,
                                  batch_sampler=test_sampler,
                                  collate_fn=collate_fn)

# Print the data of the first batch in the training set
print(next(iter(train_loader)))
```

The output result is:

```
[Tensor(shape=[64, 64], dtype=int64, place=Place(gpu:0), stop_gradient=True,
    [[431671, 118703, 459477, ..., 0     , 0     , 0     ],
     [198205, 53616 , 123738, ..., 0     , 0     , 0     ],
     [340164, 341100, 18254 , ..., 0     , 0     , 0     ],
     ...,
     [2336  , 316092, 422048, ..., 0     , 0     , 0     ],
     [45069 , 133773, 415330, ..., 0     , 0     , 0     ],
     [431671, 345062, 162709, ..., 0     , 0     , 0     ]]), Tensor(shape=[64], dtype=int64, place=Place(g
pu:0), stop_gradient=True,
    [5 , 4 , 11, 2 , 5 , 5 , 8 , 8 , 7 , 8 , 6 , 5 , 4 , 9 , 5 , 4 , 5 , 5 ,
     4 , 19, 8 , 9 , 6 , 6 , 4 , 6 , 6 , 5 , 5 , 4 , 19, 8 , 5 , 11, 7 , 5 ,
     5 , 8 , 3 , 18, 5 , 4 , 8 , 5 , 6 , 6 , 10, 6 , 8 , 5 , 6 , 7 , 6 , 5 ,
     5 , 11, 14, 3 , 5 , 10, 5 , 5 , 6 , 4 ]), Tensor(shape=[64, 64], dtype=int64, place=Place(gpu:0), stop
_gradient=True,
    [[431671, 118703, 459477, ..., 0     , 0     , 0     ],
     [123738, 198205, 53616 , ..., 0     , 0     , 0     ],
     [340164, 397067, 423468, ..., 0     , 0     , 0     ],
     ...,
     [390293, 316092, 391947, ..., 0     , 0     , 0     ],
     [445681, 423468, 439085, ..., 0     , 0     , 0     ],
     [431671, 345062, 162709, ..., 0     , 0     , 0     ]]), Tensor(shape=[64], dtype=int64, place=Place(g
pu:0), stop_gradient=True,
    [6, 5, 8, 4, 3, 6, 9, 6, 9, 8, 6, 6, 6, 10, 4, 3, 5, 4, 3, 22, 8, 10, 8, 7,
     3, 12, 6, 6, 4, 8, 20, 9, 6, 10, 6, 4, 6, 4, 4, 17, 6, 6, 7, 4, 7, 3, 12, 5,
     7, 7, 5, 6, 6, 3, 6, 13, 16, 5, 6, 5, 6, 5, 6, 6]), Tensor(shape=[64], dtype=int64, place=Place(gpu:
0), stop_gradient=True,
    [1, 1, 0, 0, 1, 0, 0, 1, 0, 1, 1, 1, 0, 1, 1, 1, 0, 1, 0, 0, 0, 0, 0, 1,
     1, 1, 1, 1, 1, 1, 0, 0, 0, 0, 0, 1, 1, 0, 1, 1, 0, 0, 0, 1, 0, 1, 0, 1,
     1, 1, 1, 1, 1, 0, 1, 1, 0, 0, 1, 1, 1, 0, 1, 1])]
```

From the output results, we can see there are five tensors. The first tensor represents the query. The second tensor represents the length of the query. The third tensor represents the title. The fourth tensor represents the length of the title, and the fifth tensor represents the label. It can be seen that each tensor contains 64 pieces of data, forming a minibatch.

10.1.3 Model Construction

SimNet is a framework for calculating the similarity of two short texts. It consists of three parts: the input layer, the representation layer, and the matching layer. These three layers process the input minibatch data in sequence. First, all the words in the text sequences are passed into SimNet, in which the input layer encodes them and outputs the corresponding vector sequences. These vector sequences are then passed into the representation layer. In this practice, we use the GRU model. The hidden state output by the GRU model can be treated as a state vector that integrates all previous words. After encoding all the words, the hidden vector of the GRU model can be regarded as the semantic vector representation of the text sequence. These semantic vectors are passed into the matching layer and the semantic vectors of the two text sequences are integrated. Finally, after softmax processing, the probabilities of semantic matching categories (similar or dissimilar) for the two text sequences are obtained. The implementation process is shown in Fig. 10.2.

In this practice, Paddlenlp.nn.GRU, Paddle.nn embedding and other APIs are used to implement the SimNet framework (GRU model), with simpler codes and higher running efficiency. Codes are implemented as follows:

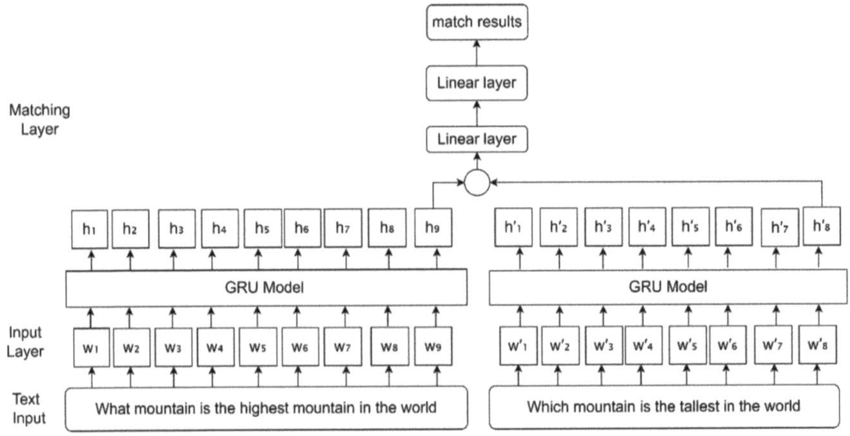

Fig. 10.2 The sketch map of the SimNet framework

```
import paddle
import paddle.nn as nn
import paddle.nn.functional as F

# Define the GRU Model, including the input layer, the presentation layer, and the matching layer
class GRUModel(nn.Layer):

    def __init__(self,
                 vocab_size,
                 num_classes,
                 emb_dim = 128,
                 padding_idx = 0,
                 gru_hidden_size = 128,
                 direction = 'forward',
                 gru_layers = 1,
                 dropout_rate = 0.0,
                 fc_hidden_size = 96):
        """

        The GRU Model, including the input layer, the presentation layer, and the matching layer

        Input:

            - vocab_size: the size of vocabulary

            - num_classes: the number of categories

            - emb_dim: the dimensions of word vectors

            - padding_idx: enter the ID value after length alignment

            - gru_hidden_size: the dimensions of input unit for GRU

            - direction: network updating direction. "Foward" refers to the unidirectional GRU network direction from the beginning of the sequences to the end of the sequences

            - gru_layers: the dimension of the GRU network

            - dropout_rate: the values of Dropout parameters in GRU

            - fc_hidden_size: the dimension of the fully connected layer in the output layer
        """
```

```
super().__init__()
        self.embedder = nn.Embedding(num_embeddings=vocab_size,
                                     embedding_dim=emb_dim,
                                     padding_idx=padding_idx)


        # Define GRU
        self.gru_encoder = nn.GRU(input_size=emb_dim,
                                  hidden_size=gru_hidden_size,
                                  num_layers=gru_layers,
                                  direction=direction,
                                  dropout=dropout_rate)
        self.fc = nn.Linear(emb_dim * 2,
                            fc_hidden_size)
        self.output_layer = nn.Linear(fc_hidden_size, num_classes)

    def forward(self, query, title, query_seq_len, title_seq_len):


        # The input layer, Shape: (batch_size, num_tokens, embedding_dim)
        embedded_query = self.embedder(query)
        embedded_title = self.embedder(title)


        # The representation layer： Shape: (batch_size, gru_hidden_size)
        encoded_query, last_hidden_query =
self.gru_encoder(embedded_query,
                                      sequence_length=query_seq_len)
        encoded_query, last_hidden_title=
self.gru_encoder(embedded_title,
                                      sequence_length=title_seq_len)


        # Concat two semantic vectors， Shape: (batch_size, 2*gru_hidden_size)
        contacted = paddle.concat([last_hidden_query[-1, :, :],
last_hidden_title[-1, :, :]], axis=-1)


        # Use the tanh activation function， Shape: (batch_size, fc_hidden_size)
        fc_out = paddle.tanh(self.fc(contacted))


        # The matching layer， Shape: (batch_size, num_classes)
        logits = self.output_layer(fc_out)
        return logits
```

And then the SimNet framework is instantiated and the model is specified by setting the network parameters. Codes are implemented as follows:

```
class SimNet(nn.Layer):

    def __init__(self,
                 network,
                 vocab_size,
                 num_classes,
                 emb_dim=128,
                 pad_token_id=0):
        """


        Implement the SimNet

        Input:
         - network: the structure of the network
        - vocab_size: the size of the word table
        - emb_dim: the dimensions of word vectors
        - pad_token_id: enter the ID value after length alignment
    """
        super().__init__()
        # Change the input strings to the lowercase
        network = network.lower()
        # SimNet's representation layer chooses GRU
        if network == 'gru':
            self.model = GRUModel(vocab_size,
                                  num_classes,
                                  emb_dim,
                                  direction='forward',
                                  padding_idx=pad_token_id)
        else:
            raise ValueError(
                "Unknown network: %s, it must be gru."
                % network)

    def forward(self, query, title, query_seq_len=None, title_seq_len=N
one):

        # Output labels for text semantic matching
        logits = self.model(query, title, query_seq_len, title_seq_len)
        return logits
```

10.1.4 Training Configuration

Next, we need to determine the computing resources, model, optimizer, loss function, and optimization indicators used during model training.

(1) Model: Use the GRU model as the backbone network for SimNet.
(2) Optimizer: Adam optimizer.
(3) Loss function: Cross-entropy.
(4) Evaluation indicators: Accuracy.

Codes are implemented as follows:

```
# Set the rounds of training
num_epochs = 3
# Set the learning rate
learning_rate = 5e-4
# Set the intervals for model evaluation in the validation set every certain number of steps
eval_steps = 50
# Set the interval for printing the log every certain number of steps
log_steps = 10
save_dir = "./checkpoints"
# Build SimNet network with GRU network as the backbone network
network = 'gru'
model = SimNet(
        network=network,
        vocab_size=len(vocab),
        num_classes=2)
# Set the optimizer
optimizer = paddle.optimizer.Adam(
        parameters=model.parameters(), learning_rate=learning_rate)

# The cross-entropy function
loss_fn = paddle.nn.CrossEntropyLoss()

# Define the calculation method for evaluation indicators
metric = paddle.metric.Accuracy()
```

10.1.5 Model Training

During the model training process, a training log is printed every certain number of log_steps, and the model is evaluated on the validation set every certain number of eval_steps. The model with the best evaluation performance during the training process is saved.

The codes implemented for evaluation on the validation set are as follows:

```
def evaluate(model, data_loader, metric):


    # Set the model to the evaluation mode
    model.eval()
    # Reset evaluations
    metric.reset()

    # Traverse each minibatch of the validation set
    for batch_id, data in enumerate(dev_loader):
        query_ids, query_seq_len, title_ids, title_seq_len, labels = data
        # Compute the output of models
        logits = model(query_ids, title_ids, query_seq_len, title_seq_len)
        # Accumulate evaluations
        correct = metric.compute(logits, labels)
        metric.update(correct)

    dev_score = metric.accumulate()

    return dev_score
```

Codes for model training are implemented as follows:

```
    for epoch in range(num_epochs):
        for step, data in enumerate(train_loader):
            query_ids,query_seq_len,title_ids,title_seq_len,labels = data
            # Obtain models' predictions
            logits = model(query_ids, title_ids, query_seq_len, title_seq_
len)

            # Compute the average of the loss functions
            loss = loss_fn(logits, labels)
            train_loss_record.append((global_step, loss.item()))

            # Gradient backpropagation
            loss.backward()
            optimizer.step()
            optimizer.clear_grad()

            if global_step % log_steps == 0:
                print(f"[Train] epoch: {epoch}/{num_epochs}, step: {global
_step}/{num_training_steps}, loss: {loss.item():.5f}")
```

```
            if global_step != 0 and (global_step % eval_steps == 0 or glob
al_step == (num_training_steps-1)):
                dev_score = evaluate(model, dev_loader, metric)
                train_score_record.append((global_step, dev_score))
                print(f"[Evaluate]  dev score: {dev_score:.5f}")
                model.train()

                # If the current indicator is the optimal indicator, save the model
                if dev_score > best_score:
                    save_path = os.path.join(save_dir, "best.pdparams")
                    paddle.save(model.state_dict(), save_path)
                    print(f"[Evaluate] best accuracy performence has been
updated: {best_score:.5f} --> {dev_score:.5f}")
                    best_score = dev_score

            global_step += 1

    save_path = os.path.join(save_dir, "final.pdparams")
    paddle.save(model.state_dict(), save_path)
    print("[Train] Training done!")

    return train_loss_record, train_score_record

train_loss_record, train_score_record = train(model)

def train(model):
    # Open the model's training mode
    model.train()
    global_step = 0
    best_score = 0.
    # Record the changes in the loss function during the training process and the accuracy on the

validation set
    train_loss_record = []
    train_score_record = []
    num_training_steps = len(train_loader) * num_epochs
    # Perform num_epochs rounds of training
```

The output result is:

[Train] epoch: 0/3, step: 3080/11193, loss: 0.47233

[Train] epoch: 0/3, step: 3090/11193, loss: 0.54170

[Train] epoch: 0/3, step: 3100/11193, loss: 0.47317

[Evaluate] dev score: 0.66530

[Train] epoch: 0/3, step: 3110/11193, loss: 0.53198

[Train] epoch: 0/3, step: 3120/11193, loss: 0.50744

[Train] epoch: 0/3, step: 3130/11193, loss: 0.54013

[Train] epoch: 0/3, step: 3140/11193, loss: 0.42351

[Train] epoch: 0/3, step: 3150/11193, loss: 0.48109

```
[Evaluate] dev score: 0.67416
[Train] epoch: 0/3, step: 3160/11193, loss: 0.42198
[Train] epoch: 0/3, step: 3170/11193, loss: 0.45638
[Train] epoch: 0/3, step: 3180/11193, loss: 0.42803
[Train] epoch: 0/3, step: 3190/11193, loss: 0.38815
[Train] epoch: 0/3, step: 3200/11193, loss: 0.42144
[Evaluate] dev score: 0.66576
[Train] epoch: 0/3, step: 3210/11193, loss: 0.44274
[Train] epoch: 0/3, step: 3220/11193, loss: 0.30856
[Train] epoch: 0/3, step: 3230/11193, loss: 0.56580
[Train] epoch: 0/3, step: 3240/11193, loss: 0.40450
[Train] epoch: 0/3, step: 3250/11193, loss: 0.40077
[Evaluate] dev score: 0.67962
[Evaluate] best accuracy performence has been updated: 0.67462 --> 0.67962
……
[Train] epoch: 2/3, step: 10660/11193, loss: 0.28567
[Train] epoch: 2/3, step: 10670/11193, loss: 0.35929
[Train] epoch: 2/3, step: 10680/11193, loss: 0.16474
[Train] epoch: 2/3, step: 10690/11193, loss: 0.27437
[Train] epoch: 2/3, step: 10700/11193, loss: 0.29359
[Evaluate] dev score: 0.72234
[Train] epoch: 2/3, step: 10710/11193, loss: 0.34276
[Train] epoch: 2/3, step: 10720/11193, loss: 0.28217
[Train] epoch: 2/3, step: 10730/11193, loss: 0.28727
[Train] epoch: 2/3, step: 10740/11193, loss: 0.17787
[Train] epoch: 2/3, step: 10750/11193, loss: 0.28457
[Evaluate] dev score: 0.74029
[Train] epoch: 2/3, step: 10760/11193, loss: 0.22572
[Train] epoch: 2/3, step: 10770/11193, loss: 0.29225
[Train] epoch: 2/3, step: 10780/11193, loss: 0.29588
```

```
[Train] epoch: 2/3, step: 10790/11193, loss: 0.22411
[Train] epoch: 2/3, step: 10800/11193, loss: 0.22890
[Evaluate] dev score: 0.73870
[Train] epoch: 2/3, step: 10810/11193, loss: 0.21487
[Train] epoch: 2/3, step: 10820/11193, loss: 0.23997
[Train] epoch: 2/3, step: 10830/11193, loss: 0.31584
[Train] epoch: 2/3, step: 10840/11193, loss: 0.19126
[Train] epoch: 2/3, step: 10850/11193, loss: 0.29022
[Evaluate] dev score: 0.74233
[Train] epoch: 2/3, step: 10860/11193, loss: 0.30609
[Train] epoch: 2/3, step: 10870/11193, loss: 0.25796
[Train] epoch: 2/3, step: 10880/11193, loss: 0.19873
[Train] epoch: 2/3, step: 10890/11193, loss: 0.32246
[Train] epoch: 2/3, step: 10900/11193, loss: 0.20712
[Evaluate] dev score: 0.74233
[Train] epoch: 2/3, step: 10910/11193, loss: 0.22017
[Train] epoch: 2/3, step: 10920/11193, loss: 0.32539
[Train] epoch: 2/3, step: 10930/11193, loss: 0.29813
[Train] epoch: 2/3, step: 10940/11193, loss: 0.38045
[Train] epoch: 2/3, step: 10950/11193, loss: 0.31829
[Evaluate] dev score: 0.74415
[Train] epoch: 2/3, step: 10960/11193, loss: 0.30175
[Train] epoch: 2/3, step: 10970/11193, loss: 0.33339
[Train] epoch: 2/3, step: 10980/11193, loss: 0.25792
[Train] epoch: 2/3, step: 10990/11193, loss: 0.29822
[Train] epoch: 2/3, step: 11000/11193, loss: 0.19026
[Evaluate] dev score: 0.75369
[Evaluate] best accuracy performence has been updated: 0.74778 --> 0.75369
[Train] epoch: 2/3, step: 11010/11193, loss: 0.30189
[Train] epoch: 2/3, step: 11020/11193, loss: 0.20098
```

[Train] epoch: 2/3, step: 11030/11193, loss: 0.28482

[Train] epoch: 2/3, step: 11040/11193, loss: 0.24817

[Train] epoch: 2/3, step: 11050/11193, loss: 0.18661

[Evaluate] dev score: 0.74733

[Train] epoch: 2/3, step: 11060/11193, loss: 0.32549

[Train] epoch: 2/3, step: 11070/11193, loss: 0.26358

[Train] epoch: 2/3, step: 11080/11193, loss: 0.29076

[Train] epoch: 2/3, step: 11090/11193, loss: 0.22489

[Train] epoch: 2/3, step: 11100/11193, loss: 0.21053

[Evaluate] dev score: 0.73847

[Train] epoch: 2/3, step: 11110/11193, loss: 0.22301

[Train] epoch: 2/3, step: 11120/11193, loss: 0.35071

[Train] epoch: 2/3, step: 11130/11193, loss: 0.18207

[Train] epoch: 2/3, step: 11140/11193, loss: 0.31599

[Train] epoch: 2/3, step: 11150/11193, loss: 0.33996

[Evaluate] dev score: 0.74960

[Train] epoch: 2/3, step: 11160/11193, loss: 0.28782

[Train] epoch: 2/3, step: 11170/11193, loss: 0.26186

[Train] epoch: 2/3, step: 11180/11193, loss: 0.39324

[Train] epoch: 2/3, step: 11190/11193, loss: 0.20669

[Evaluate] dev score: 0.74847

[Train] Training done!

Then the loss changes during the training process (train_loss_record) and the score changes on the validation set (train_score_record) are saved, and the training process is visualized. Codes are implemented as follows:

```
%matplotlib inline
from tools import plot_training_acc, plot_training_loss

fig_path = "./checkpoints/chapter9_loss.pdf"
plot_training_loss(train_loss_record, fig_path, loss_legend_loc="upper
right", sample_step=100)

fig_path = "./checkpoints/chapter9_acc.pdf"
plot_training_acc(train_score_record, fig_path, sample_step=1,
loss_legend_loc="lower right")
```

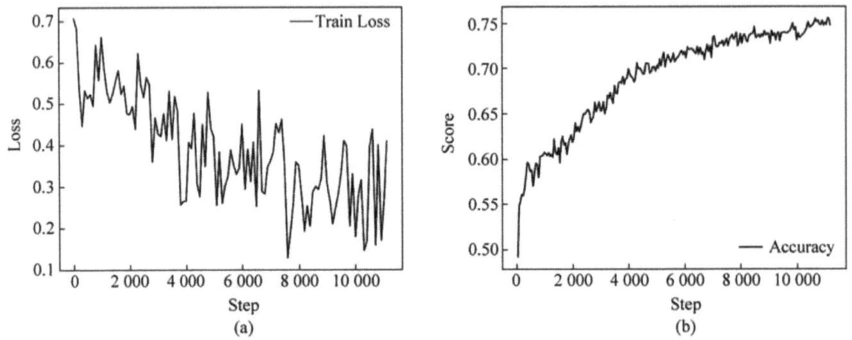

Fig. 10.3 The trend of the loss function on the training set and the trend of the accuracy on the validation set **a** The trend of the loss function's changes on the training set; **b** The trend of the accuracy on the validation set

The output result is shown in Fig. 10.3.

The output results show that as the training progresses, the loss function value on the training set continuously decreases and gradually converges, while the accuracy score on the validation set continuously increases and stabilizes after the model converges.

10.1.6 Model Evaluation

Then, the model is evaluated with the best performance during the training process and the training effect of the model is verified by using the test set. Codes are implemented as follows:

```
# Load the trained model for evaluation, re-instantiate the model, and then load the trained model's
parameters into the new model
saved_state = paddle.load("./checkpoints/best.pdparams")
model = SimNet(
        network=network,
        vocab_size=len(vocab),
        num_classes=2)
model.load_dict(saved_state)

# Model evaluation
evaluate(model, test_loader, metric)
```

The output result is: *0.7555101113383322*.

10.1.7 Model Prediction

After completing all the steps above, the model has been successfully established. We can input any two sentences, such as "What are the athletics events at the Olympic Games?" and "What are the track and field events in the Olympic Games?", and the model will provide the semantic matching results for these two sentences after inference. Codes are implemented as follows:

```
def infer(model,tokenizer,query,title):
    label_map = {0: 'dissimilar', 1: 'similar'}

    # Code is mapped to ID
    query_id = tokenizer.encode(query)
    title_id = tokenizer.encode(title)

    # Convert data into tensor
    query_ids = paddle.to_tensor([query_id])
    title_ids = paddle.to_tensor([title_id])
    query_seq_lens = paddle.to_tensor(len(query_id))
    title_seq_lens = paddle.to_tensor(len(title_id))

    # Model evaluation
    model.eval()
    logits = model(query_ids, title_ids, query_seq_lens, title_seq_lens)
    probs = F.softmax(logits, axis=1)
    idx = paddle.argmax(probs, axis=1).numpy()
    idx = idx.tolist()
    pred_label=label_map[idx[0]]
    print('Label: {}'.format(pred_label))


# Input a sample
query='What are the athletics events at the Olympic Games?'
title='What are the track and field events in the Olympic Games?'
infer(model,tokenizer,query,title)
```

The output result is: Label: similar.

10.2 The Text Semantic Matching Based on RocketQA

10.2.1 Task Objective and Implementation Processes

The task objective introduced in this sub-section is the same as those described in Sub-Sect. 10.1, which is to implement the semantic matching task (similar or dissimilar) between two short texts. Although the Chinese semantic matching dataset LCQMC from Baidu Zhidao (the world's leading Chinese Q&A interactive platform) is used, the method adopted in this sub-section is different from that used in Sub-Sect. 10.1. In this sub-section, we use the pre-training model RocketQA (Qu et al., 2021).

RocketQA is a pre-training model independently developed by Baidu for dense vector retrieval in the open-domain Q&A. It has achieved excellent performance on the industrial-level open-domain Q&A datasets such as Microsoft's MSMARCO, Google's Natural Questions, and the large-scale Chinese Q&A reading comprehension dataset DuReader. RocketQA, based on the ERNIE model, adopts three training optimization strategies based on the ERNIE model, including cross-minibatch negatives, denoised hard negatives, and data augmentation, to enhance the generalization of trained models, especially to improve the ability to sort questions and paragraphs. In this way, models are more suitable for search, sorting, and matching scenarios.

(1) Cross-batch Negatives: Use the data in the form of sentences from multiple GPUs as negatives for training. This is no longer restricted to a single minibatch. Instead, we use more negatives for model training.
(2) Denoised Hard Negatives: On the basis of increasing the number of negatives, the cross-encoder scoring strategy is adopted to score the results retrieved by the model. Only strong negatives with high confidence are retained, and then they are sent into the model for training using the sampling method. In this way, the trained model's generalization will be improved.
(3) Data Augmentation: The purpose of data augmentation is to effectively utilize a vast amount of unlabeled data and use the trained cross encoder to automatically label questions and paragraphs. Then, positive and negative samples are selected based on the model's threshold, thereby increasing the amount of data for model training.

Figure 10.4 shows the semantic matching network framework of RocketQA. The framework uses the cross encoder to concatenate two input texts, and converts them into the corresponding encoding format. Then they are passed into RockeQA to obtain the semantic vector representation of the concatenated sentence. Finally, the probability of text similarity is calculated through the linear layer. Unlike the SimNet method, RocketQA first concatenates the two texts, and then calculates the similarity of the entire sentence. It fully utilizes the attention mechanism to get the deep semantic relationship between the two sentences. The model achieves higher prediction accuracy.

The idea of implementing text semantic matching based on RocketQA is shown in Fig. 10.5. The input of the model is two sentences, and the output is the result of text semantic matching (similar or not similar). In the modeling process, data processing is first carried out, then the semantic vector representation of the text is obtained using RocketQA, and finally the result of text semantic matching is obtained after processing through a fully connected layer and softmax.

10.2.2 Data Processing

As previously mentioned, this practice still uses the LCQMC dataset for text semantic matching. The introduction to the dataset and the method of loading the dataset are

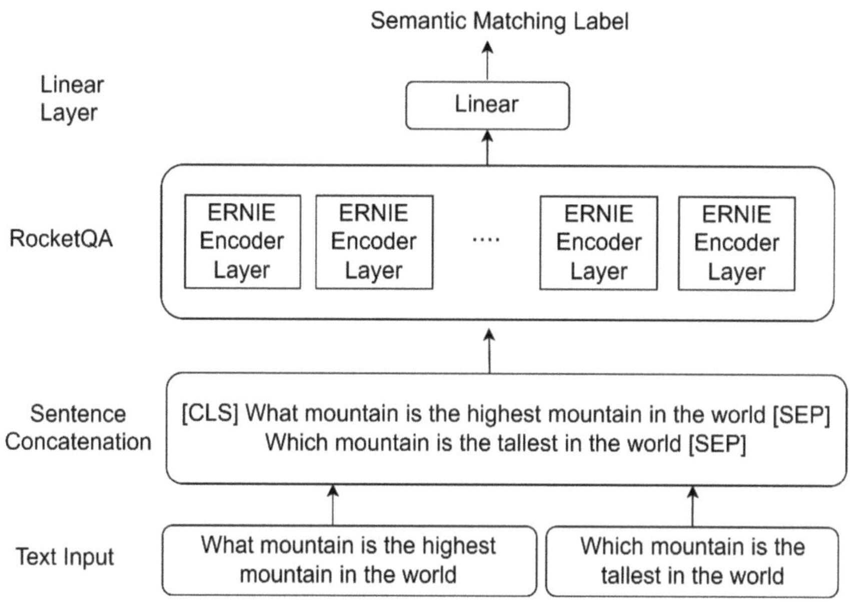

Fig. 10.4 An example of text semantic matching using RocketQA

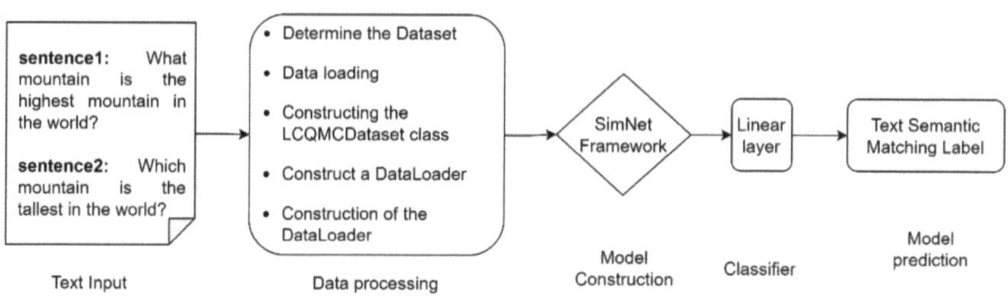

Fig. 10.5 The text semantic matching process based on RocketQA

exactly the same as those described in Sub-Sect. 10.1.2, so we will not repeat them here. The following will first introduce the implementation steps such as converting data into the feature form and constructing the DataLoader, and then the same batch of data is processed into feature sequences with equal length, and the DataLoader is used to input them into the RocketQA model in batches.

1. **Transform Data into the Feature Form**

The original data is in the form of texts, which cannot be directly read by the RocketQA model. It is necessary to first convert the text into the corresponding feature data, including input coding input_ids and text type coding token_type_ids. For example, for the following two texts:

{'text_a': 'What kind of girls do boys who like to play basketball like?', 'text_b': 'What kind of girls do boys who love playing basketball like?' }

RocketQA converts the problem of matching two texts into the problem of classification of concatenated texts. The model first concatenates the two texts into one sentence, "[CLS]What kind of girls do boys who like to play basketball like?[SEP]What kind of girls do boys who love playing basketball like?[SEP]," in which "[CLS]" represents the semantic coding of the entire sentence and "[SEP]" represents the segmentation boundary between the two sentences. Then, the concatenated sentence is mapped into the encoding form through vocabulary and obtains the input_id:

[1, 692, 811, 445, 2001, 497, 5, 654, 21, 692, 811, 614, 356, 314, 5, 291, 2, 329, 445, 2001, 497, 5, 654, 21, 692, 811, 614, 356, 314, 5, 291, 2]

For the token_type_ids of text type, the characters of text_a are usually encoded as "0," and the characters of text_b are encoded as "1," and so on. The token_type_ids corresponding to the above example are:

[0, 0, 0, 0, 0, 0, 0, 0, 0, 0, 0, 0, 0, 0, 0, 0, 0, 1, 1, 1, 1, 1, 1, 1, 1, 1, 1, 1, 1, 1, 1]

The above feature encoding can be quickly implemented using rocketqa-zh-base-query-encoder of tokenizer. codes are implemented as follows:

```
from paddlenlp.transformers import AutoTokenizer

tokenizer = AutoTokenizer.from_pretrained("rocketqa-zh-base-query-encoder
")

encoded_inputs = tokenizer(text="What kind of girls do boys who like to play basketball
like? ",
                           text_pair="What kind of girls do boys who love playi
ng basketball like?",
                           max_seq_len=32)
print(encoded_inputs)
```

The output result is:

{'input_ids': [1, 692, 811, 445, 2001, 497, 5, 654, 21, 692, 811, 614, 356, 314, 5, 291, 2, 329, 445, 2001, 497, 5, 654, 21, 692, 811, 614, 356, 314, 5, 291, 2], 'token_type_ids': [0, 0, 0, 0, 0, 0, 0, 0, 0, 0, 0, 0, 0, 0, 0, 0, 0, 1, 1, 1, 1, 1, 1, 1, 1, 1, 1, 1, 1, 1, 1]}.

From the output results, it can be seen that by default, the tokenizer returns input_ids and token_type_ids in the form of a list.

Next we will convert the loaded training set, validation set, and test set into the corresponding feature representations based on the convert_example_to_feature function. Codes are implemented as follows:

```
import numpy as np

def convert_example_to_feature(example,
                    tokenizer,
                    max_seq_length=512,
                    is_test=False,
                    is_pair=False):
    if is_pair:
        # Sentence a
        text = example["text_a"]
        # Sentence b
        text_pair = example["text_b"]
    else:
        text = example["text"]
        text_pair = None

    # Use the tokenizer to convert sentence a and sentence b into the feature form
    encoded_inputs = tokenizer(text=text,
                         text_pair=text_pair,
                         max_seq_len=max_seq_length)

    if is_test:
        return encoded_inputs

    # Convert label into the Numpy format
    encoded_inputs["label"] = np.array([example["label"]], dtype="int64")
    return encoded_inputs

from functools import partial
max_seq_length= 128

# Set the default value of the convert_example_to_feature function
trans_func = partial(convert_example_to_feature,
                tokenizer=tokenizer,
                max_seq_length=max_seq_length,
                is_pair=True)

# Convert the input training set, validation set, and test set data into the feature form
train_dataset = train_dataset.map(trans_func)
dev_dataset = dev_dataset.map(trans_func)
test_dataset = test_dataset.map(trans_func)

# Output the samples of the training set, validation set, and test set
print("Number of samples of the training set:", len(train_dataset))
print("Number of samples of the validation set:", len(dev_dataset))
print("Number of samples of the test set:", len(test_dataset))
```

The output result is:

Number of samples of the training set: 238766

Number of samples of the validation set: 4401

Number of samples of the test set: 4401

2. **Construction of the DataLoader**

Due to the limitation of computer memory, it is necessary to train the data in batches in the process of model training, that is, to select a minibatch of data for training each time. We achieve this using the DataLoader. However, since the length of the data in the same batch may not be the same, it is necessary to change the data in the same batch into the same length. As mentioned earlier, there are two common methods: text truncation and text padding, the convert_example_to_feature function can truncate the excessively long text, and the text padding uses the DataCollatorWithPadding function. Codes are implemented as follows:

```
from paddle.io import BatchSampler, DataLoader
from paddlenlp.data import DataCollatorWithPadding


# Use DataCollatorWithPadding for text padding
data_collator = DataCollatorWithPadding(tokenizer)
batch_size = 64


# Set shuffle to True for the training set to randomize data operation
train_sampler = BatchSampler(
        train_dataset, batch_size=batch_size, shuffle=True)
train_loader = DataLoader(dataset=train_dataset,
                                   batch_sampler=train_sampler,
                                   collate_fn=data_collator)

# The validation set and test set do not need to randomize data, and set shuffle to False
dev_sampler = BatchSampler(
        dev_dataset, batch_size=batch_size, shuffle=False)
dev_loader = DataLoader(dataset=dev_dataset,
                                   batch_sampler=dev_sampler,
                                   collate_fn=data_collator)

test_sampler = BatchSampler(
        test_dataset, batch_size=batch_size, shuffle=False)
test_loader = DataLoader(dataset=test_dataset,
                                   batch_sampler=test_sampler,
                                   collate_fn=data_collator)

# Print the data of the first minibatch in the training set

for batch in train_loader:
    print(batch)
    break
```

The output result is:

```
{'input_ids': Tensor(shape=[64, 46], dtype=int64, place=Place(gpu:0), stop
_gradient=True,
       [[1   , 422 , 1189, ..., 0   , 0   , 0   ],
        [1   , 429 , 433 , ..., 0   , 0   , 0   ],
        [1   , 75  , 1173, ..., 0   , 0   , 0   ],
        ...,
        [1   , 416 , 1480, ..., 0   , 0   , 0   ],
        [1   , 3586, 936 , ..., 0   , 0   , 0   ],
        [1   , 1183, 4016, ..., 0   , 0   , 0   ]]), 'token_type_ids': Ten
sor(shape=[64, 46], dtype=int64, place=Place(gpu:0), stop_gradient=True,
       [[0, 0, 0, ..., 0, 0, 0],
        [0, 0, 0, ..., 0, 0, 0],
        [0, 0, 0, ..., 0, 0, 0],
        ...,
        [0, 0, 0, ..., 0, 0, 0],
        [0, 0, 0, ..., 0, 0, 0],
        [0, 0, 0, ..., 0, 0, 0]]), 'labels': Tensor(shape=[64, 1], dtype=i
nt64, place=Place(gpu:0), stop_gradient=True,
       [[1], [1], [1], [1], [1], [0], [1], [0], [1], [1], [1], [0], [1],
[1], [1], [0], [0], [0], [1], [1], [0], [1], [0], [0], [0], [1], [0],
[1], [1], [1], [0], [0], [1], [1], [1], [1], [1], [1], [1], [1], [0], [1],
 [1], [1], [0], [0], [0], [1], [1], [0], [0], [0], [1], [0], [0], [0],
[0], [1], [0], [1], [1], [1], [0], [1]])}
```

From the output results, the generated minibatch data contains three parts: input_ids, token_type_ids, and labels. The size of input_ids and token_type_ids is [64, 58], indicating that there are 64 pieces of data, and the length of each piece of data is 58. The size of labels is [64, 1], indicating that there are 64 labels, and the length of each label is 1.

10.2.3 Model Construction

The RocketQA model processes each input minibatch of data in turn. First, the processed text feature data is sent into the RocketQA model, which contains multiple ERNIE encoding layers. Each layer encodes the input data, and the output vector sequences from the first and last encoding layers are averaged and aggregated. Then, the two vectors are further averaged after convergence, and then the final vector sequence after convergence is passed to the linear classification layer (vector sequence multiplied by weight, plus bias). After softmax processing, the final classification result (either "similar" or "dissimilar") is obtained. The model's architecture is shown in Fig. 10.6.

When implementing codes, readers can use avg_pool1d average pooling method to implement the average pooling operation. Set the kernel_size to the length of the sequences. Here are the codes:

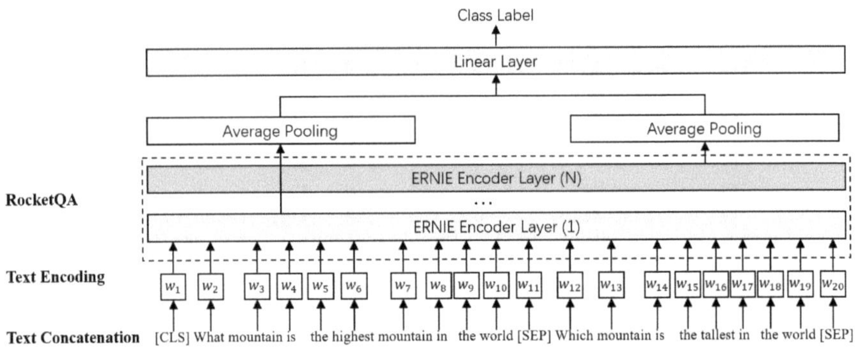

Fig. 10.6 Semantic matching based on RocketQA in the form of text classification

```
class CrossEncoder(nn.Layer):

    def __init__(self, pretrained_model, dropout=None, num_classes=
2):
        """

        The implementation of CrossEncoder

        Input:

            - pretrained_model: pre-training language models

            - dropout: parameters of dropout

            - num_classes: the number of categories for semantic matching tasks is
generally binary classification.
        """
        super().__init__()
        self.ernie = pretrained_model
        self.dropout = nn.Dropout(dropout if dropout is not None els
e self.
                                  ernie.config["hidden_dropout_prob
"])
        self.classifier = nn.Linear(self.ernie.config["hidden_size
"],
                                    num_classes,
                                    weight_attr=nn.initializer.Trunc
atedNormal(
                                        mean=0.0, std=0.02),
                                    bias_attr=nn.initializer.Constan
t(value=0))

    def forward(self,
                input_ids,
                token_type_ids=None,
                encoder_type='first-last-avg'):
        """
```

```
        Input:
            - input_ids: input_ids parameter
            - token_type_ids: token_type_ids parameter
            - encoder_type: types of encoding, 'first-last-avg' represents the average of
the outputs of the first and last layers
        """
        sequence_output, pooled_output, hidden_output = self.ernie(i
nput_ids,
                                                token_type_ids=token_type_ids,
                                                output_hidden_states=True)
        # The average of the first and last encoding layers

        if encoder_type == 'first-last-avg':

            # The hidden_output list has 13 hidden_states. The first is the embeddings,
and the second is the hidden_states of the first layer

            first = hidden_output[1]
            last = hidden_output[-1]
            seq_length = first.shape[1]
            first_avg = F.avg_pool1d(first.transpose([0, 2, 1]), ker
nel_size=seq_length).squeeze(-1)
            last_avg = F.avg_pool1d(last.transpose([0,2, 1]), kernel
_size=seq_length).squeeze(-1)
            final_encoding = F.avg_pool1d(paddle.concat([first_avg.u
nsqueeze(1),last_avg.unsqueeze(1)],axis=1).transpose([0,2, 1]), kern
el_size=2).squeeze(-1)
        logits = self.classifier(final_encoding)
        return logits
```

10.2.4 Training Configuration

Determine the computing resources, model, optimizer, loss function, and optimization indicators used during model training.

(1) Model: RocketQA.
(2) Optimizer: The AdamW optimizer.
(3) Loss function: Cross-entropy loss.
(4) Evaluation indicators: Accuracy.

Codes are implemented as follows:

```
from paddlenlp.transformers import AutoModel
import paddle

# Set training rounds
num_epochs = 1
# Set the learning rate
learning_rate = 0.0001
# Set the interval for model evaluation on the validation set every certain number of steps
eval_steps = 50
# Set the interval for printing the log every certain number of steps
log_steps = 10
# Set the AdamW optimizer's weight decay coefficient
weight_decay = 0.0
save_dir = "./checkpoints"

# The name of models
model_name_or_path = "rocketqa-zh-base-query-encoder"

# The number of categories
num_classes=2

# Load the RocketQA model
pretrained_model = AutoModel.from_pretrained(model_name_or_path)

# Instantiate CrossEncoder
model = CrossEncoder(pretrained_model, num_classes=num_classes)


# All the LayerNorm and bias' parameters do not set weight decay
decay_params = [
        p.name for n, p in model.named_parameters()
        if not any(nd in n for nd in ["bias", "norm"])
    ]

# Set the optimizer
optimizer = paddle.optimizer.AdamW(
        parameters=model.parameters(),
        weight_decay=weight_decay,
        apply_decay_param_fun=lambda x: x in decay_params,
        learning_rate=learning_rate)
# Define the loss function: cross-entropy loss
loss_fn = paddle.nn.CrossEntropyLoss()

# Define the calculation method for evaluation indicators
metric = paddle.metric.Accuracy()
```

10.2.5 Model Training

During the model's training process, a training log is printed every certain log_steps, and the model is evaluated on the validation set every certain eval_steps. The model with the best evaluation performance during the training process is saved.

The codes for evaluation on the validation set are implemented as follows:

```
def evaluate(model, data_loader, metric):
    """

    The model evaluation function


    Input:
        - model: model samples to be evaluated
        - data_loader: datasets to be evaluated
        - metric: Class instances used for analyzing evaluation indicators
  """

    # Set the model into the evaluation mode
    model.eval()

    # Reset metric
    metric.reset()

    # Traverse each minibatch of the validation set
    for batch_id, data in enumerate(dev_loader):
        input_ids, token_type_ids, labels = data["input_ids"], data["toke
n_type_ids"], data["labels"]

        # Compute the output of models
        logits = model(input_ids, token_type_ids)

        # Accumulate evaluation
        correct = metric.compute(logits, labels)
        metric.update(correct)

    dev_score = metric.accumulate()

    return dev_score
```

The codes for model training are implemented as follows:

```
import os

def train(model):

    # Start the training mode of models
    model.train()
    global_step = 0
    best_score = 0.

    # Record the loss function value during training and the score of the model on the
    validation set
    train_loss_record = []
    train_score_record = []
    num_training_steps = len(train_loader) * num_epochs

    # Perform num_epochs rounds of training
    for epoch in range(num_epochs):
        for step, data in enumerate(train_loader):
            input_ids, token_type_ids, labels = data["input_ids"],
data["token_type_ids"], data["labels"]

            # Obtain model prediction
            logits = model(input_ids, token_type_ids)

            # Compute the average of the loss function
            loss = loss_fn(logits, labels)
            train_loss_record.append((global_step, loss.item()))


            # Gradient backpropagation
            loss.backward()
            optimizer.step()
            optimizer.clear_grad()

            if global_step % log_steps == 0:
                print(f"[Train] epoch: {epoch}/{num_epochs}, step:
{global_step}/{num_training_steps}, loss: {loss.item():.5f}")

            if global_step != 0 and (global_step % eval_steps == 0 or
global_step == (num_training_steps-1)):
                dev_score = evaluate(model, dev_loader, metric)
                train_score_record.append((global_step, dev_score))
                print(f"[Evaluate]  dev score: {dev_score:.5f}")
                model.train()

                # If the current indicator is the optimal indicator, save the model
                if dev_score > best_score:
                    save_path = os.path.join(save_dir, "best.pdparams")
                    paddle.save(model.state_dict(), save_path)
                    print(f"[Evaluate] best accuracy performence has been
updated: {best_score:.5f} --> {dev_score:.5f}")
                    best_score = dev_score

            global_step += 1

    save_path = os.path.join(save_dir, "final.pdparams")
    paddle.save(model.state_dict(), save_path)
    print("[Train] Training done!")

    return train_loss_record, train_score_record
```

The output result is:

```
train_loss_record, train_score_record = train(model)
[Train] epoch: 0/1, step: 0/3731, loss: 0.71879
[Train] epoch: 0/1, step: 10/3731, loss: 0.62647
[Train] epoch: 0/1, step: 20/3731, loss: 0.52406
[Train] epoch: 0/1, step: 30/3731, loss: 0.34259
[Train] epoch: 0/1, step: 40/3731, loss: 0.33003
[Train] epoch: 0/1, step: 50/3731, loss: 0.28793
[Evaluate] dev score: 0.84640
[Evaluate] best accuracy performence has been updated: 0.00000 --> 0.84640
[Train] epoch: 0/1, step: 60/3731, loss: 0.22907
[Train] epoch: 0/1, step: 70/3731, loss: 0.33072
[Train] epoch: 0/1, step: 80/3731, loss: 0.26247
[Train] epoch: 0/1, step: 90/3731, loss: 0.42596
[Train] epoch: 0/1, step: 100/3731, loss: 0.30369
[Evaluate] dev score: 0.82163
[Train] epoch: 0/1, step: 110/3731, loss: 0.32494
[Train] epoch: 0/1, step: 120/3731, loss: 0.18590
[Train] epoch: 0/1, step: 130/3731, loss: 0.20824
[Train] epoch: 0/1, step: 140/3731, loss: 0.18202
[Train] epoch: 0/1, step: 150/3731, loss: 0.28326
[Evaluate] dev score: 0.85526
[Evaluate] best accuracy performence has been updated: 0.84640 --> 0.85526
……
[Train] epoch: 0/1, step: 3510/3731, loss: 0.15867
[Train] epoch: 0/1, step: 3520/3731, loss: 0.10175
[Train] epoch: 0/1, step: 3530/3731, loss: 0.12430
[Train] epoch: 0/1, step: 3540/3731, loss: 0.22932
[Train] epoch: 0/1, step: 3550/3731, loss: 0.13192
[Evaluate] dev score: 0.89593
```

[Evaluate] best accuracy performence has been updated: 0.89275 --> 0.89593

[Train] epoch: 0/1, step: 3560/3731, loss: 0.27284

[Train] epoch: 0/1, step: 3570/3731, loss: 0.20879

[Train] epoch: 0/1, step: 3580/3731, loss: 0.18811

[Train] epoch: 0/1, step: 3590/3731, loss: 0.18174

[Train] epoch: 0/1, step: 3600/3731, loss: 0.23823

[Evaluate] dev score: 0.88775

[Train] epoch: 0/1, step: 3610/3731, loss: 0.15869

[Train] epoch: 0/1, step: 3620/3731, loss: 0.18524

[Train] epoch: 0/1, step: 3630/3731, loss: 0.19346

[Train] epoch: 0/1, step: 3640/3731, loss: 0.22400

[Train] epoch: 0/1, step: 3650/3731, loss: 0.13313

[Evaluate] dev score: 0.89298

[Train] epoch: 0/1, step: 3660/3731, loss: 0.13891

[Train] epoch: 0/1, step: 3670/3731, loss: 0.30161

[Train] epoch: 0/1, step: 3680/3731, loss: 0.16025

[Train] epoch: 0/1, step: 3690/3731, loss: 0.12038

[Train] epoch: 0/1, step: 3700/3731, loss: 0.21716

[Evaluate] dev score: 0.87321

[Train] epoch: 0/1, step: 3710/3731, loss: 0.18570

[Train] epoch: 0/1, step: 3720/3731, loss: 0.17670

[Train] epoch: 0/1, step: 3730/3731, loss: 0.21010

[Evaluate] dev score: 0.89025

The loss changes (train_loss_record) during the training process and the score changes (train_score_record) on the validation set are saved, and the training process is visualized. Codes are implemented as follows:

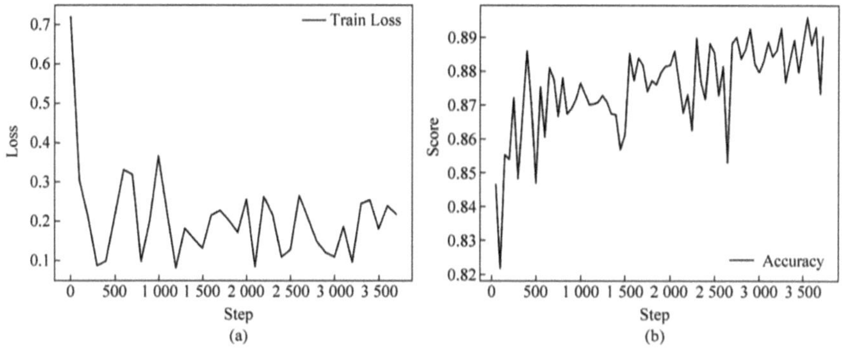

Fig. 10.7 The trend of the loss function on the training set and the trend of the accuracy on the validation set **a** The trend of the loss function on the training set; **b** The trend of the accuracy on the validation set

```
%matplotlib inline
from tool import plot_training_acc, plot_training_loss

fig_path = "./checkpoints/chapter9_loss.pdf"
plot_training_loss(train_loss_record, fig_path, loss_legend_loc="upper
right", sample_step=100)

fig_path = "./checkpoints/chapter9_acc.pdf"
plot_training_acc(train_score_record, fig_path, sample_step=1,
loss_legend_loc="lower right")
```

The output result is shown in Fig. 10.7.

From the output results, it can be seen that as the training progresses, the loss value on the training set continues to decrease and gradually converges. The accuracy on the validation set initially increases and then stabilizes after the model converges, and eventually converges to an accuracy of about 89%.

10.2.6 Model Evaluation

The model with the best performance is evaluated by using the test set during the training process to verify the training effect of the model. Codes are implemented as follows:

```
# Load the trained model for prediction, re-instantiate a model, and then load the trained model's
parameters into the new model
saved_state = paddle.load("./checkpoints/best.pdparams")
model = CrossEncoder(pretrained_model)
model.load_dict(saved_state)


# Model evaluation
evaluate(model, test_loader, metric)
```

The output result is: 0.8959327425585094.

It can be seen that with only one round of training, the model's accuracy reached 89%, while the SimNet network in Sect. 10.1 was trained for three rounds and the accuracy was only 75.55%.

10.2.7 Model Prediction

Now let's input any two sentences, such as "What are the athletics events at the Olympic Games?" and "What are the track and field events in the Olympic Games?", and the model will infer the semantic matching results between the two sentences. Codes are implemented as follows:

```
def infer(model, tokenizer, text_a, text_b):
    label_map = {0: 'dissimilar', 1: 'similar'}


    # Convert text data into the ID format
    encoded_inputs = tokenizer(text_a, text_pair = text_b)

    # Convert text encoding into tensor
    input_ids=paddle.to_tensor([encoded_inputs["input_ids"]])
    token_type_ids=paddle.to_tensor([encoded_inputs["token_type_ids"]])


    # Model evaluation
    model.eval()
    logits = model(input_ids, token_type_ids)
    probs = F.softmax(logits, axis=1)
    idx = paddle.argmax(probs, axis=1).numpy()
    idx = idx.tolist()
    pred_label=label_map[idx[0]]
    print('Label: {}'.format(pred_label))


# Input a sample
text_a = 'What are the athletics events at the Olympic Games?'
text_b =' What are the track and field events in the Olympic Games?'
infer(model, tokenizer, text_a, text_b)
```

10.3 Exercises

In the experiment introduced in this chapter, the GRU network is used as the representation layer of SimNet to obtain the semantic vector representation of texts. Readers are encouraged to try replacing the representation layer of SimNet with the LSTM network and observe the effect of model training.

In this chapter, we employ RocketQA as the feature extraction network, and take the outputs of its first and last layers as classification vectors. Readers are encouraged to compare the experiments to see what results would be obtained if only the output of the last layer is selected for classification?

Chapter 11
PEGASUS-Based Text Summarization

Abstract This chapter introduces the implementation of Chinese text summarization using the PEGASUS pre-trained model within PaddleNLP. It begins with an overview of extractive and generative summarization methods, highlighting the advantages of generative approaches enabled by encoder–decoder architectures. The chapter then details the use of the LCSTS dataset, data preprocessing steps, feature construction, and DataLoader preparation. The PEGASUS model is trained with AdamW optimization and evaluated using BLEU and ROUGE metrics, demonstrating steadily improving performance with training. Experimental results show strong summarization quality, with high ROUGE and BLEU scores on the test set. Finally, the chapter presents model inference examples, practical applications, and exercises for extending and improving summarization approaches.

The text summarization means automatically extracting key information from a specified single document or multiple documents, and generating a summary of a specified length that represents the main points of the document. The generated summary can be regenerated based on the content of the document. It can also be a selection of key sentences extracted from the given document(s).

Automatic summarization from different perspectives can be divided into different types, such as single-document summarization and multi-document summarization; single language summary and cross-language summary; ordinary summary and user-oriented query summary; and extractive summary and generative summary. Extractive summary only needs to extract key sentences from the document(s), usually using methods such as Lead-3, TextRank, and clustering. The task of generative summary is more complex and it requires understanding of the document content to generate summary texts of a specified length.

In recent years, with the development of deep learning, sequence-to-sequence models based on the encoder–decoder have been applied to text summarization tasks, such as MASS, UniLM, T5, BART, Unimo-Text, and PEGASUS, and they all achieve good results. The basic idea of these models is to first use an encoder to encode the semantic meanings of the input text, and then use a decoder to generate the corresponding summary.

C. Zong et al., *Natural Language Processing and Large Language Models*,
https://doi.org/10.1007/978-981-92-0682-7_11

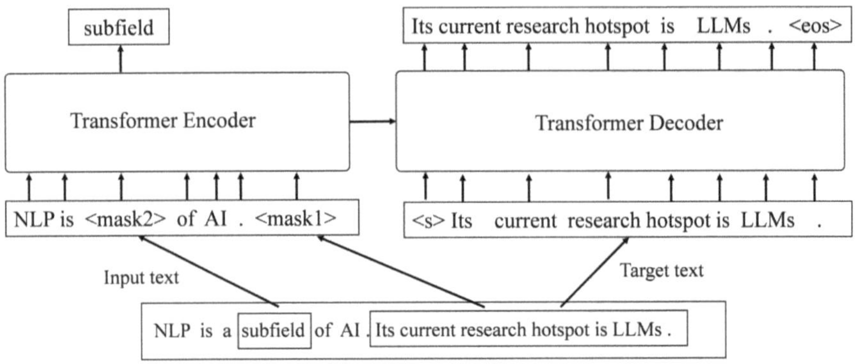

Fig. 11.1 PEGASUS's architecture (Zhang et al. 2020)

In this chapter, we use the pre-training model PEGASUS supported by PaddleNLP to implement automatic summarization of Chinese texts. PEGASUS is a model architecture proposed by Google at the 2020 ICML conference. Its network structure is shown in Fig. 11.1, which is the encoder–decoder structure. It designs an unsupervised pre-training task for the text summarization task, namely, gap sentence generation (GSG). PEGASUS uses the principal strategy, first selects the top M sentences based on their importance, and covers the selected sentences to allow the model to generate the covered sentences. This pre-training task can be well matched with the text summarization task, and the pre-training model can generate good summarization results after simple fine-tuning. Currently, the two Chinese pre-trained models (Randeng-Pegasus-238 M-Summary-Chinese and Randeng-Pegasus-523 M-Summary-Chinese) are proposed by the International Digital Economy Academy and perform well in the LCSTS dataset (a large-scale Chinese short text summarization dataset). The PaddlePaddle team implemented these two pre-training models.

11.1 Task Objectives and Implementation Processes

In this practice, we use generative modeling method to complete the text summarization task based on the PEGASUS pre-training model, as shown in Fig. 11.2. For an input text, it automatically generates a summary of a specified length. When generating the summary, the model first needs to extract the semantic information of the text, and then decodes these semantic information to generate a summary. Currently, the technique of text summarization has been widely used in news recommendation, voice broadcasting, hot-discussed news aggregation, and other businesses.

For example, given an input paragraph as follows:

Gu Ailing won the silver medal in the women's slopestyle event

Fig. 11.2 A Chinese text summarization task

In the women's slopestyle final of freestyle skiing at the Beijing Winter Olympics, Chinese athlete Gu Ailing won the silver metal. Congratulations to Gu Ailing! This morning, the women's slopestyle final of freestyle skiing was held. The final round will be held in three rounds, and the MEDALS will be determined by ranking the best performances of the contestants. In the first jump, Chinese athlete Gu Ailing scored 69.90 points. Ranked third among the 12 contestants. After completing the action, Gu Ailing made a funny face again, which was very cute. In the second round, Gu Ailing made a mistake at the third obstacle in the props area and fell down when she landed. Get 16.98 points. Netizen: It doesn't matter if you fell down. Keep going! Despite making a mistake and falling in the second jump, Gu Ailing withstood the pressure and performed steadily in the third jump, landing smoothly! Got 86.23 points! In this round of the competition, a total of 12 contestants participated, and Gu Ailing was the 10th to take the stage. Netizen: I was more nervous than Gu Ailing when watching the game. Come on!

The output summary information:

The process of implementing text summarization based on PEGASUS is shown in Fig. 11.3. The input of the model is a longer text sequence and the output is the summary of the given text. During the modeling process, for the input long text sequence x, it is necessary to carry out data processing first. Then use the PEGASUS encoder to encode the given text sequences to obtain its semantic vector representation and use the PEGASUS decoder to generate the corresponding summary y.

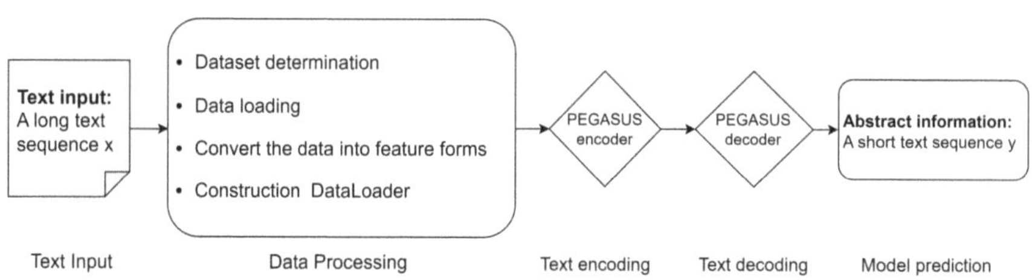

Fig. 11.3 The implementation flowchart for Chinese text summarization tasks

11.2 Data Processing

Data processing includes steps such as determining the dataset, data loading, converting data into the feature form, and constructing the DataLoader. Finally, the same batch of data is processed into feature sequences with the equal length, and the data is sent into the PEGASUS model iteratively in batches using the DataLoader.

11.2.1 Determination of the Dataset

In this practice, we use the LCSTSdataset for text summarization. This dataset comes from Sina Weibo and it is a large-scale Chinese short text summarization dataset, consisting of more than two million real Chinese short texts, with each text having a brief summary provided by the author. In this practice, we use a part of the corpus from the LCSTS dataset, including 8,000 text–summary pairs in the training set, 800 text–summary pairs in the validation set, and 100 text–summary pairs in the test set.

The following is a sample text–summary pair from the training set, consisting of two parts: the given text and its summary.

{“title”: “The World’s First Smart Toothbrush Kolibree: Keeping You Away from Tooth Decay”, “content”: In the era of the explosion of the Internet of Things, almost everything is “connected”. Yesterday, the world’s first electric toothbrush was unveiled at CES. It can automatically analyze users’ brushing habits through built-in sensors and record relevant brushing data every time they brush their teeth, reducing tooth decay for everyone! ”}.

In this example, the sentence after “title” is the summary of the text following “content.”

11.2.2 Data Loading

First, we need to read the training set, validation set, and test set data into memory. Codes are implemented as follows:

```
from datasets import load_dataset
from paddlenlp.transformers import AutoTokenizer
from functools import partial

# Load the training set, validation set, and test set through load_dataset
train_dataset = load_dataset("json", data_files='train.json', split="trai
n")
dev_dataset = load_dataset("json", data_files='dev.json', split="train")
test_dataset = load_dataset("json", data_files='test.json',
split="train")
# Print and view the first data in the training set
print(load_dataset["train"][:2])
```

The output result is:

{'title': ['A 63-year-old retired teacher, Xie Shuhua, pulled a human-powered cart and spent a year traveling 24,000 miles, taking her 90-year-old mother on a tour around China, fulfilling her mother's wish of "spending her entire life by the stove and also wanting to go out for a walk". She said: If Mom is willing to go out for a walk, I'm willing to hold her. Filial piety can't wait. We should go as far as we can.']}.

From the output we can see that each data item consists of "title" and "content," which represent the summary written by human and text information, respectively.

11.2.3 Convert Data into the Feature Form

The loaded text data is a sequence, which cannot be directly read by the PEGASUS model. It is necessary to first tokenize the text data and then convert it into the corresponding feature data. As we mentioned above that the PEGASUS model is an encoder–decoder structure, and the input data needs to follow a fixed format, similar to the Transformer model introduced in Sub-Sect. 4.3. The differences between the data before and after conversion are shown in Table 11.1.

Table 11.1 Corresponding feature forms of title and content

Original data	The converted data format
Title	labels: The id after title word segmentation, that is, the label of the model labels: The ids of the segmented title, which is the label of the model
Content	input _ids:The ids of the segmented content, which is the input of encoder attention_mask: padding the ids after content segmentation, which needs to be removed during model training decoder_input_ids: The ids of the title that is segmented and shifted as input to the decoder

The convert_example_to_feature function is defined below, and the input text is converted by loading the tokenizer, while constructing labels.

```
def convert_example(example, text_column, summary_column, tokenizer,
                    max_source_length, max_target_length):
    """
    # Construct the input of models
    Example: input text data
    text_column: the text column(content column)
    summary_column: the summary column (title column)
    max_source_length: the maximum text length, the exceeded part will be truncated
    max_target_length: the maximum summary length, the exceeded part will be truncated
    """
    inputs = example[text_column]
    targets = example[summary_column]
    # Use the tokenizer to convert the content into the corresponding feature format
    model_inputs = tokenizer(inputs,
                             max_length=max_source_length,
                             padding=False,
                             truncation=True,
                             return_attention_mask=True)
    # Use the tokenizer to convert the title into the corresponding feature format
    summary_inputs = tokenizer(targets,
                         max_length=max_target_length,
                         padding=False,
                         truncation=True)
    # Labels are the ids of title after being tokenized, and it is necessary to shift to the right
when inputting the decoder
    model_inputs["labels"] = summary_inputs["input_ids"]
    return model_inputs
```

The following codes convert the loaded training set, validation set, and test set into corresponding feature forms based on the convert_example_to_feature function. The partial function is needed here to set the parameters in the convert_example_to_feature function, then the conversion is performed based on the map function. The map function provided by datasets supports data conversion one by one or batch by batch. Since the convert_example_to_feature function processes data one by one, the parameter “batched” is set to “False” in the map function. Codes are implemented as follows:

```
# Initialize tokenizer
tokenizer = AutoTokenizer.from_pretrained('IDEA-CCNL/Randeng-Pegasus-
238M-Summary-Chinese')
# Remove original columns
remove_columns = ['content', 'title']
# Set the maximum length of the text
max_source_length = 128
# Set the maximum length of the summary
max_target_length = 64
# Define the converter to convert content and title into text_column and summary_column
respectively
trans_func = partial(convert_example,
                     text_column='content',
                     summary_column='title',
                     tokenizer=tokenizer,
                     max_source_length=max_source_length,
                     max_target_length=max_target_length)

# Convert the data in the training set, validation set, and test set into the feature forms
train_dataset = train_dataset.map(trans_func,
                                  batched=True,
                                  load_from_cache_file=True,
                                   remove_columns=remove_columns)
dev_dataset = dev_dataset.map(trans_func,
                              batched=True,
                              load_from_cache_file=True,
                               remove_columns=remove_columns)

test_dataset = test_dataset.map(trans_func,
                                batched=True,
                                load_from_cache_file=True,
                                 remove_columns=remove_columns)

# Output the previous sample of the training set
for idx, example in enumerate(train_dataset):
    if idx < 1:
        print(example)
```

The output result is:

{'input_ids': [9356, 30,226, 12,756, 16,670, 5661, 2333, 22,062, 3399, 1477, 13,863, 31,272, 8334, 5661, 22,746, 5230, 3399, 8209, 5661, 22,062, 3399, 13,863, 8534, 26,138, 873, 8534, 30,451, 179, 2651, 2238, 1082, 3848, 45,396, 2274, 297, 201, 297, 3399, 13,653, 5661, 16,987, 7709, 178, 28,289, 7709, 178, 28,289, 223, 3399, 6957, 28,290, 7709, 5661, 5034, 4566, 486, 25,191, 2651, 2238, 1082, 3848, 179, 1], 'attention_mask': [1, 1, 1, 1, 1, 1, 1, 1, 1, 1, 1, 1, 1, 1, 1, 1, 1, 1, 1, 1, 1, 1, 1, 1, 1, 1, 1, 1, 1, 1, 1, 1, 1, 1, 1, 1, 1, 1, 1, 1, 1, 1, 1, 1, 1, 1, 1, 1, 1, 1, 1, 1, 1, 1, 1, 1, 1, 1,

1, 1, 1, 1, 1, 1], 'labels': [44907, 4785, 7423, 24,916, 30,346, 22,746, 33,335, 3399, 22,451, 32,511, 1]}.

From the data results, the context and title of the original data have been converted into the corresponding feature forms, which are input_ids, attention_mask, and labels, respectively.

11.2.4 Construction of the DataLoader

Next, we use DataCollatorForSeq2Seq to automatically change a minibatch of data into the same length. DataCollatorForSeq2Seq takes tokenizer as a parameter, including input_ids, attention_mask, and labels. Codes are implemented as follows:

```
from paddlenlp.transformers import AutoModelForConditionalGeneration
from paddlenlp.data import DataCollatorForSeq2Seq

# Initialize the model. AutoModelForConditionalGeneration encapsulation achieves the model structure used by PEGASUS to generate tasks
model = AutoModelForConditionalGeneration.from_pretrained('IDEA-CCNL/Randeng-Pegasus-238M-Summary-Chinese')
# Assemble the minibatch data, and use the DataCollatorForSeqSeq function to automatically change the data into the same length
batchify_fn = DataCollatorForSeq2Seq(tokenizer=tokenizer, model=model)
```

And then, we need to construct the DataLoader for iteration of data in batches and specify the batch size. During data iteration, we need to specify whether to perform random sample sorting by the parameter shuffle. If shuffle is set to False, the DataLoader iterates a batch of data in order; if it is set to True, the DataLoader iterates a batch of data randomly. Codes are implemented as follows:

```
from paddle.io import BatchSampler, DistributedBatchSampler, DataLoader
# Distributed minibatch sampler for multi-GPU distributed training
train_batch_sampler = DistributedBatchSampler(
    train_dataset, batch_size=12, shuffle=True)

# Construct the Dataloader for the training set, validation set and test set, and iterate data in batches according to the batch size
train_data_loader = DataLoader(dataset=train_dataset,
                               batch_sampler=train_batch_sampler,
                               num_workers=0,
                               collate_fn=batchify_fn,
                               return_list=True)

dev_batch_sampler = BatchSampler(dev_dataset,
                                 batch_size=12,
                                 shuffle=False)
dev_data_loader = DataLoader(dataset=dev_dataset,
                             batch_sampler=dev_batch_sampler,
                             num_workers=0,
                             collate_fn=batchify_fn,
                             return_list=True)
test_batch_sampler = BatchSampler(test_dataset,
                                  batch_size=1,
                                  shuffle=False)
test_data_loader = DataLoader(dataset=test_dataset,
                              batch_sampler=test_batch_sampler,
                              num_workers=0,
                              collate_fn=batchify_fn,
                              return_list=True)

# Print the data of the first minibatch in the training set
for i in test_data_loader:
    print(i)
    break
```

The output result is:

The output result is:

{'input_ids': Tensor(shape=[1, 67], dtype=int64, place=Place(gpu:0), stop_gradient=True,

[[7423 , 297 , 17002, 46011, 4566 , 2029 , 129 , 3280 , 2669 , 179 ,
207 , 723 , 27785, 14495, 30402, 5661 , 15249, 7423 , 200 , 19684,
39275, 24227, 5661 , 1463 , 1426 , 266 , 37022, 179 , 20755, 5661 ,
18583, 175 , 787 , 198 , 2333 , 218 , 1969 , 322 , 176 , 10567,
5661 , 8536 , 34705, 3399 , 39275, 1463 , 7034 , 516 , 33296, 14288,
179 , 23185, 1463 , 16389, 10567, 39275, 10736, 26027, 5661 , 1463 ,
11234, 19063, 29521, 7540 , 25961, 179 , 1]]), 'attention_mask': Tensor(shape=[1, 67], dtype=int64, place=Place(gpu:0), stop_gradient=True,

[[1, 1, 1, 1, 1, 1, 1, 1, 1, 1, 1, 1, 1, 1, 1, 1, 1, 1, 1, 1, 1, 1, 1, 1,
1, 1, 1, 1, 1, 1, 1, 1, 1, 1, 1, 1, 1, 1, 1, 1, 1, 1, 1, 1, 1, 1, 1, 1,
1, 1, 1, 1, 1, 1, 1, 1, 1, 1, 1, 1, 1, 1, 1, 1, 1, 1, 1]]), 'labels': Tensor(shape=[1, 11], dtype=int64, place=Place(gpu:0), stop_gradient=True,

[[23185, 1463 , 16389, 10567, 39275, 10736, 26027, 7371 , 39275, 24227,
1]]), 'decoder_input_ids': Tensor(shape=[1, 11], dtype=int64, place=Place(gpu:0), stop_gradient=True,

[[0 , 23185, 1463 , 16389, 10567, 39275, 10736, 26027, 7371 , 39275, 24227]])}

From the output results, it is clear that the data length of the same minibatch is the same.

11.3 Model Construction

As mentioned above, the PEGASUS model adopts the encoder–decoder structure of the sequence-to-sequence model, to process each input minibatch of data in turn. The processed text data is first sent into the PEGASUS model, and the encoding layer of the PEGASUS model encodes it and outputs the corresponding vector sequence. Then, the summary shifted one position to the right is passed into the decoding layer, and the corresponding decoding results are obtained one by one, which is the summary information corresponding to the input text. The first vector in the decoding layer is < start > , representing the beginning of decoding; the last vector of the decoding layer is < eos > , which means the end of decoding, as shown in Fig. 11.4.

Based on the modeling approach shown in Fig. 11.4, codes are implemented as follows:

```
# Initialize the model, AutoModelForConditionalGeneration
model = AutoModelForConditionalGeneration.from_pretrained('IDEA-
CCNL/Randeng-Pegasus-238M-Summary-Chinese')
```

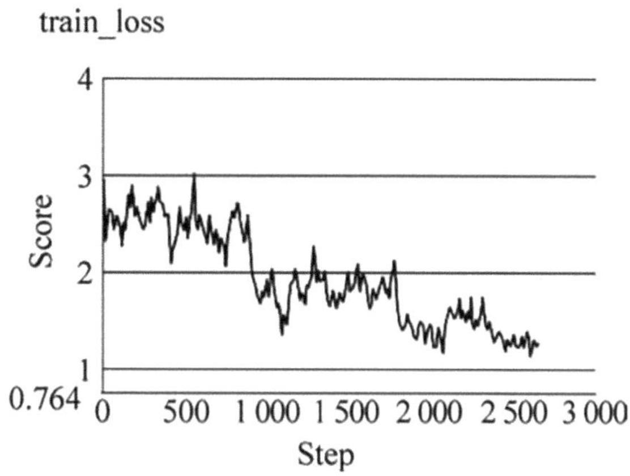

Fig. 11.4 A generative modeling approach used for Chinese text summarization based on PEGASUS

11.4 Training Configuration

Define hyperparameters, the optimizer, the loss function, evaluation indicators, and so on during model training.

(1) Model: The PEGASUS model.
(2) Optimizer: The AdamW optimizer.
(3) Loss function: The cross-entropy loss function (defined inside the model).
(4) Evaluation indicators: BLEU and ROUGE are commonly used evaluation indicators in sequence generation tasks.

11.4.1 The Metric of BLEU

BLEU (BiLingual Evaluation Understudy) is an index for measuring the overlap of n-grams between a generated sequence and a reference sequence. It was proposed for evaluating the quality of machine translation and has been widely used in various sequence generation tasks including automatic summarization.

Given an input sequence $\boldsymbol{s}$ and its corresponding reference sequence $\boldsymbol{r}$,l_s, and/ or l_r represent the lengths of the generated sequence and the reference sequence, respectively. Let $\boldsymbol{x}$ be a candidate sequence generated by the model based on $\boldsymbol{s}$, and w be the set of all N-grams extracted from the generated candidate sequence. The BLEU score can be gotten by calculating the precision of N-grams with different lengths $P_N(\boldsymbol{x}), N = 1,\ 2,\ \cdots$ ... and performing a weighted geometric mean on it. The calculation formula is:

$$\text{BLEU-N(x)} = b(\boldsymbol{x}) \cdot \exp\left(\sum_{N=1}^{N\prime} \alpha_N \log P_N\right) \tag{11.1}$$

$$b(\boldsymbol{x}) = \begin{cases} 1 & l_x > l_r \\ \exp(1 - l_s/l_r) & l_s \le l_r \end{cases} \tag{11.2}$$

$$P_N(\boldsymbol{x}) = \frac{\sum_{w \in W} \min\left(c_w(x), \max_{k=1}^{n} c_w(r_k)\right)}{\sum_{w \in W} c_w(x)}. \tag{11.3}$$

In Formula (11.1), $N\prime$ is the length of the longest n-gram and α_N is the weight of different n-grams, which is generally set to $\frac{1}{N\prime}$. The value range of the BLEU algorithm is [0,1].The bigger the value is, the closer the generated sequence is to the reference sequence. The core idea of $P_N(\boldsymbol{x})$ is to measure the proportion of n-grams appearing in the reference sequence and ones in the generated candidate sequence $\boldsymbol{x}$. In Formula (11.3), $c_w(\boldsymbol{x})$ indicates the number of occurrences.

11.4.2 The Metric of ROUGE

The Recall-Oriented Understudy for Gisting Evaluation (ROUGE) is a method to evaluate the text summarization proposed by Chin-Yew Lin et al. (2004), which measures the number of overlapping units between the summaries generated by the model and the manually labeled reference summaries. Similar to BLEU's calculation method, ROUGE calculates the recall rate of n-grams between the generated summaries and the reference summaries. The ROUGE method includes ROUGE-N, ROUGE-L, ROUGE-S, ROUGE-W, and other types. In this practice, we use the ROUGE-N and ROUGE-L scores for model evaluation.

1. The ROUGE-N Algorithm

The calculation formula for ROUGE-N is:

$$\mathrm{ROUGE-N} = \frac{\sum_{S \in \{\mathrm{ReferenceSummaries}\}} \sum_{\mathrm{gram}_n \in S} \mathrm{Count}_{\mathrm{match}}\left(\mathrm{gram}_n\right)}{\sum_{S \in \{\mathrm{ReferenceSummaries}\}} \sum_{\mathrm{gram}_n \in S} \mathrm{Count}\left(\mathrm{gram}_n\right)}. \tag{11.4}$$

In Formula (11.4), ReferenceSummaries represents the reference summaries, which are the manually labeled "standard answers," and $gram_n$ represents the n-grams in the reference summary. When $n = 1$, $gram_n$ is the individual word in the reference summary; when $n = 2$, $gram_n$ is bi-grams in each summary. $Count_{match}(gram_n)$ represents the number of n-grams that appear in both the generated summary and the reference summary; $Count(gram_n)$ indicates the number of n-grams appearing in the reference summary.

The following is an example of the calculation method for ROUGE-N. Let's assume the generated summary and the reference summary are as follows:

> Generated summary：It is raining.
>
> Reference (golden standard)：It is raining outside.

(1) When calculating ROUGE-1, the 1-g of the summary can be expressed as follows:

> Generated(uni-gram)：It、is、raining
>
> Reference (uni-gram)：It、is、raining、outside

It is clear that the number of overlapping uni-grams between the generated summary and the reference summary is 3, while the number of uni-grams in the reference summary is 4. The value of ROUGE-1 is:

$$ROUGE - 1 = \frac{3}{4} = 0.75$$

(2) When calculating ROUGE-2, the 2-g of the summary can be expressed as follows:

> Generated summary (2-gram)：It is、is raining
>
> Reference summary (2-gram)：It is、is raining、raining outside

It can be seen that the number of overlapping 2-g between the generated summary and the reference summary is 2, while the number of 2-g in the reference summary is 3. The score of ROUGE-2 is:

$$ROUGE - 2 = \frac{2}{3} \approx 0.667$$

When n takes other values, the calculation method of ROUGE-N is similar.

11.5 The ROUGE-L Algorithm

In ROUGE-L, L stands for the longest common sub-sequence (LCS). Assuming that X represents the reference summary, Y represents the generated summary, and m and n are the lengths of the summaries, respectively. $LCS(X, Y)$ represents the maximum common sub-sequence in the reference summary and the generated summary. The calculation formulas for recall rate R_{cls} and precision rate P_{cls} are shown as follows:

$$R_{lcs} = \frac{LCS(XY)}{m} \times 100\% \tag{11.5}$$

$$P_{lcs} = \frac{LCS(XY)}{n} \times 100\% \tag{11.6}$$

The calculation formula of ROUGE-L is:

$$F_{lcs} = \frac{(1 + \beta^2) R_{lcs} P_{lcs}}{R_{lcs} + \beta^2 P_{lcs}}. \tag{11.7}$$

In the evaluation tasks organized by the DUC (Document Understanding Conference), β is usually set to a large number, so ROUGE-L basically only considers the recall rate.

Codes implemented for the training configuration are as follows:

```
# Install the evaluation indicators ROUGE base for text summarization tasks
!pip install rouge==1.0.1 -i https://pypi.tuna.tsinghua.edu.cn/simple

import paddle
from paddlenlp.transformers import LinearDecayWithWarmup
# Load the PaddlePaddle's visualization tool visualdl's LogWriter to customize a logger
from visualdl import LogWriter
from rouge import Rouge

# Set the proportion of warm-up training during the training process
warmup_proportion = 0.02
# Set the learning rate
learning_rate = 5e-5
# Set the rounds of training
num_epochs = 3
# Set needed training steps in total
num_training_steps = len(train_data_loader) * num_epochs
# Set the AdamW optimizer parameter epsilon
adamw_epsilon = 1e-6
# Set the weight decay coefficient during training
weight_decay=0.01
# Set the intervals for printing the log every certain number of steps
log_steps = 10
# Set the intervals for model evaluation in the validation set every certain number of steps
eval_steps = 100
# Set the shortest length of summaries
min_target_length = 0
# Set the model's saving path and automatically save the model with the best performance
during the training process
save_dir = 'checkpoints'

# Set the visualization of evaluation indicators
log_writer = LogWriter('visualdl_log_dir')
lr_scheduler = LinearDecayWithWarmup(learning_rate,
num_training_steps, warmup_proportion)

# Except for the parameters of bias and LayerNorm, other parameters perform decay operations
during training
decay_params = [
    p.name for n, p in model.named_parameters()
    if not any(nd in n for nd in ["bias", "norm"])
]
# Define the optimizer AdamW
optimizer = paddle.optimizer.AdamW(
    learning_rate=lr_scheduler,
    beta1=0.9,
    beta2=0.999,
    epsilon=adamw_epsilon,
    parameters=model.parameters(),
    weight_decay=weight_decay,
    apply_decay_param_fun=lambda x: x in decay_params)
```

During the training process, the evaluation indicators Rouge-1, Rouge-2, Rouge-L, and BLEU-4 are calculated. Codes are implemented as follows:

```
def compute_metrics(preds, targets):
    assert len(preds) == len(targets), (
        'The length of pred_responses should be equal to the length of '
        'target_responses. But received {} and {}.'.format(
            len(preds), len(targets)))
    rouge = Rouge()
    bleu4 = BLEU(n_size=4)
    scores = []
    for pred, target in zip(preds, targets):
        try:
            score = rouge.get_scores(' '.join(pred), ' '.join(target))
            scores.append([
                score[0]['rouge-1']['f'], score[0]['rouge-2']['f'],
                score[0]['rouge-l']['f']
            ])
        except ValueError:
            scores.append([0, 0, 0])
        bleu4.add_inst(pred, [target])
    rouge1 = np.mean([i[0] for i in scores])
    rouge2 = np.mean([i[1] for i in scores])
    rougel = np.mean([i[2] for i in scores])
    bleu4 = bleu4.score()
   print('[Evaluate] BLEU-4:', round(bleu4*100, 2), 'rouge-1:',
round(rouge1*100, 2), 'rouge-2:', round(rouge2*100, 2), 'rouge-L:',
round(rougel*100, 2), 'BLEU-4:', round(bleu4*100, 2))
    return rouge1, rouge2, rougel, bleu4
```

11.6 Model Training

During the model training process, a training log is printed every certain number of log_steps, and the model is evaluated on the validation set every certain number of eval_steps. The model with the best evaluation performance during the training process is saved.

The codes implemented for evaluation on the validation set are as follows:

```
import time
from paddlenlp.utils.log import logger
from paddlenlp.metrics import BLEU
from tqdm import tqdm
import numpy as np
import os

@paddle.no_grad()
def evaluate(model, data_loader, tokenizer, min_target_length,
             max_target_length):
    model.eval()
    all_preds = []
    all_labels = []
    model = model._layers if isinstance(model, paddle.DataParallel) else
model
    for batch in tqdm(data_loader, total=len(data_loader), desc="Eval
step"):
        labels = batch.pop('labels').numpy()

        # Generate models
        preds = model.generate(input_ids=batch['input_ids'],
                               attention_mask=batch['attention_mask'],
                               min_length=min_target_length,
                               max_length=max_target_length,
                               use_cache=True)[0]
        # tokenizer converts id into the form of strings
        all_preds.extend(
            tokenizer.batch_decode(preds.numpy(),
                                   skip_special_tokens=True,
                                   clean_up_tokenization_spaces=False))
        labels = np.where(labels != -100, labels, tokenizer.pad_token_id)
        all_labels.extend(
            tokenizer.batch_decode(labels,
                                   skip_special_tokens=True,
                                   clean_up_tokenization_spaces=False))
    rouge1, rouge2, rougel, bleu4 = compute_metrics(all_preds,
all_labels)
    model.train()
    return rouge1, rouge2, rougel, bleu4
```

Codes implemented for the model training are as follows:

```
def train(model, train_data_loader):
    global_step = 0
    best_rougel = 0
    tic_train = time.time()
    for epoch in range(num_epochs):
        for step, batch in enumerate(train_data_loader):
            global_step += 1

            # Forward computation, calculate the loss function
            lm_logits, _, loss = model(**batch)
            loss.backward()
            optimizer.step()
            lr_scheduler.step()
            optimizer.clear_grad()
            if global_step % log_steps == 0:
                logger.info(
                    "[Train] global step %d/%d, epoch: %d, batch: %d,
rank_id: %s, loss: %f, lr: %.10f, speed: %.4f step/s"
                    % (global_step, num_training_steps, epoch, step,
                        paddle.distributed.get_rank(), loss,
optimizer.get_lr(),
                        log_steps / (time.time() - tic_train)))
                log_writer.add_scalar("train_loss", loss.numpy(),
global_step)
                tic_train = time.time()
            if global_step % eval_steps== 0 or global_step ==
num_training_steps:
                tic_eval = time.time()
                rouge1, rouge2, rougel, bleu4 = evaluate(model,
dev_data_loader, tokenizer,
                            min_target_length, max_target_length)
                logger.info("eval done total : %s s" % (time.time() -
tic_eval))
                log_writer.add_scalar("eval/ROUGE-1",
round(rouge1*100, 2), global_step)
                log_writer.add_scalar("eval/ROUGE-2",
round(rouge2*100, 2), global_step)
                log_writer.add_scalar("eval/ROUGE-L",
round(rougel*100, 2), global_step)
                log_writer.add_scalar("eval/BLEU-4", round(bleu4*100,
2), global_step)
                if best_rougel < rougel:
                    best_rougel = rougel
                    if paddle.distributed.get_rank() == 0:
                        if not os.path.exists(save_dir):
                            os.makedirs(save_dir)

                        model_to_save = model._layers if isinstance(
                            model, paddle.DataParallel) else model
                        model_to_save.save_pretrained(save_dir)
                        tokenizer.save_pretrained(save_dir)

# Start the model's training
train(model, train_data_loader)
```

The output result is:

```
[Train] step 10/2667, epoch: 0, loss: 2.963048
[Train] step 20/2667, epoch: 0, loss: 1.921782
[Train] step 30/2667, epoch: 0, loss: 2.666176
[Train] step 40/2667, epoch: 0, loss: 2.847842
[Train] step 50/2667, epoch: 0, loss: 2.615268
[Train] step 60/2667, epoch: 0, loss: 2.603299
[Train] step 70/2667, epoch: 0, loss: 2.174151
[Train] step 80/2667, epoch: 0, loss: 2.478480
[Train] step 90/2667, epoch: 0, loss: 2.782897
[Train] step 100/2667, epoch: 0, loss: 2.417608
[Evaluate]
rouge-1: 34.47
rouge-2: 20.5
rouge-L: 31.06
BLEU-4: 15.39
[Train] step 110/2667, epoch: 0, loss: 2.373282
[Train] step 120/2667, epoch: 0, loss: 1.972023
[Train] step 130/2667, epoch: 0, loss: 2.894804
[Train] step 140/2667, epoch: 0, loss: 2.311194
[Train] step 150/2667, epoch: 0, loss: 2.941457
[Train] step 160/2667, epoch: 0, loss: 3.054342
[Train] step 170/2667, epoch: 0, loss: 2.458297
[Train] step 180/2667, epoch: 0, loss: 3.266744
[Train] step 190/2667, epoch: 0, loss: 2.919276
```

[Train] step 200/2667, epoch: 0, loss: 2.064613

[Evaluate]

rouge-1: 33.78

rouge-2: 20.23

rouge-L: 30.64

BLEU-4: 14.98

……

[Train] step 2410/2667, epoch: 2, loss: 1.455715

[Train] step 2420/2667, epoch: 2, loss: 1.368634

[Train] step 2430/2667, epoch: 2, loss: 1.243881

[Train] step 2440/2667, epoch: 2, loss: 1.309806

[Train] step 2450/2667, epoch: 2, loss: 1.003384

[Train] step 2460/2667, epoch: 2, loss: 1.509695

[Train] step 2470/2667, epoch: 2, loss: 1.384019

[Train] step 2480/2667, epoch: 2, loss: 1.107425

[Train] step 2490/2667, epoch: 2, loss: 1.393112

[Train] step 2500/2667, epoch: 2, loss: 1.470335

[Evaluate]

rouge-1: 36.36

rouge-2: 22.45

rouge-L: 32.77

BLEU-4: 17.82

The visualized training results are shown in Figs. 11.5, 11.6 and 11.7.

From the output results, it is easy to see that the loss value on the training set is continuously decreasing, and the values of ROUGE and BLUE on the validation set are steadily increasing.

11.7 Model Evaluation

The model is evaluated with the best performance during the training process using the test set to verify the effectiveness of the model training. Codes are implemented as follows:

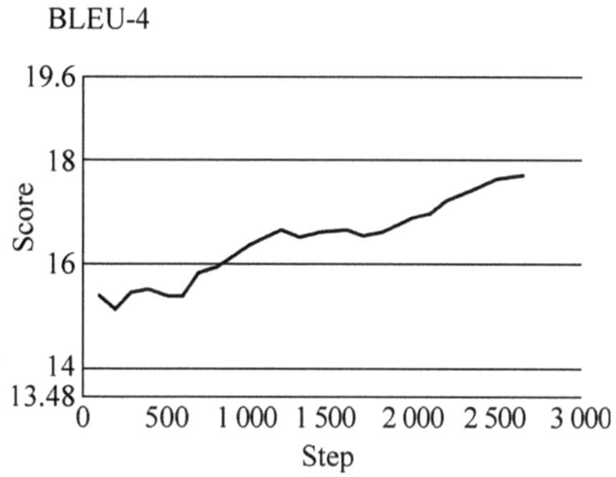

Fig. 11.5 The trend of the loss value on the training set

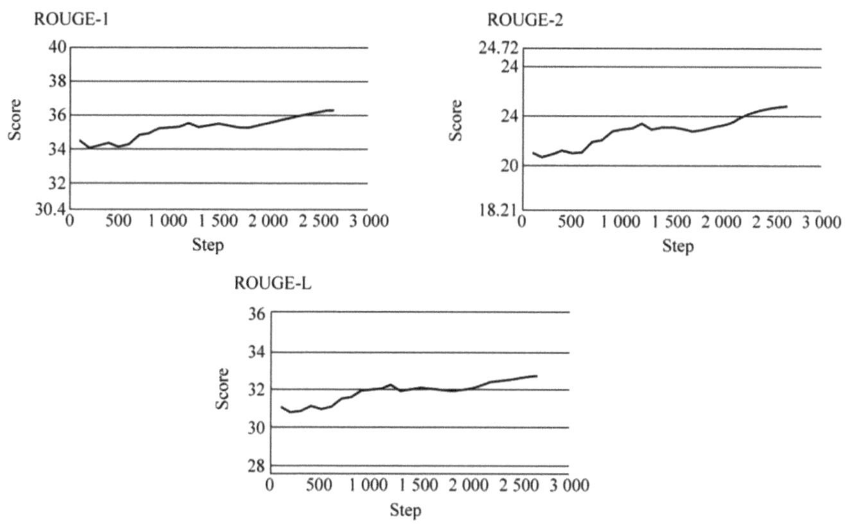

Fig. 11.6 The trend of BLEU indicators on the validation set

```
evaluate(model, test_data_loader, tokenizer, min_target_length,
max_target_length)
```

The output result is:

The auto evaluation result is:

rouge-1: 57.38

rouge-2: 44.19

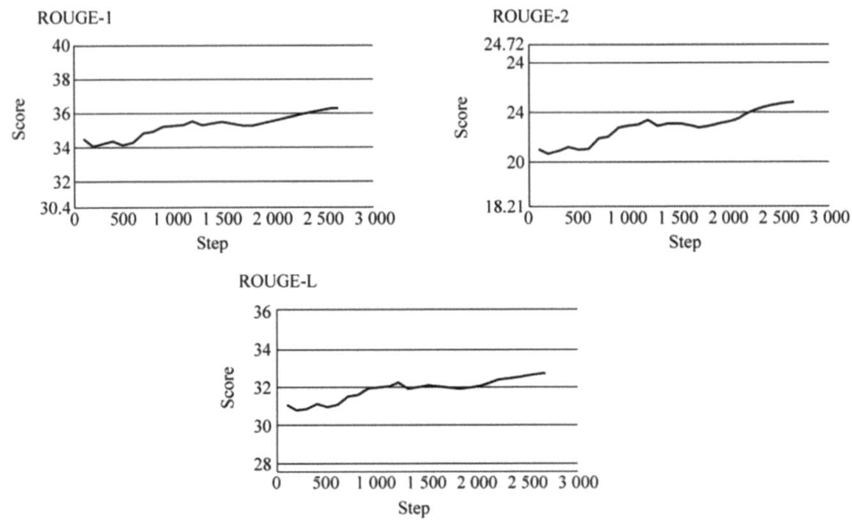

Fig. 11.7 The trend of ROUGE indicators on the validation set

rouge-L: 54.72

Bleu-4: 38.51

11.8 Model Prediction

After the model training is completed, we may input any text information and obtain its summary through model prediction. Codes are implemented as follows:

```
# Model inference, generate summaries for individual texts
def infer(text, model, tokenizer):
    tokenized = tokenizer(text,
                          truncation=True,
                          max_length=max_source_length,
                          return_tensors='pd')
    preds, _ = model.generate(input_ids=tokenized['input_ids'],
                              max_length=max_target_length,
                              min_length=min_target_length,
                              decode_strategy='beam_search',
                              num_beams=4)
    print(tokenizer.decode(preds[0], skip_special_tokens=True,
clean_up_tokenization_spaces=False))

# Load trained models
model = AutoModelForConditionalGeneration.from_pretrained('checkpoints')
tokenizer = AutoTokenizer.from_pretrained('checkpoints')

# Model prediction
text = ''In the women's slopestyle final of freestyle skiing at the
Beijing Winter Olympics, Chinese athlete Gu Ailing won the silver medal.
Congratulations to Gu Ailing! This morning, the women's slopestyle final
of freestyle skiing was held. The final round will be held in three
rounds, and the MEDALS will be determined by ranking the best
performances of the contestants. In the first jump, Chinese athlete Gu
Ailing scored 69.90 points. Ranked third among the 12 contestants. After
completing the action, Gu Ailing made a funny face again, which was very
cute. In the second round, Gu Ailing made a mistake at the third obstacle
in the props area and fell down when she landed. Get 16.98 points.
Netizen: It doesn't matter if you fall down. Keep going! Despite making a
mistake and falling in the second jump, Gu Ailing withstood the pressure
and performed steadily in the third jump, landing smoothly! Got 86.23
points! In this round of the competition, a total of 12 contestants
participated, and Gu Ailing was the 10th to take the stage. Netizen: I
was more nervous than Gu Ailing when watching the game. Come on!''
infer(text, model, tokenizer)
```

The output result is:

title:Gu Ailing won the silver medal in the women's slopestyle event of freestyle skiing at the Winter Olympics.

11.9 Exercises

(1) Apart from some common indicators such as ROUGE and BLEU, are there any better methods to evaluate text summarization? For example, human evaluation?
(2) If you are asked to improve the PEGASUS model, what ideas do you have? For example, can you improve it from the perspectives of data or model's architecture?
(3) The PEGASUS model introduced in this chapter is a method based on the encoder–decoder. Please try to use the pre-training language model based on the Transformer's decoder to implement text summarization tasks.

Chapter 12
ERNIE 3.0-Based Intent Recognition

Abstract This chapter presents the implementation of an intent recognition system based on ERNIE 3.0, framing the task as a multi-label text classification problem. Using the CrossWOZ dataset, the study demonstrates how ERNIE 3.0 leverages large-scale pre-training with text and knowledge graphs to effectively capture semantic representations. The workflow covers data preprocessing, feature conversion, model construction, and training with Adam optimization and binary cross-entropy loss, evaluated using Macro F1 and Micro F1 metrics. Experimental results show that the model achieves strong performance, with stable convergence and high F1 scores on validation and test sets. The chapter concludes with prediction examples and exercises, illustrating the practical application of ERNIE 3.0 to improve the accuracy and robustness of intent recognition in real-world dialogue and search scenarios.

Intent recognition refers to the task that the machine correctly analyzes the text given by the user and returns the corresponding intent results. This is a hot topic in the field of NLP in recent years and is widely used in scenarios such as search engines, dialogue systems, and robots. For example, in a search engine, if a user searches "Is it cold today?", the system will recognize the user is concerned about the "weather." It then directly returns the weather forecast as the result to the user, thereby reducing the number of users' clicks and enhancing the user experience, as shown in Fig. 12.1. In a human–computer dialogue system, if the user says "My iPhone never experiences lag issues," then the model can determine through intent recognition that "iPhone" at this moment refers to an electronic device rather than a fruit, so the dialogue can proceed smoothly. The accuracy of intent recognition can greatly improve the matching of search and the intelligence of human–computer dialogue systems.

Intent recognition can be seen as a problem of classification, in which the classification model calculates the probability of each candidate intent based on the input texts, and ultimately gives a judgment on the user's intent. The classification model can realize one of different classifications, binary, multiple, or multi-label.[1]

[1] In some dialogue systems, intent recognition includes two sub-tasks: domain recognition and intent slot recognition. Domain recognition can be seen as a task of text classification, while intent

C. Zong et al., *Natural Language Processing and Large Language Models*,
https://doi.org/10.1007/978-981-92-0682-7_12

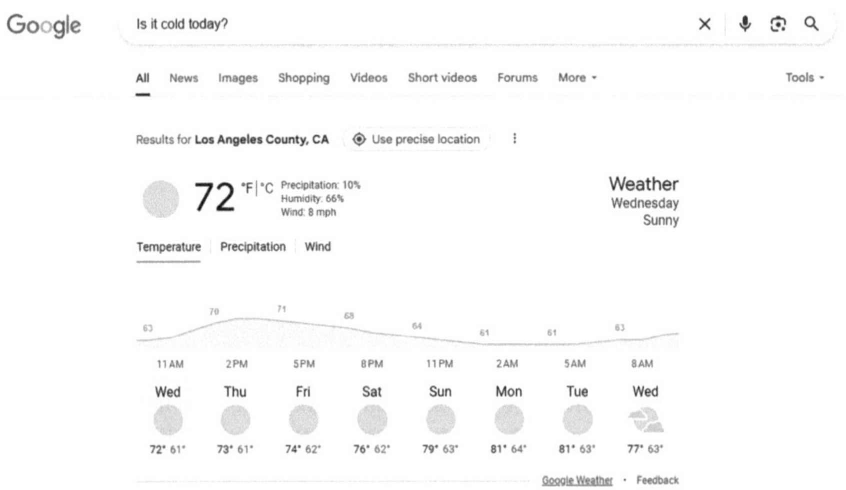

Fig. 12.1 The application effect of intent analysis in search engines

This chapter introduces the intention recognition method of the ERNIE 3.0 based on pre-training models and adopts the multi-label classification strategy.

12.1 Task Objectives and Implementation Processes

As previously mentioned, the goal of this task is to analyze the true intent of the user's input query. The experiment introduced in this chapter regards intent recognition as a classification task. As shown in Fig. 12.2, when a user inputs the query "Resident Evil," there are games, TV series, news, pictures, etc. It is clear that the user wants to watch TV series through user intent recognition, so we hope to see the TV series as the first result to the user.

As mentioned above, this experiment based on the pre-training model ERNIE 3.0 is designed for intent recognition tasks using text classification methods. ERNIE 3.0 introduces a large-scale knowledge graph into the pre-training model with a 10

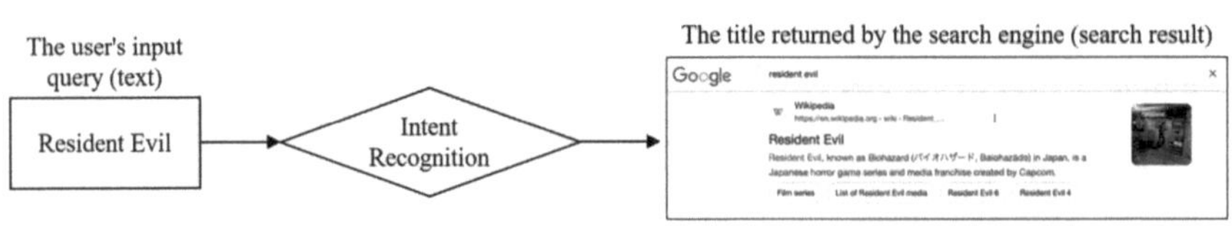

Fig. 12.2 The application of the intent recognition technology in retrieval

---

slot recognition can be seen as a sequence labeling task. Here, intent recognition is treated as a problem of classification.

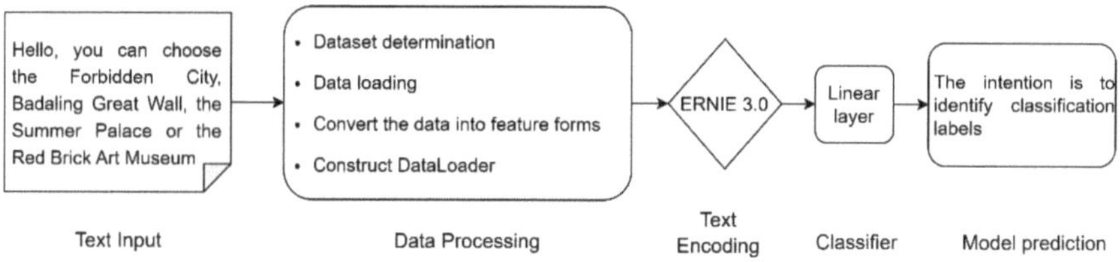

Fig. 12.3 The implementation process of the intention recognition task

billion-scale and proposes a pre-training method based on massive unsupervised text and large-scale knowledge graph. For building ERNIE 3.0, the 50 million triples obtained from the knowledge graph mining algorithm and the 4 TB large-scale corpus are simultaneously input into the pre-training models for joint masking training. This approach enables efficient information sharing between structured knowledge and unstructured texts, which significantly improves the model's knowledge memorization and inference ability.

The process of implementing intent recognition based on ERNIE 3.0 is shown in Fig. 12.3. The input of the model is text and the output is the intent classification label of the text. During the modeling process, the input text is first processed and then the pre-training model ERNIE 3.0 is used to perform text sequence encoding to obtain the semantic vector representation of the text. Finally, it is processed through a linear layer (fully connected layer) and softmax to obtain the corresponding intent recognition classification label for the input text.

12.2 Data Processing

As described in other tasks, data processing includes steps such as determining the dataset, dataset loading, converting data into feature representations, and constructing the DataLoader. Ultimately, the same batch of data is processed into feature sequences with the equal length, and the DataLoader is used to iterate and send data into the ERNIE 3.0 model batch by batch.

12.2.1 Determine the Dataset

In this practice, we use the CrossWOZ (Cross-Domain Wizard-of-Oz task-oriented) dataset (Zhu et al., 2020). This is a large-scale, task-oriented conversation dataset in Chinese, constructed by the CoAI group of the Department of Computer Science and Technology and the Institute for Artificial Intelligence of Tsinghua University. It contains 6,000 dialogues and 102,000 sentences, covering 5 domains (scenic spots sightseeing, booking hotels or restaurants, talking on subway or in a taxi). This dataset simulates users' conversations of booking hotels or restaurants, scenic spots sightseeing, or chatting with the system, and all data are labeled. To facilitate readers' use, in this practice, we only use the dialogue data about booking hotels and scenic spots sightseeing in the CrossWOZ dataset, which is a multi-label dataset containing 158 categories. The training set contains 44,409 text–intent pairs, the validation set contains 4,935 text–intent pairs, and the test set contains 4,909 text–intent pairs. Sample data are as follows:

An example: {'sentence': 'Hello, could you recommend a hotel with a price range of 400–500 Yuan and a rating of 4.5 or above? Thank you!', 'labels': ['General + greet + none + none', 'General + thank + none + none', 'Request + hotel + name + ']}.

From the example, it can be seen that each text–intent pair contains two parts: text and its corresponding intent classification labels. This example is multi-label data.

12.2.2 Data Loading

The purpose of dataset loading is to read the training set, validation set, test set, and labels into memory. Codes are implemented as follows:

```
from paddlenlp.datasets import load_dataset

# Load labels of the dataset
label_list = {}
label_map = {}
with open('data/data174910/label.txt', 'r', encoding='utf-8') as f:
    for i, line in enumerate(f):
        l = line.strip()
        label_list[l] = i
        label_map[i] = l
f.close()

# Define data and load the function
def load_data(path, label_list=None, is_test=False):
    with open(path, 'r', encoding='utf-8') as f:
        for line in f:
            if is_test:
                items = line.strip().split('\t')
                sentence = ''.join(items)
                yield {'sentence': sentence}
            else:
                items = line.strip().split('\t')
                if len(items) == 0:
                    continue
                elif len(items) == 1:
                    sentence = items[0]
                    labels = []
                else:
                    sentence = ''.join(items[:-1])
                    label = items[-1]
                    labels = [label_list[l] for l in label.split(',')]
                yield {'sentence': sentence, 'label': labels}
    f.close()

# Load the training set, test set, and validation set
train_dataset = load_dataset(load_data,
path='data/data174910/train.txt', label_list=label_list, lazy=False)
dev_dataset = load_dataset(load_data, path='data/data174910/dev.txt',
label_list=label_list, lazy=False)
test_dataset = load_dataset(load_data,
path='data/data174910/test.txt', label_list=label_list, lazy=False)

print('Number of samples of the training set: ', len(train_dataset))
print('Number of samples of the validation set: ', len(dev_dataset))
print('Number of samples of the test set: ', len(test_dataset))
print('Samples: ', train_dataset[0])
```

The output result is:

Number of samples of the training set: 44,409.

Number of samples of the validation set: 4935.

Number of samples of the test set: 4909.

Sample: {'sentence': 'Hello, could you recommend a hotel with a price range of 400–500 yuan and a rating of 4.5 or above? Thank you!)', 'labels': ['General + greet + none + none', 'General + thank + none + none', 'Request + hotel + name + ']}.

In this example, "General + greet + none + none" indicates that the sample contains the intention of "hello"; "General + thank + none + none" indicates that the sample contains the intention of "thanks"; and "Request + hotel + name + " indicates that the sample includes the intention to ask for the name of a hotel.

The output results include two parts of data: the user's question "sentence" and the labels corresponding to the question "labels." This example is a multi-label data.

12.2.3 Convert Data into the Feature Form

The data, after being loaded, is in the form of a text. The ERNIE 3.0 model cannot read directly, and it is necessary to convert the text data into corresponding feature data. The input data of the ERNIE 3.0 model needs to follow a fixed format, including information such as input_ids and token_type_ids. Therefore, it is necessary to convert the text data into feature data that meets the requirements of the ENIRE format.

Next, we use PaddleNLP's tokenizer to perform word segmentation and convert the segmented words and labels into feature representations. We need to define the 'convert_example_to_feature' function and convert the input text data into the feature data in batches by loading the tokenizer. Codes are implemented as follows:

```
import functools
from paddlenlp.transformers import AutoTokenizer

# Use the tokenizer to convert the input data into the feature form
def preprocess_function(examples, tokenizer, max_seq_len,
num_classes, is_test=False):
    encoded_inputs = tokenizer(text=examples["sentence"],
max_seq_len=max_seq_len)
    # Change labels into the one-hot representation
    if not is_test:
        encoded_inputs["labels"] = [float(1) if i in
examples["label"] else float(0) for i in range(num_classes)]
    return encoded_inputs

tokenizer = AutoTokenizer.from_pretrained("ernie-3.0-medium-zh")

# Define the function partial to set the parameters in the convert_example_to_feature function,
and then convert based on the map function
convert_example_to_feature = functools.partial(preprocess_function,
                                tokenizer=tokenizer,
                                max_seq_len=128,
                                num_classes=len(label_list))

# Unify the input training set, validation set, and test set data into the feature form
train_dataset = train_dataset.map(convert_example_to_feature)
dev_dataset = dev_dataset.map(convert_example_to_feature)
test_dataset = test_dataset.map(convert_example_to_feature)

# Print the data samples processed ahead
print('An example: ', train_dataset[0])
```

The output result is:

Samples: {‘input_ids’: [1, 226, 170, 4, 836, 75, 426, 1645, 7, 27, 463, 144, 11, 2125, 12,051, 1657, 183, 46, 143, 5, 661, 737, 4, 480, 59, 41, 11, 397, 42, 317, 59, 22, 28, 5, 4, 1183, 1183, 12,043, 2], ‘token_type_ids’: [0, 0, 0, 0, 0, 0, 0, 0, 0, 0, 0, 0, 0, 0, 0, 0, 0, 0, 0, 0, 0, 0, 0, 0, 0, 0, 0, 0, 0, 0, 0, 0, 0, 0, 0, 0, 0, 0, 0], ‘labels’: [1.0, 0.0, 0.0, 0.0, 0.0, 0.0, 1.0, 0.0, 1.0, 0.0]}

From the output results, the text sequence has been converted into a form that ERNIE 3.0 can recognize, including input_ids and token_type_ids, and the labels have been converted into the one-hot encoding form.

12.2.4 Construction of the DataLoader

And then, we construct a DataLoader to iterate data in batches during model training and specify the batch size, that is, select a minibatch of data for training each time. To ensure that the length of data in the same batch is consistent, text truncation or text padding methods are needed. The convert_example_to_feature function in the previous section has already truncated the overly long text. Next, DataCollatorWithPadding is used to perform text padding on the data of the same minibatch. When constructing the DataLoader, we need to specify whether to shuffle the samples by using the parameter shuffle. Codes are implemented as follows:

```
from paddle.io import DataLoader, BatchSampler
from paddlenlp.data import DataCollatorWithPadding
# Text padding
collate_fn = DataCollatorWithPadding(tokenizer)
# Construct the batch sampler
train_batch_sampler = BatchSampler(train_dataset, batch_size=32,
shuffle=True)
dev_batch_sampler = BatchSampler(dev_dataset, batch_size=32,
shuffle=False)
test_batch_sampler = BatchSampler(test_dataset, batch_size=32,
shuffle=False)

# Construct the DataLoader, and iterate through the training set, validation set, and test set
data in batches based on the batch size
train_loader = DataLoader(dataset=train_dataset,
batch_sampler=train_batch_sampler, collate_fn=collate_fn)
dev_loader = DataLoader(dataset=dev_dataset,
batch_sampler=dev_batch_sampler, collate_fn=collate_fn)
test_loader = DataLoader(dataset=test_dataset,
batch_sampler=test_batch_sampler, collate_fn=collate_fn)

# Print the data of the first minibatch in the training set
print(next(iter(train_loader)))
```

The output result is:

```
mini batch : {'input_ids': Tensor(shape=[32, 60], dtype=int64, place=Place(gpu:0), stop_gradient=True,
       [[1  , 226, 170, ..., 0  , 0  , 0  ],
        [1  , 81 , 454, ..., 0  , 0  , 0  ],
        [1  , 384, 10 , ..., 0  , 0  , 0  ],
        ...,
        [1  , 160, 353, ..., 0  , 0  , 0  ],
        [1  , 374, 86 , ..., 0  , 0  , 0  ],
        [1  , 673, 176, ..., 0  , 0  , 0  ]]), 'token_type_ids': Tensor(shape=[32, 60], dtype=int64, place=Place(gpu:0), stop_gradient=True,
       [[0, 0, 0, ..., 0, 0, 0],
        [0, 0, 0, ..., 0, 0, 0],
        [0, 0, 0, ..., 0, 0, 0],
        ...,
        [0, 0, 0, ..., 0, 0, 0],
        [0, 0, 0, ..., 0, 0, 0],
        [0, 0, 0, ..., 0, 0, 0]]), 'labels': Tensor(shape=[32, 158], dtype=float32, place=Place(gpu:0), stop_gradient=True,
       [[1., 0., 0., ..., 0., 0., 0.],
        [0., 0., 0., ..., 0., 0., 0.],
        [0., 0., 0., ..., 0., 0., 0.],
        ...,
        [0., 0., 0., ..., 0., 0., 0.],
        [0., 0., 0., ..., 0., 0., 0.],
        [0., 0., 0., ..., 0., 0., 0.]])}
```

12.3 Model Construction

The ERNIE 3.0 model processes each incoming minibatch of data in turn. First, the processed text features are passed into the ERNIE 3.0 model, which encodes them and outputs the corresponding vector sequences. Then, these vector sequences are passed into the linear layer (the vector sequences are multiplied by weights plus biases), and after softmax processing, the classification labels corresponding to the

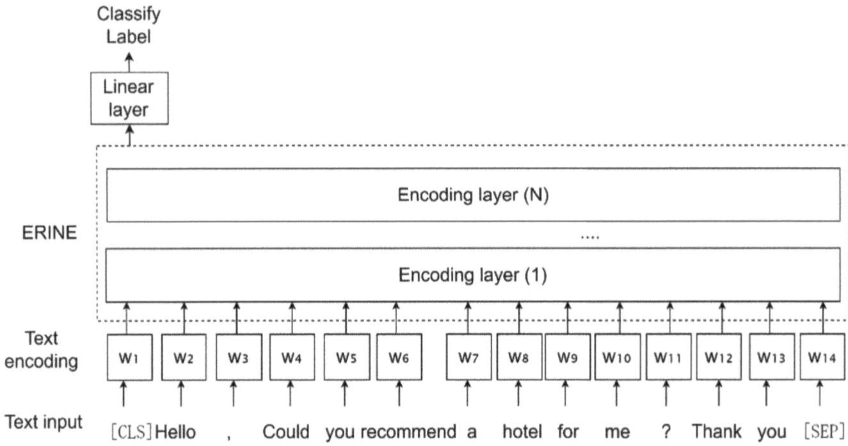

Fig. 12.4 Intent recognition using text classification based on ERNIE 3.0

text are obtained. After decoding, the intent recognition results can be obtained. The basic idea is shown in Fig. 12.4. In the input text, "[CLS]" represents the features of the entire sentence, and "[SEP]" represents the split point between two sentences. Here, there is only one sentence, so [SEP] represents the end marker of the sentence. In this way, we can transform the task of intent recognition into a work of sentence classification.

Based on the modeling approach shown in Fig. 12.4, codes are implemented as follows:

```
# Use the Auto module to load the ERNIE 3.0 Medium pre-training model and word segmenter
from paddlenlp.transformers import
AutoModelForSequenceClassification, AutoTokenizer

model =
AutoModelForSequenceClassification.from_pretrained("ernie-3.0-
medium-zh", num_classes=158)
tokenizer = AutoTokenizer.from_pretrained("ernie-3.0-medium-zh")
```

Here, we use PaddleNLP's AutoModelForSequenceClassification to call the classification model. In this way, the implementing codes are designed more concisely and conveniently.

12.4 Training Configuration

The following gives the definitions of the computing resources, models, the optimizer, the loss function, and optimization indicators used during model training.

(1) Model: ERNIE 3.0.
(2) Optimizer: The Adam optimizer.
(3) Loss function: Binary cross-entropy.
(4) Evaluation indicators: Macro F1 and Micro F1.

There are four possible outcomes for data prediction results, represented by TP, FP, FN, and TN. T and F represent the correctness and incorrectness of the prediction results, respectively, and P and N represent the positive and negative prediction samples. The calculation methods for the precision, recall rate, and F1-score are as follows:

$$Precision = \frac{TP}{TP + FP} \times 100\%$$

$$\text{Recall} = \frac{TP}{TP + FN} \times 100\%$$

$$\text{F1} = \frac{2 \times \text{Precision} \times \text{Recall}}{\text{Precision} + \text{Recall}} \times 100\%$$

In this experiment, we use Macro F1 and Micro F1, and their calculation methods are as follows:

(1) Macro F1: We first compute True Positives (TP), False Positives (FP), False Negatives (FN) and True Negatives (TN) for each category, then calculate the corresponding precision and recall. Then, the respective F1 score is gotten, and the average is taken to obtain the Macro F1. Macro F1 does not consider the data volume of each category and treats the calculation indicators of each category equally.
(2) Micro F1: Similar to Macro F1, we first compute True Positives (TP), False Positives (FP), False Negatives (FN), and True Negatives (TN) for each category, then calculate the corresponding precision and recall. Finally, the micro-precision and micro-recall are calculated to obtain the Micro F1. Micro F1 takes into account the data volume of each category, making it more suitable for situations with imbalanced data distribution. In the case of extreme imbalanced data, the categories with a larger data volume will affect the value of Micro F1. In order to know more about macro F1 and micro F1, please refer to (Zong Chengqing et al., 2022).

Codes implemented for training configuration are as follows:

```
import os
import paddle
from metric import MetricReport

# Set the rounds of training
num_epochs = 30
# Set the interval for printing the log every certain number of steps
log_steps = 50
# Set the intervals for model evaluation in the validation set every certain number of steps
eval_steps = 500
# Set the learning rate
learning_rate = 0.00003
# Set the intervals for model evaluation in the validation set every certain number of steps
save_dir = "./checkpoints"
if not os.path.exists(save_dir):
    os.makedirs(save_dir)

# Define the optimizer
optimizer = paddle.optimizer.Adam(learning_rate=learning_rate,
parameters= model.parameters())

# Define the loss function and use the cross-entropy
loss_fn = paddle.nn.BCEWithLogitsLoss()

# Define the calculation method for evaluation indicators
metric = MetricReport()
```

12.5 Model Training

In the model training phase, training logs will be output at each log_steps interval, and validation evaluation is conducted every eval_steps. The model with the best evaluation performance during the training process is saved.

The codes implemented for evaluation on the validation set are as follows:

```
import paddle.nn.functional as F

@paddle.no_grad()
def evaluate(model, data_loader, metric):
    """
    The model evaluation function

    Input:
        -model: model
        -data_loader: data in minibatches
        -metric: the evaluation indicator function
    Output:
        -micro_f1_score: Micro F1
        -macro_f1_score: Macro F1
     """
    # Set models into the evaluation mode
    model.eval()
    # Reset evaluations
    metric.reset()

    # Traverse every batch of the validation set
    for batch in data_loader:
        labels = batch.pop("labels")

        # Compute the output of models
        logits = model(**batch)
        probs = F.sigmoid(logits)

        # Accumulate evaluation
        metric.update(probs, labels)

    micro_f1_score, macro_f1_score = metric.accumulate()
    model.train()
    metric.reset()

    return micro_f1_score, macro_f1_score
```

During the training process, the changes in the loss function are saved using "train_loss_record," and the evaluation scores on the validation set are saved using "train_score_record." The codes implemented for model training are as follows:

```
def train(model):
    """
    Train the function
    Input:
        -model: model
    Output:
        -train_loss_record: the loss list during the training process
        -train_score_record: the accuracy of the development set during the training
process
    """
    # Start the training mode of models
    model.train()
    global_step = 0
    best_score = 0.

    # Record the loss function value during training and the Macro F1 and Micro F1 on the
validation set
    train_loss_record = []
    train_score_record = []
    num_training_steps = len(train_loader) * num_epochs

    # Perform num_epochs rounds of training
    for epoch in range(num_epochs):
        for step, batch in enumerate(train_loader):

            labels = batch.pop("labels")
            logits = model(**batch)
            loss = loss_fn(logits, labels)
            train_loss_record.append((global_step, loss.item()))


            # Gradient backpropagation
            loss.backward()
            optimizer.step()
            optimizer.clear_grad()

            if global_step > 0 and global_step % log_steps == 0:
                print(f"- epoch: {epoch} - global_step:
{global_step}/{num_training_steps} -loss:
{loss.numpy().item():.6f}")
            if global_step > 0 and global_step % eval_steps == 0:
                micro_f1_score, macro_f1_score = evaluate(model,
dev_loader, metric)
                train_score_record.append((global_step,
macro_f1_score))
            # If the current Macro F1 indicator is the optimal indicator, save the model
                if macro_f1_score > best_score:
                    print(f"best F1 performance has been updated:
{best_score :.5f} --> {macro_f1_score :.5f}")
                    best_score = macro_f1_score
                    paddle.save(model.state_dict(),
f"{save_dir}/best.pdparams")
```

```
f"{save_dir}/best.pdparams")
                print(f"evaluation result: Micro F1 score:
{micro_f1_score:.5f}, Macro F1 score: {macro_f1_score:.5f}")
            global_step += 1
    paddle.save(model.state_dict(), f"{save_dir}/final.pdparams")
    return train_loss_record, train_score_record

train_loss_record, train_score_record = train(model)
```

The output result is:

[Train] epoch: 1/30 step: 50/41640 loss: 0.232961

[Train] epoch: 1/30 step: 100/41640 loss: 0.109148

[Train] epoch: 1/30 step: 150/41640 loss: 0.074335

[Train] epoch: 1/30 step: 200/41640 loss: 0.061572

[Train] epoch: 1/30 step: 250/41640 loss: 0.049068

[Train] epoch: 1/30 step: 300/41640 loss: 0.049263

[Train] epoch: 1/30 step: 350/41640 loss: 0.046645

[Train] epoch: 1/30 step: 400/41640 loss: 0.042554

[Train] epoch: 1/30 step: 450/41640 loss: 0.041090

[Train] epoch: 1/30 step: 500/41640 loss: 0.040871

[Evaluate]: Micro F1 score: 0.00000, Macro F1 score: 0.00000

[Train] epoch: 1/30 step: 550/41640 loss: 0.043147

[Train] epoch: 1/30 step: 600/41640 loss: 0.035055

[Train] epoch: 1/30 step: 650/41640 loss: 0.037858

[Train] epoch: 1/30 step: 700/41640 loss: 0.035718

[Train] epoch: 1/30 step: 750/41640 loss: 0.033693

[Train] epoch: 1/30 step: 800/41640 loss: 0.033031

[Train] epoch: 1/30 step: 850/41640 loss: 0.033343

[Train] epoch: 1/30 step: 900/41640 loss: 0.040167

[Train] epoch: 1/30 step: 950/41640 loss: 0.028225

[Train] epoch: 1/30 step: 1000/41640 loss: 0.030494

Best Macro F1 performance has been updated: 0.00000 – > 0.02536

[Evaluate]: Micro F1 score: 0.93796, Macro F1 score: 0.66886

[Train] epoch: 29/30 step: 39,050/41640 loss: 0.000537

[Train] epoch: 29/30 step: 39,100/41640 loss: 0.000595

[Train] epoch: 29/30 step: 39,150/41640 loss: 0.000787

[Train] epoch: 29/30 step: 39,200/41640 loss: 0.000938

[Train] epoch: 29/30 step: 39,250/41640 loss: 0.001806

[Train] epoch: 29/30 step: 39,300/41640 loss: 0.000069

[Train] epoch: 29/30 step: 39,350/41640 loss: 0.001138

[Train] epoch: 29/30 step: 39,400/41640 loss: 0.002254

[Train] epoch: 29/30 step: 39,450/41640 loss: 0.000789

[Train] epoch: 29/30 step: 39,500/41640 loss: 0.000202

[Evaluate]: Micro F1 score: 0.93956, Macro F1 score: 0.68682

[Train] epoch: 29/30 step: 39,550/41640 loss: 0.000819

[Train] epoch: 29/30 step: 39,600/41640 loss: 0.000162

[Train] epoch: 29/30 step: 39,650/41640 loss: 0.000154

[Train] epoch: 29/30 step: 39,700/41640 loss: 0.001387

[Train] epoch: 29/30 step: 39,750/41640 loss: 0.000454

[Train] epoch: 29/30 step: 39,800/41640 loss: 0.000784

[Train] epoch: 29/30 step: 39,850/41640 loss: 0.000717

[Train] epoch: 29/30 step: 39,900/41640 loss: 0.000097

[Train] epoch: 29/30 step: 39,950/41640 loss: 0.001101

[Train] epoch: 29/30 step: 40,000/41640 loss: 0.001130

Best Macro F1 performance has been updated: 0.69926 – > 0.71424

Then, the loss "train_loss_record" during the training process and the score "train_score_record" on the validation set are saved, and the training process is visualized. Codes are implemented as follows:

```
from tools import plot_training_loss, plot_training_f1

# Visualize the training process
plot_training_loss(train_loss_record, 'loss.jpg')
plot_training_f1(train_score_record, 'f1.jpg')
```

The output result is shown in Fig. 12.5.

From the output results, it can be seen that as the training progresses, the loss value on the training set continuously decreases and then converges towards 0. Meanwhile, the scores of Macro F1 and Micro F1 on the validation set initially continue to

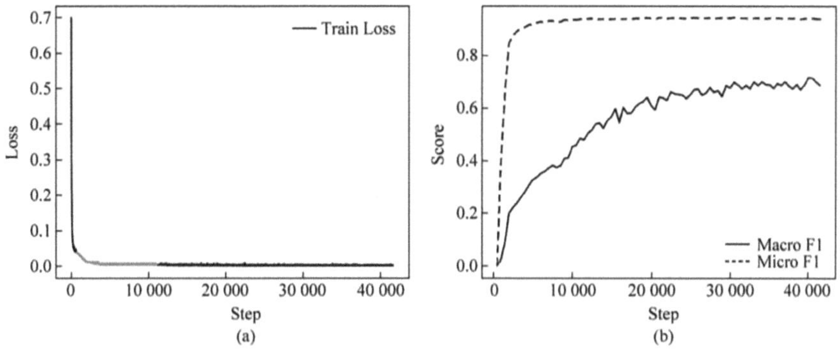

Fig. 12.5 The trend of the loss function on the training set and the trend of Macro F1 and Micro F1 on the validation set **a** The trend of the loss function on the training set; **b** The trend of Macro F1 and Micro F1 on the validation set

increase and then stabilize after the model converges. The model with the highest Macro F1 value is finally selected as the optimal model for subsequent evaluation and prediction.

12.6 Model Evaluation

The model with the best performance during the training process is evaluated using the test set to verify the performance of the model training. Codes are implemented as follows:

```
# Load the trained model for evaluation, re-instantiate the model, and then load the trained
model's parameters into the new model
saved_state = paddle.load("./checkpoints/best.pdparams")
model = AutoModelForSequenceClassification.from_pretrained("ernie-
3.0-medium-zh", num_classes=158)
model.load_dict(saved_state)
# Model evaluation
micro_f1_score, macro_f1_score = evaluate(model, test_loader,
metric)
# Output the evaluation results of models
print(f"test dataset result: Micro F1 score: {micro_f1_score:.5f},
Macro F1 score: {macro_f1_score:.5f}")
```

The output result is:

[Evaluate] Micro F1 score: 0.94019, Macro F1 score: 0.70192

12.7 Model Prediction

After completing all the above tasks, the model has been successfully developed. We can input any text, such as “Hello, could you recommend a hotel with a rating of 5 and a price range of 100–200 Yuan?”, and the model predicts the classification result of intent recognition. Codes are implemented as follows:

```
def infer(model, text):
    """
    Predict the inference function

    Input:
        -model: model
        -text: text
    """

    model.eval()
    # Data processing
    encoded_inputs = tokenizer(text=text, max_seq_len=128)

    # Construct the data of the input models
    input_ids =
paddle.to_tensor(encoded_inputs["input_ids"]).unsqueeze(0)
    token_type_ids =
paddle.to_tensor(encoded_inputs["token_type_ids"]).unsqueeze(0)

    # Calculate the log probability of the output classification of the model
    logits = model(input_ids, token_type_ids)
    probs = F.sigmoid(logits).numpy()
    # Analyze the label with the highest score
    labels = []
    for i, p in enumerate(probs[0]):
        if p > 0.5:
            labels.append(label_map[i])
    print("Text: ", text)
    print("Label: ", ','.join(labels))

text = "Hello, could you recommend a hotel with a rating of 5 and a price range of 100 -
200 Yuan?"
infer(model, text)
```

The output result is:

Text: Hello, could you recommend a hotel with a rating of 5 and a price range of 100–200 Yuan?

Label: General+greet+none+none, Request+hotel+name+

12.8 Exercises

(1) In this practice, we use Macro F1 and Micro F1 as the evaluation indicators for the model. Readers are encouraged to think that in the multi-label classification tasks, which kind of data distribution is more suitable for Macro F1 and Micro F1?

(2) Please repeat the work described in this chapter by setting different batch sizes and learning rates to observe their impact on model training.

Chapter 13
Machine Reading Comprehension

Abstract This chapter provides an overview of machine reading comprehension (MRC), a natural language processing task where a computer system analyzes an article to answer questions about it. The document focuses on implementing a Chinese MRC system using the ERNIE 3.0 pre-training model and the extractive Q&A method, which involves identifying the exact start and end positions of an answer within a given text. It details the entire process, including data processing using the DuReaderrobust dataset, model construction with a linear layer to predict answer positions, training configuration with the AdamW optimizer and cross-entropy loss, and model evaluation using F1 score and overall exact match as metrics. The chapter concludes with a demonstration of how the trained model can predict answers for new input texts.

Machine reading comprehension (MRC) refers to a processing task that a computer system uses relevant algorithms to analyze the content of a given article and obtain the correct answers according to questions. In this task, the types of questions include but are not limited to multiple-choice questions, fill-in-the-blank questions, and essay questions. MRC is a challenging yet very popular research topic in the field of NLP and related technologies are widely used in fields such as customer service robots, search engines, and Q&A systems.

In most tasks of MRC, the question-answering (Q&A) evaluation method is adopted to design natural language description questions related to the content of the article, allowing the model to analyze the questions and provide answers based on the content of the article. Common types of questions are as follows:

(1) Multiple-choice: The model needs to select the correct answers from several given options. This type of task is relatively simple and has a low evaluation difficulty.
(2) Extractive Q&A: In this task type, the candidate answers are text fragments within the given article, requiring the model to indicate the correct starting and ending positions of the answers in the article. The difficulty level of this task is moderate, and during evaluation, it is necessary to calculate the overlap between the answer output by the model and the labeled results.

C. Zong et al., *Natural Language Processing and Large Language Models*,
https://doi.org/10.1007/978-981-92-0682-7_13

(3) Free Q&A: This type does not restrict the form of the answers generated by the model, allowing the model to freely generate sentences. This task is more difficult, and the matching between the output answers and the labeled results needs to be calculated during evaluation.
(4) Fill-in-the-blank: Several words or phrases are removed from the original texts, and the model is required to fill in the correct answers. This task is relatively simple, and the exact matching method is used for evaluation.

In addition, in some datasets the questions "without answers" are also designed, that is, a question may not have a suitable answer in the article, and the model is required to output the unanswerable conclusion.

The case introduced in this chapter is based on ERNIE3.0 to achieve Chinese reading comprehension. The extractive Q&A method is used to accurately identify the starting and end positions of the answers in the given article.

13.1 Task Objectives

For a given piece of text, the MRC system should accurately analyze the semantics of the text and quickly find the corresponding answers according to the question. As shown in Fig. 13.1, for the input text and question, the model analyzes their semantics and outputs the correct answers.

Input text: Water resistance has become a standard feature of high-end mobile phones at present. Especially after Apple also supports water resistance, most high-end flagship domestic mobile phones have already supported water resistance. Although we rarely don't deliberately put our phones in water, with water resistance, users will have an extra sense of security in their hearts. So, is the recently most popular Xiaomi 6 waterproof? What is the water resistance level of the Xiaomi 6? The editor has looked up a lot of information and found that the Xiaomi 6 is indeed waterproof. However, to keep a low profile and avoid being accused of having insufficient water resistance, many materials do not indicate whether Xiaomi is waterproof or not. According to the evaluation data, the Xiaomi 6 supports IP68-level water resistance and is definitely capable of meeting the water resistance requirements in daily life.

Question 1: What is the water resistance level of Xiaomi 6?

Answer 1: IP68 level

Question 2: Does the Xiaomi 6 support water resistance?

Answer 2: The Xiaomi 6 is indeed waterproof

Fig. 13.1 Machine reading comprehension

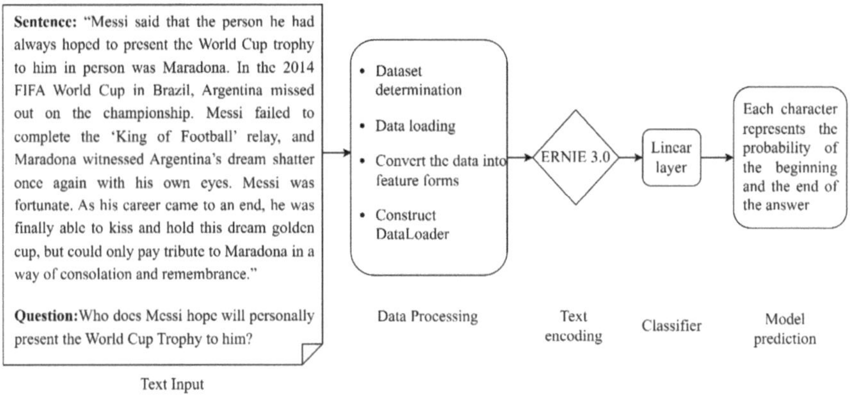

Fig. 13.2 The implementation flowchart of the Chinese machine reading comprehension task

13.2 Implementation Processes

The processes of implementing MRC task based on the pre-training model ERNIE 3.0 are shown in Fig. 13.2. The input to the model consists of a given text and a question, and the output is the corresponding answers. During the modeling process, the input text and question are first concatenated, followed by data processing. The ERNIE 3.0 model is then used to perform text sequence encoding to obtain the semantic vector representation of the text. Finally, a linear layer (the fully connected layer and softmax) is applied to calculate the probabilities of each character as the start and end of an answer in the text, which are then decoded into the corresponding answers.

13.3 Data Processing

Data processing includes steps such as determining the dataset, data loading, converting data into the feature form, and constructing the DataLoader. The same batch of data is processed into feature sequences with equal length, and the DataLoader is used to iterate in batches and send data into the ERNIE 3.0 model.

13.3.1 Determination of the Dataset

The robustness of MRC models is one of the key indicators to measure whether the technology can be implemented on a large scale in practical applications. With the advancement of technology, although models can achieve good performance on some reading comprehension test sets, the robustness demonstrated by the models in practical applications is still unsatisfactory. In this practice, we use the $DuReader_{robust}$[1]

[1] For a detailed introduction to the DuReaderrobust dataset, please refer to the reference (Tang et al. 2021).

dataset (the first Chinese dataset with its focus on the robustness of reading comprehension models), which contains a total of 65,937 data, including 14,520 data in the training set, 1,417 data in the test set, and 50,000 data in the validation set. Here, the "data" includes a given question, a passage, and the answer to the question. This is a single-passage, extractive reading comprehension dataset. The definition of this specific task is: given a question q and a passage p, the answer a to the question q should be provided based on p. For example:

Question q: How many seasons did Jordan play?
Passage p: Michael Jordan played in the NBA for 15 seasons. He entered the NBA in 1984 and retired for the first time on October 6, 1993, switching to baseball. He returned on March 18, 1995 and retired for the second time on January 13, 1999. He made a comeback on October 31, 2001 and finally retired in 2003. …
Reference answer a: ['15','15 seasons']

Question q: How many seasons did Jordan play?

Passage p: Michael Jordan played in the NBA for 15 seasons. He entered the NBA in 1984 and retired for the first time on October 6, 1993, switching to baseball. He returned on March 18, 1995 and retired for the second time on January 13, 1999. He made a comeback on October 31, 2001 and finally retired in 2003. …

Reference answer a: ['15','15 seasons']

13.3.2 Data Loading

This module loads the training set, test set, validation set, and labels into memory. Codes are implemented as follows:

```
# Load the training set, test data and validation data
from datasets import load_dataset


train_dataset_raw, dev_dataset_raw, test_dataset_raw =
load_dataset('dureader_robust', split=("train", "validation",
"test"))
train_dataset_raw, dev_dataset_raw, test_dataset_raw =
train_dataset_raw, dev_dataset_raw, test_dataset_raw
print('Number of samples of the training set： ', len(train_dataset_raw))
print('Number of samples of the validation set： ', len(dev_dataset_raw))
```

The output result is:
Number of samples of the training set: 14,520.
Number of samples of the validation set: 1417.
Number of samples of the test set: 50,000.

Samples: {'id': 'b9e74d4b9228399b03701d1fe6d52940', 'title': '', 'context': 'Michael. Jordan played in the NBA for 15 seasons. He entered the NBA in 1984 and retired for the first time on October 6, 1993, switching to baseball. He returned on March 18, 1995 and retired for the second time on January 13, 1999. He returned on October 31, 2001 and finally retired in 2003. Michael Jordan, born on February 17, 1963 in Brooklyn, New York, is a famous American basketball player who plays as a shooting guard and is the greatest basketball player in history. In the 1984 NBA Draft, Jordan was selected by the Chicago Bulls with the third overall pick in the first round. In the 1986–1987 season, Jordan averaged 37.1 points per game and won his first scoring title. In the 1990–1991 season, Jordan won both the regular season MVP and NBA Finals MVP titles, leading the Chicago Bulls to their first NBA championship. In the 1997–1991 season, Jordan won his 10th career scoring championship and led the Bulls to their sixth championship. On September 11, 2009, Jordan was officially inducted into the NBA Hall of Fame.', 'question': 'How many seasons did Jordan play?', 'answers': {'text': ['15 seasons'], 'answer_start': [12]}}.

From the output result we can see that the SQuAD data format is used in the DuReader$_{robust}$ dataset. Each data contains six parts: 'id', 'title', 'context', 'question', 'answers', and 'answer_start', representing the sample ID, the title of the article, the content of the article, the question, the answers and the starting position of the answer, respectively.

13.3.3 Convert Data into the Feature Form

Since the length of the context plus question for each piece of data may be greater than max_seq_length, and the answers may appear at the end of the context, we should not simply truncate the context. For overly long contexts, the sliding window approach is used to divide them into multiple paragraphs, which are then combined with the question to generate new input texts. The implementation process is shown in Fig. 13.3.

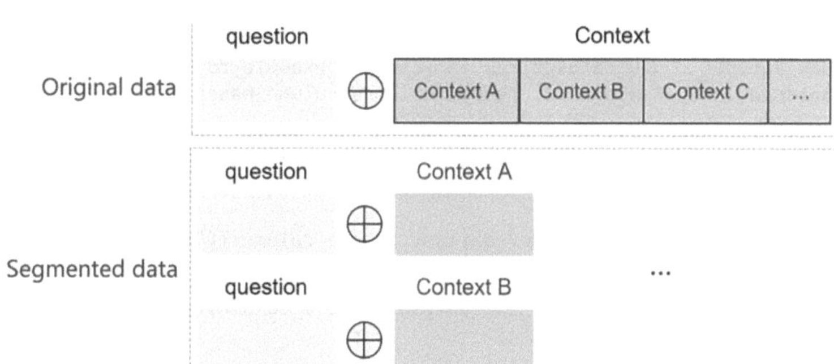

Fig. 13.3 The process of generating input texts using the sliding window method

The loaded text data is in the form of strings, which cannot be directly read by the ERNIE 3.0 model. It is necessary to convert the text data into corresponding feature data. We use PaddleNLP's tokenizer to perform word segmentation on the input text and map it to the index ID in the dictionary, so that the model can easily find the vector representation corresponding to the word based on this ID. The convert_example_to_feature function is first defined and the input text data is converted into the feature data by loading the tokenizer. The doc_stride parameter is used to control the distance of each slide. Codes are implemented as follows:

```
# Data processing ahead
from utils import preprocess_function_train, preprocess_function_dev
from functools import partial
# Set the largest length of texts
max_seq_length = 512
# Set the distance for each slide
doc_stride = 128
convert_example_to_feature_train =
partial(preprocess_function_train,
                            max_seq_length=max_seq_length,
                            doc_stride=doc_stride,
                            tokenizer=tokenizer)

# Use convert_example_to_feature to convert text data into the feature form
convert_example_to_feature_dev = partial(preprocess_function_dev,
                            max_seq_length=max_seq_length,
                            doc_stride=doc_stride,
                            tokenizer=tokenizer)


column_names = train_dataset_raw.column_names
train_dataset =
train_dataset_raw.map(convert_example_to_feature_train,
batched=True, num_proc=4, remove_columns=column_names)
dev_dataset = dev_dataset_raw.map(convert_example_to_feature_dev,
batched=True, num_proc=4, remove_columns=column_names)
test_dataset = test_dataset_raw.map(convert_example_to_feature_dev,
batched=True, num_proc=4, remove_columns=column_names)


dev_dataset_for_model = dev_dataset.remove_columns(["example_id",
"offset_mapping"])
test_dataset_for_model = test_dataset.remove_columns(["example_id",
"offset_mapping"])
# Print the processed data samples
print('Sample: ', train_dataset[0])
```

The output result is:

Sample:{'input_ids': [1, 1034, 1189, 734, 2003, 241, 284, 131, 553, 271, 28, 125, 280, 2, 131, 1773, 271, 1097, 373, 1427, 1427, 501, 88, 662, 1906, 4, 561, 125, 311, 1168, 311, 692, 46, 430, 4, 84, 2073, 14, 1264, 3967, 5, 1034, 1020, 1829, 268, 4, 373, 539, 8, 154, 5210, 4, 105, 167, 59, 69, 685, 12043, 539, 8, 883, 1020, 4, 29, 720, 95, 90, 427, 67, 262, 5, 384, 266, 14, 101, 59, 789, 416, 237, 12043, 1097, 373, 616, 37, 1519, 93, 61, 15, 4, 255, 535, 7, 1529, 619, 187, 4, 62, 154, 451, 149, 12043, 539, 8, 253, 223, 3679, 323, 523, 4, 535, 34, 87, 8, 203, 280, 1186, 340, 9, 1097, 373, 5, 262, 203, 623, 704, 12043, 84, 2073, 1137, 358, 334, 702, 5, 262, 203, 4, 334, 702, 405, 360, 653, 129, 178, 7, 568, 28, 15, 125, 280, 518, 9, 1179, 487, 12043, 84, 2073, 1621, 1829, 1034, 1020, 4, 539, 8, 448, 91, 202, 466, 70, 262, 4, 638, 125, 280, 83, 299, 12043, 539, 8, 61, 45, 7, 1537, 176, 4, 84, 2073, 288, 39, 4, 889, 280, 14, 125, 280, 156, 538, 12043, 190, 889, 280, 71, 109, 124, 93, 292, 889, 46, 1248, 4, 518, 48, 883, 125, 12043, 539, 8, 268, 889, 280, 109, 270, 4, 1586, 845, 7, 669, 199, 5, 3964, 3740, 1084, 4, 255, 440, 616, 154, 72, 71, 109, 12043, 49, 61, 283, 3591, 34, 87, 297, 41, 9, 1993, 2602, 518, 52, 706, 109, 12043, 37, 10, 561, 125, 43, 8, 445, 86, 576, 65, 1448, 2969, 4, 469, 1586, 118, 776, 5, 1993, 2602, 4, 108, 25, 179, 51, 1993, 2602, 498, 1052, 122, 12043, 1082, 1994, 1616, 11, 262, 4, 518, 171, 813, 109, 1084, 270, 12043, 539, 8, 3006, 580, 11, 31, 4, 2473, 306, 34, 87, 889, 280, 846, 573, 12043, 561, 125, 14, 539, 889, 810, 276, 182, 4, 67, 351, 14, 889, 1182, 118, 776, 156, 952, 4, 539, 889, 16, 38, 4, 445, 15, 200, 61, 12043, 2], 'token_type_ids': [0, 0, 0, 0, 0, 0, 0, 0, 0, 0, 0, 0, 0, 0, 1], 'start_positions': 14, 'end_positions': 16}

From the output results, it can be seen that the text sequences in the dataset have been converted into the feature forms, including input_ids, token_type_ids, start_positions, and end_positions.

(1) input_ids: representing the token ID of the input text.
(2) token_type_ids: indicating whether the corresponding token belongs to the input question or answer. "0" represents questions, and "1" represents articles.
(3) start_positions: the starting positions of answers.
(4) end_positions: answers' end positions.

13.3.4 Construction of the DataLoader

Next, we construct a DataLoader for the iteration of data in batches and specify the batch size. When the model iterates data for training, it uses DataCollatorWithPadding to pad each minibatch to the equal length and specify whether to perform sample rotation by setting the shuffle parameter. If shuffle is set to False, the DataLoader iterates a batch of data in order; if it is set to True, the DataLoader iterates a batch of data randomly. Codes are implemented as follows:

```
import paddle
from paddle.io import DataLoader, BatchSampler
from paddlenlp.data import DataCollatorWithPadding

batch_size = 12

# Define BatchSampler
train_batch_sampler = BatchSampler(train_dataset,
batch_size=batch_size, shuffle=True)
dev_batch_sampler = BatchSampler(dev_dataset,
batch_size=batch_size, shuffle=False)
test_batch_sampler = BatchSampler(test_dataset,
batch_size=batch_size, shuffle=False)

# Text padding
collate_fn = DataCollatorWithPadding(tokenizer)

# Construct the DataLoader
train_loader = DataLoader(dataset=train_dataset,
batch_sampler=train_batch_sampler, collate_fn=collate_fn,
return_list=True)
dev_loader = DataLoader(dataset=dev_dataset_for_model,
batch_sampler=dev_batch_sampler, collate_fn=collate_fn,
return_list=True)
test_loader = DataLoader(dataset=test_dataset_for_model,
batch_sampler=test_batch_sampler, collate_fn=collate_fn,
return_list=True)
```

The output result is:

mini batch: {'input_ids': Tensor(shape = [12, 512], dtype = int64, place = Place(gpu:0), stop_gradient = True,

[[1, 931, 40, ..., 0, 0, 0],

[1, 395, 141, ..., 0, 0, 0],

[1, 102, 837, ..., 0, 0, 0],

...,

[1, 3147, 5095, ..., 0, 0, 0],

[1, 536, 1400, ..., 0, 0, 0],

[1, 252, 560, …, 0, 0, 0]]), ‘token_type_ids’: Tensor(shape = [12, 512], dtype = int64, place = Place(gpu:0), stop_gradient = True,
[[0, 0, 0, …, 0, 0, 0],
[0, 0, 0, …, 0, 0, 0],
[0, 0, 0, …, 0, 0, 0],
…,
[0, 0, 0, …, 0, 0, 0],
[0, 0, 0, …, 0, 0, 0],
[0, 0, 0, …, 0, 0, 0]]), ‘start_positions’: Tensor(shape = [12], dtype = int64, place = Place(gpu:0), stop_gradient = True,
[28, 17, 27, 15, 31, 72, 174, 14, 91, 86, 68, 388]), ‘end_positions’: Tensor(shape = [12], dtype = int64, place = Place(gpu:0), stop_gradient = True,
[29, 21, 29, 17, 34, 75, 178, 17, 92, 88, 69, 390])}

From the output results we can see the data length of the same minibatch is the same.

13.4 Model Construction

The ERNIE 3.0 model processes the incoming minibatch data in sequence. The processed text data (including questions and articles) is passed into the ERNIE 3.0 model. Then they are encoded, and the corresponding feature vectors are output. Then, the feature vectors of each word are passed into the linear layer (the vector sequence is multiplied by the weight plus bias). There are two linear layers, which are used to identify the starting position “Start token” of the answers and the end position “End token” of the answers, respectively. Finally, the probabilities of each token as the start and end of the answers in the text are obtained through softmax processing. After decoding, the corresponding answers to the questions will be obtained. The implementation process is shown in Fig. 13.4.

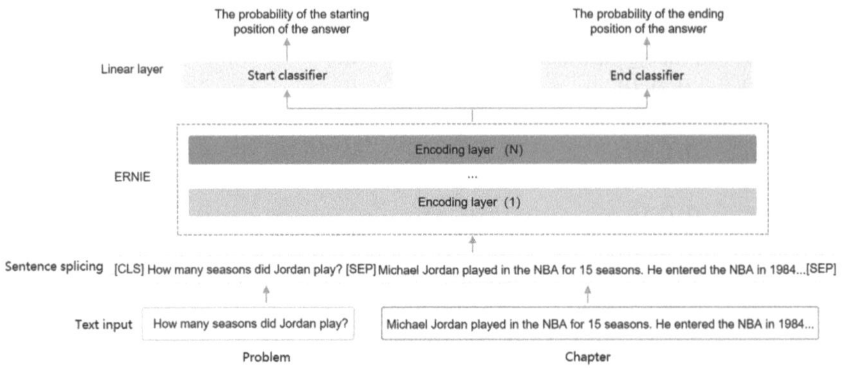

Fig. 13.4 Chinese reading comprehension using the extractive Q&A method based on RENIE 3.0

Based on the modeling approach shown in Fig. 13.4, codes are implemented as follows:

```
# Use the Auto module to load the pre-training model ERNIE and the word segmenter

from paddlenlp.transformers import AutoModelForQuestionAnswering,
AutoTokenizer

model = AutoModelForQuestionAnswering.from_pretrained("ernie-3.0-
medium-zh")

tokenizer = AutoTokenizer.from_pretrained("ernie-3.0-medium-zh")
```

In this paragraph of codes, the AutoModelForQuestionAnswering of PaddleNLP is used to quickly implement reading comprehension tasks.

13.5 Training Configuration

Net, the computing resources, models, the optimizer, the loss function, and optimization indicators used during model training should be defined.

(1) Model: the ERNIE 3.0 model
(2) Optimizer: the AdamW optimizer.
(3) Loss function: the cross-entropy loss. The AutoModelForQuestionAnswering of PaddleNLP splits the semantic vector into start_logits and end_logits outputs. Hence in this practice, the loss function is the sum of the start loss and end loss.

(4) Evaluation indicators: the overall exact match and character-level F1 score.

%1 Overall exact match: calculating the overall accuracy rate of the answers. If the model's response matches any one of the standard answers, it is counted as "1", otherwise it is counted as "0".

%1 Character-level F1 score: the model's answers and the standard answers are treated as character sets. The precision rate of character matching is calculated by using the following formula $P = \frac{TP}{TP+FP} \times 100\%$ and the recall rate is calculated by $R = \frac{TP}{TP+FN} \times 100\%$ and then the average F1 of all questions is calculated by formula $F1 = \frac{2PR}{P+R}$. The character-level F1 is calculated as a more reliable evaluation method.

The codes implemented for defining the loss function are as follows:

```
# Define the loss function, which is the sum of start_loss and end_loss
class CrossEntropyLossForRobust(paddle.nn.Layer):
    def __init__(self):
        super(CrossEntropyLossForRobust, self).__init__()

    def forward(self, y, label):
        start_logits, end_logits = y
        start_position, end_position = label
        start_position = paddle.unsqueeze(start_position, axis=-1)
        end_position = paddle.unsqueeze(end_position, axis=-1)
        start_loss = paddle.nn.functional.cross_entropy(
            input=start_logits, label=start_position)
        end_loss = paddle.nn.functional.cross_entropy(
            input=end_logits, label=end_position)
        loss = (start_loss + end_loss) / 2
        return loss
```

The codes implemented for the training configuration are as follows:

```
import os

import paddlenlp
# Set the rounds of training
num_epochs = 3
# Set the learning rate
learning_rate = 3e-5
# Set the proportion of warm-up training during the training process
warmup_proportion = 0.1
# Set the weight decay coefficient during training
weight_decay = 0.01
# Set training steps needed in total
num_training_steps = len(train_loader) * num_epochs
lr_scheduler =
paddlenlp.transformers.LinearDecayWithWarmup(learning_rate,
num_training_steps, warmup_proportion)
save_dir = "./checkpoints"
log_steps = 10
eval_steps = 100
if not os.path.exists(save_dir):
    os.makedirs(save_dir)

# Specify the optimizer
lr_scheduler =
paddlenlp.transformers.LinearDecayWithWarmup(learning_rate,
num_training_steps, warmup_proportion)

# layer_norm and bias variables do not use weight decay
decay_params = [
    p.name for n, p in model.named_parameters()
    if not any(nd in n for nd in ["bias", "norm"])
]

# Define the optimizer
optimizer = paddle.optimizer.AdamW(
    learning_rate=lr_scheduler,
    parameters=model.parameters(),
    weight_decay=weight_decay,
    apply_decay_param_fun=lambda x: x in decay_params)

# Specify the loss function
loss_fn = CrossEntropyLossForRobust()
```

13.6 Model Training

During the model training process, a training log is printed every certain number of log_steps, and the model is evaluated on the validation set every certain number of eval_steps. The model with the best evaluation performance during the training process is saved.

Codes are implemented as follows:

```
from paddlenlp.metrics.squad import squad_evaluate, compute_prediction

# Define the evaluation function
@paddle.no_grad()
def evaluate(model, raw_dataset, dataset, data_loader, is_test=False):
    """
    The evaluation function
    Input:
        -model: models
        -raw_dataset: Raw data that has not been processed ahead by the tokenizer
        -dataset: Data that has been processed ahead by the tokenizer
        -data_loader: Data in batches
        -is_test: Whether the dataset is a test set or not
    Output:
        -results: Evaluate the output dictionary
    """
    model.eval()

    all_start_logits = []
    all_end_logits = []

    for batch in data_loader:
        start_logits_tensor, end_logits_tensor =
model(batch["input_ids"], batch["token_type_ids"])
        for idx in range(start_logits_tensor.shape[0]):
            all_start_logits.append(start_logits_tensor.numpy()[idx])
            all_end_logits.append(end_logits_tensor.numpy()[idx])
    all_predictions, _, _ = compute_prediction(raw_dataset, dataset,
(all_start_logits, all_end_logits), False, 20, 30)
    model.train()
    if is_test:
        return all_predictions
    else:
        results = squad_evaluate(examples=[raw_data for raw_data in
raw_dataset], preds=all_predictions, is_whitespace_splited=False)
        return results
```

The codes for model training are given as follows:

```
def train(model):
    """
    Train the function

    Input:
        -model: models
    Output:
        -train_loss_record: the loss list during training
        -train_score_record: the accuracy list of the development set during training

    """
    # Start the training mode of models
    model.train()
    global_step = 0
    best_score = 0.

    # Record the loss function value during training, the F1 score on the validation set, and the
overall accurate match value
    train_loss_record = []
    f1_score_record = []
    exact_score_record = []

    # Perform num_epochs rounds of training
    for epoch in range(num_epochs):
        for step, batch in enumerate(train_loader):

            logits = model(input_ids=batch["input_ids"],
token_type_ids=batch["token_type_ids"])
            loss = loss_fn(logits, (batch["start_positions"],
batch["end_positions"]))
            train_loss_record.append((global_step, loss.item()))


            # Gradient backpropagation
            loss.backward()
            optimizer.step()
            lr_scheduler.step()
            optimizer.clear_grad()
```

```
            if global_step > 0 and global_step % log_steps == 0:
                print(f"[Train] epoch: {epoch+1}/{num_epochs} step:
{global_step}/{num_training_steps} loss: {loss.numpy().item():.6f}")
            if global_step > 0 and global_step % eval_steps == 0:
                results = evaluate(model, dev_dataset_raw,
dev_dataset, dev_loader)
                f1_score = results['f1']
                exact_score = results['exact']
                print(f"[Evaluate]: F1 score: {f1_score:.5f}, Exact
Match: {exact_score:.5f}")

                # Record the loss changes during training
                train_score_record.append((global_step, f1_score))
                if f1_score > best_score:
                    print(f"best F1 performance has been updated:
{best_score :.5f} --> {f1_score :.5f}")
                    best_score = f1_score

                    # Save the best model's parameters
                    paddle.save(model.state_dict(),
f"{save_dir}/best.pdparams")
            global_step += 1

    # Save the models with the best performance during training
    paddle.save(model.state_dict(), f"{save_dir}/final.pdparams")
    return train_loss_record, train_score_record
# Start training models
train_loss_record, train_score_record = train(model)
```

The output result is:
[Train] epoch: 1/4 step: 10/5884 loss: 6.191539
[Train] epoch: 1/4 step: 20/5884 loss: 6.072257
[Train] epoch: 1/4 step: 30/5884 loss: 5.872962
[Train] epoch: 1/4 step: 40/5884 loss: 5.795141
[Train] epoch: 1/4 step: 50/5884 loss: 5.675653
[Train] epoch: 1/4 step: 60/5884 loss: 5.305289
[Train] epoch: 1/4 step: 70/5884 loss: 4.892242
[Train] epoch: 1/4 step: 80/5884 loss: 4.349128
[Train] epoch: 1/4 step: 90/5884 loss: 4.234649
[Train] epoch: 1/4 step: 100/5884 loss: 3.371776
[Evaluate]: F1 score: 58.50232, Exact Match: 38.32040
best F1 performance has been updated: 0.00000 --> 58.50232
[Train] epoch: 1/4 step: 110/5884 loss: 3.276013
[Train] epoch: 1/4 step: 120/5884 loss: 2.369620
[Train] epoch: 1/4 step: 130/5884 loss: 2.000223

[Train] epoch: 1/4 step: 140/5884 loss: 2.416104
[Train] epoch: 1/4 step: 150/5884 loss: 1.414138
[Train] epoch: 1/4 step: 160/5884 loss: 1.975554
[Train] epoch: 1/4 step: 170/5884 loss: 2.008070
[Train] epoch: 1/4 step: 180/5884 loss: 1.398116
[Train] epoch: 1/4 step: 190/5884 loss: 1.509223
[Train] epoch: 1/4 step: 200/5884 loss: 1.442685
[Evaluate]: F1 score: 79.16693, Exact Match: 61.89132
best F1 performance has been updated: 58.50232 --> 79.16693
……
[Train] epoch: 2/4 step: 2210/5884 loss: 0.935110
[Train] epoch: 2/4 step: 2220/5884 loss: 0.359446
[Train] epoch: 2/4 step: 2230/5884 loss: 0.629553
[Train] epoch: 2/4 step: 2240/5884 loss: 1.196259
[Train] epoch: 2/4 step: 2250/5884 loss: 0.922623
[Train] epoch: 2/4 step: 2260/5884 loss: 0.728656
[Train] epoch: 2/4 step: 2270/5884 loss: 1.049675
[Train] epoch: 2/4 step: 2280/5884 loss: 0.893290
[Train] epoch: 2/4 step: 2290/5884 loss: 0.719477
[Train] epoch: 2/4 step: 2300/5884 loss: 0.663813
[Evaluate]: F1 score: 86.34958, Exact Match: 72.54764
[Train] epoch: 2/4 step: 2310/5884 loss: 0.614171
[Train] epoch: 2/4 step: 2320/5884 loss: 0.819235
[Train] epoch: 2/4 step: 2330/5884 loss: 0.385778
[Train] epoch: 2/4 step: 2340/5884 loss: 1.084770
[Train] epoch: 2/4 step: 2350/5884 loss: 1.025034
[Train] epoch: 2/4 step: 2360/5884 loss: 0.726824
[Train] epoch: 2/4 step: 2370/5884 loss: 0.641483
[Train] epoch: 2/4 step: 2380/5884 loss: 0.459251
[Train] epoch: 2/4 step: 2390/5884 loss: 0.534366
[Train] epoch: 2/4 step: 2400/5884 loss: 0.179930
[Evaluate]: F1 score: 87.09196, Exact Match: 74.10021
best F1 performance has been updated: 86.52995 --> 87.09196

The loss record train_loss_record during training and the score train_socre_record on the validation set are saved, and the training process is visualized. Codes are implemented as follows:

```
# Visualize the training process
from tools import plot_training_loss, plot_f1

plot_training_loss(train_loss_record, 'loss.jpg')
plot_f1(f1_score_record, exact_score_record, 'score.jpg')
```

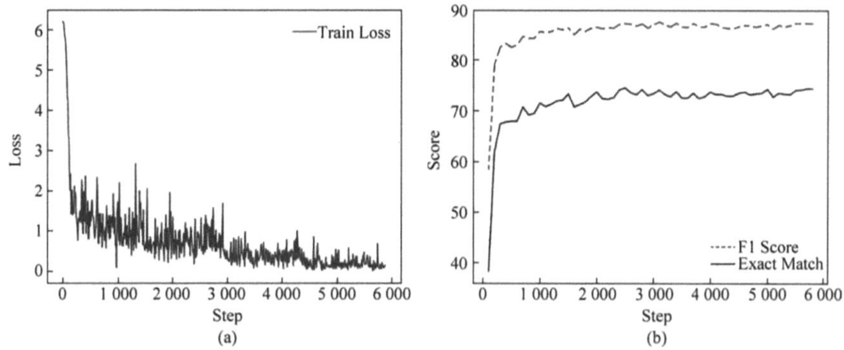

Fig. 13.5 The trend of the loss function on the training set and the trend of the F1 score and the overall exact match on the validation set. **a** The trend of the loss function on the training set; **b** The trend of the F1 score and the overall exact match on the validation set

The input results are shown in Fig. 13.5. From the output results, we can see as the training progresses, the loss value of the model on the training set continues to decline, then converges, and the value tends to 0. At the same time, the overall exact match and character-level F1 score on the validation set initially continue to rise, and gradually stabilize after the model converges.

13.7 Model Evaluation

The model with the best performance during training is evaluated by using the validation set to verify the performance of model training. Codes are implemented as follows:

```
# Load the trained model for evaluation, re-instantiate the model, and then load the trained
model's parameters into the new model
saved_state = paddle.load("./checkpoints/best.pdparams")
model = AutoModelForQuestionAnswering.from_pretrained("ernie-3.0-
medium-zh")
model.load_dict(saved_state)

# Model evaluation
results = evaluate(model, dev_dataset_raw, dev_dataset, dev_loader)
f1_score = results['f1']
exact_score = results['exact']
print(f"[Evaluate]: F1 score: {f1_score:.5f}, Exact Match:
{exact_score:.5f}")
```

The output result is:

[Evaluate]: F1 score: 87.34910, Exact Match: 74.52364

In this practice, we use the validation set as the test set for model evaluation. If readers are interested, they can try to use the test set $DuReader_{robust}$ to re-evaluate the model's performance.

13.8 Model Prediction

Given any input passage and question, the model can predict the corresponding answers. Codes are implemented as follows:

```
from datasets import Dataset

def infer(input_data, model,  batch_size=12):
    """
    The inference function
    Input:
        -input_list: input lists, including sample ID, questions and texts
        -model: models
        -batch_size: a mini batch contains the number of samples
    Output:
        -results: evaluate the output dictionary
    """
    model.eval()
    dataset_raw = Dataset.from_list(input_data)
    column_names = dataset_raw.column_names
    dataset = dataset_raw.map(convert_example_to_feature_dev,
batched=True, num_proc=2, remove_columns=column_names)
    dataset_for_model = dataset.remove_columns(["example_id",
"offset_mapping"])
    batch_sampler = BatchSampler(dataset, batch_size=batch_size,
shuffle=False)
    loader = DataLoader(dataset=dataset_for_model,
batch_sampler=batch_sampler, collate_fn=collate_fn, return_list=True)
    all_predictions = evaluate(model, dataset_raw, dataset, loader,
True)
    return all_predictions
```

```
input_data = [
    {'id':'1', 'question':'How many watts is 220V voltage and one ampere equivalent to?', 'context':'Under the state of 220v alternating current, one ampere is equal to 220 watts. The calculation method for the size of the switch based on 32 ta1.5-horsepower is: 1 horsepower =0.735 watts. 0.735*1.5*32=35.28 kilowatts. 1 kilowatt =4.5 amperes. 35.28*4.5=158.76 amperes Here is the actual current. In practical applications, a 160-ampere switch cannot be used. A single 1.5-horsepower start-up has a relatively large starting current. In actual use, multiply it by 1.5 times: 158.76*1.5=238.14 amperes. The current of the switch should be a 250A air switch.'},
    {'id':'2', 'question':'The ionic equation of copper oxide and dilute hydrochloric acid', 'context':'Chemical equation: CuO+2HCl=CuCl2+H2O. When writing ionic equations, only strong electrolytes (strong acids, strong bases, salts) are written in ionic form. Ionic equation: CuO+2H+=Cu^2+ +H2O'}
    ]
all_predictions = infer(input_data, model)
for i in range(len(input_data)):
    print('content:', input_data[i]['context'])
    print('question:', input_data[i]['question'])
    print('answer:', all_predictions[input_data[i]['id']])
    print('-'*20)
```

The output result is:

```
Content: Under the state of 220v alternating current, one ampere is equal to
 220 watts. The calculation method for the size of the switch based on 32
ta1.5-horsepower is: 1 horsepower =0.735 watts. 0.735*1.5*32=35.28 kilowat
ts. 1 kilowatt =4.5 amperes. 35.28*4.5=158.76 amperes Here is the actual c
urrent. In practical applications, a 160-ampere switch cannot be used. A s
ingle 1.5-horsepower start-up has a relatively large starting current. In
actual use, multiply it by 1.5 times: 158.76*1.5=238.14 amperes. The curre
nt of the switch should be a 250A air switch.

Question: How many watts is 220V voltage and one ampere equivalent to?

Answer: 220 watts

--------------------

Content: Chemical equation: CuO+2HCl=CuCl2+H2O. When writing ionic equation
s, only strong electrolytes (strong acids, strong bases, salts) are writte
n in ionic form. Ionic equation: CuO+2H+=Cu^2+ +H2O

Question: The ionic equation of copper oxide and dilute hydrochloric acid

Answer: CuO+2H+=Cu^2+ +H2O

--------------------
```

13.9 Exercises

(1) Please think about it: why in MRC tasks, the character-level F1 is usually the evaluation indicator, rather than overall exact match?
(2) Based on the learning and practices introduced in this chapter, readers are encouraged to think what role token_type_ids plays in reading comprehension tasks?
(3) The practices introduced in this chapter are based on the pre-training model ERNIE3.0. Two linear layers are used to identify the initial position of the "Start token" and the end position of the "End token," respectively. Please think about it, are there any other ways to better identify the initial and end positions of the answers?

Chapter 14
Machine Translation

Abstract This chapter focuses on machine translation (MT) using the Transformer model. It introduces the history and applications of MT, detailing the implementation of a Chinese-to-English translation method based on the Transformer framework. The chapter explains the entire process, including data processing (loading, tokenization, converting data to feature form), model construction (embedding layers, encoder, decoder), training configuration (optimizer, loss function, evaluation metrics), and model evaluation. It also provides code examples for each step and analyzes key concepts like positional encoding, label smoothing, and the Adam optimizer with the Noam learning rate decay schedule. The chapter concludes by demonstrating model prediction with sample Chinese texts and provides exercises for further learning.

Machine translation (MT) is a technology that uses computers to automatically translate one language into another one. The language being translated is usually called the source language, and the output translation is called the target language. Machine translation is a very important research topic in the field of natural language processing and has important and wide applications. It has been widely used in various industries and people's daily lives, such as translation machines, simultaneous interpreting, and cross-language retrieval.

The research on machine translation technology can be traced back to the 1950s. During this process, there is a series of twists and turns. Scientists have proposed different machine translation methods, such as rule-based MT, interlingua-based MT, example-based MT, and statistical MT (Zong Chengqing 2013).

With the successful applications of neural networks and deep learning technology, researchers have proposed the end-to-end neural machine translation method. In this method, discrete symbolic representations are embedded into the low-dimensional and continuous real vector spaces, alleviating the problem of data sparsity. With this method, the linguistic features can be automatically extracted without the need for the manual design of feature templates. Currently, most neural machine translation engines are based on the encoder–decoder framework of the attention mechanism, say Transformer framework. Using this framework the long text sequences can be well

C. Zong et al., *Natural Language Processing and Large Language Models*,
https://doi.org/10.1007/978-981-92-0682-7_14

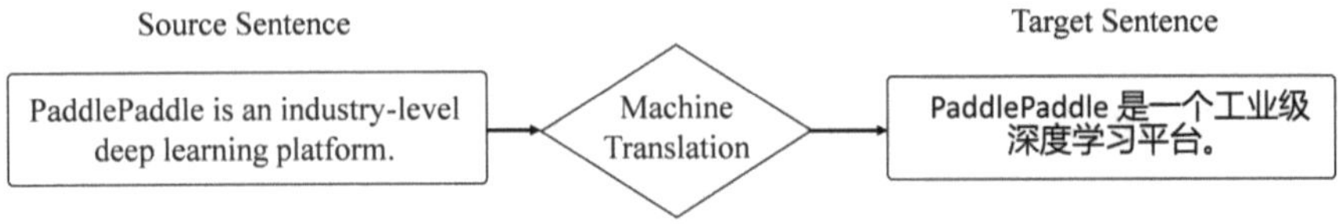

Fig. 14.1 The application of the machine translation technology in the translation system

handled, and the long-distance dependency problem of long texts can be effectively alleviated.

In this chapter, we take the translation of TED speech texts as the example to introduce the specific implementation processes of the Chinese-English translation method based on the Transformer model (Vaswani et al. 2017).

14.1 Task Objectives and Implementation Processes

As mentioned earlier, machine translation implements the process of automatically translating a text from the source language into the target language, as shown in Fig. 14.1.

The implementation process of machine translation based on the Transformer is shown in Fig. 14.2. The input of the model is the source language (Chinese sentence), and the output is the target language (English sentence). During the modeling process, the model first processes the input source language text, then encodes it using an encoder, then decodes it using a decoder, and finally performs post-processing on the generated results to obtain the translation result.

The detailed implementation processes of each step are introduced as follows.

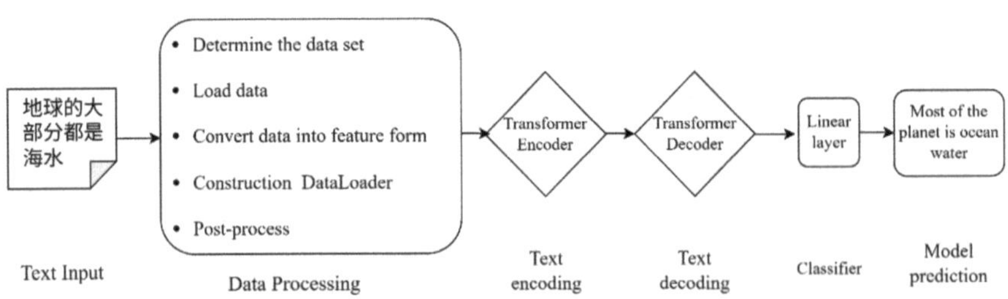

Fig. 14.2 The implementation flowchart of machine translation

14.2 Data Processing

Data processing includes steps such as determining the dataset, data loading, word segmentation, converting data into the feature form, constructing the DataLoader, and post-processing. Ultimately, the same batch of data is processed into feature sequences with an equal length, and the DataLoader is used to iterate and send data into the Transformer model in batches.

14.2.1 Determine the Dataset

This practice uses the data in the IWSLT 2015 dataset① for the English-Chinese translation, which includes a training set of 2,000 dialogues, 200,000 sentences, and 4,000,000 tokens, as well as 10–15 dialogues, 1,000–1,500 sentences, and 20,000–30,000 tokens in the validation and test sets, respectively. The texts in the IWSLT 2015 dataset come from TED talks and include translations between English and French, English and German, English and Chinese, English and Thai, English and Vietnamese, and English and Czech.

Please see a sample data below randomly selected from the training set:

```
# Chinese text
大卫·盖罗：这位是比尔·兰格，我是大卫·盖罗。
# English text
This is Bill Lange. I'm Dave Gallo.
```

To facilitate readers' learning and use, we have segmented the Chinese sequences of the original data into words, tokenized the English data, and generated a word list. The implementation process is shown in Fig. 14.3. First, the jieba word segmentation tool[1] is used to segment the Chinese texts into words, followed by the sub-word segmentation using BPE (refer to Sect. 7.2). For English sentences, the Moses word processing tool[2] is used firstly, then followed by the sub-word compression using BPE. After word segmentation and compression, the text sequence in the sample data is segmented and becomes a sub-word, in which @@ is the separator symbol between sub-words.

[1] https://github.com/fxsjy/jieba.

[2] http://www2.statmt.org/moses/?n=Main.HomePage.

Fig. 14.3 The process of word segmentation in English and Chinese texts

14.2.2 Data Loading

The codes to read datasets are implemented as follows:

```
import os
from datasets import load_dataset

# The paths of datasets
data_dir = r"./data"

# Construct data_files, and the key-value pairs correspond to the paths of the Chinese and
English text files for the training set, validation set, and test set respectively
splits = ("train", "dev", "test")
src2trg = ("zh", "en")

data_files = {
    key: [os.path.join(data_dir, f"{key}_{lang}.bpe") for lang in
src2trg]
    for key in splits
}
print("data_files struct: \n", data_files)

# Use language_pair.py to load data and read the dataset specified by split from data_files. For
display purposes, only load the training set
(dataset, ) = load_dataset("language_pair.py",
                           data_files=data_files,
                           split=["train"])
print("the sample structure of the training set: ")
dataset
```

The output result is:

The structure of data_files:

{‘train’: [‘./data/train_zh.bpe’, ‘./data/train_en.bpe’], ‘dev’: [‘./data/dev_zh.bpe’, ‘./data/dev_en.bpe’], ‘test’: [‘./data/test_zh.bpe’, ‘./data/test_en.bpe’]}.

The sample structure of the training set:

Dataset({

features: ['id', 'source', 'target'],
num_rows: 209,491.
})

From the output results, we can see that the Chinese data and the English data are separately in two different files. Each sample can be seen as a dictionary, containing two parts: features and num_rows, representing "features" and "position," respectively. Features contain three attributes, representing "feature id," "source language," and "target language."

Note In this practice, the source and target in each data are referred to as "text pairs."

14.2.3 Convert Data into the Feature Form

The Transformer model cannot directly read the loaded text data. It needs to be converted into the feature data required by the Transformer model, which mainly includes the following three steps:

(1) word2id: Each token in the text pair is converted into its index ID in the dictionary. Tokens not included in the dictionary are replaced with the ID of the special marker < UNK > (Unknown, unknown tokens).
(2) Sequence pairs whose lengths exceed the specified range are filtered out. Limited by the positional encoding of the Transformer model, the model can only accept sequences of token numbers within a certain length range as input.
(3) Special tokens < BOS > (begin of sentence) and < EOS > (end of sentence) are added to represent the beginning and end of the sentence, respectively. The specific processes of adding these tokens are as follows. The special token < EOS > is added to the end of the source language sequence to serve as the input to the encoder (see the left side of Fig. 14.4). The token < BOS > is added at the beginning of the target language sequence as the input to the decoder (see the bottom right of Fig. 14.4). The token < EOS > is added to the end of the target language sequence to represent the output of the decoder (see the top right of Fig. 14.4).

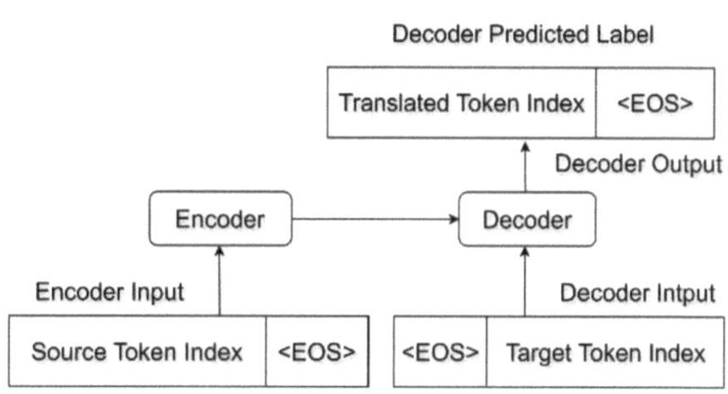

Fig. 14.4 The conversion process of data from text to adding special tokens

During the procedure of training model, the teacher forcing method is adopted. Regardless of whether the decoder decodes correctly, the correct answers are input into the encoder to calculate the loss between the model's prediction and the true result.

The following codes correspond to Step (1) mentioned above. To convert tokens into IDs, we need to first read the dictionary and obtain two dictionaries: word2idx and idx2word. Codes are implemented as follows:

```
def load_vocab(fpath, special_token_list):
    word2idx = dict()
    idx2word = dict()

    # Add the special token <UNK>
    for idx, sp_token in enumerate(special_token_list):
        word2idx.setdefault(sp_token, idx)
        idx2word.setdefault(idx, sp_token)

    start_idx = len(special_token_list)

    # Read fpath, create word2idx and idx2word dictionaries
    with open(fpath, "r", encoding="utf8") as file:
        for idx, line in enumerate(file.readlines()):

            # line = `The number of times a token appears`
            word = line.strip().split()[0]
            word2idx.setdefault(word, idx + start_idx)
            idx2word.setdefault(idx + start_idx, word)
    return word2idx, idx2word


# 词表
src_vocab_fpath = os.path.join(data_dir, "vocab.zh")
trg_vocab_fpath = os.path.join(data_dir, "vocab.en")
special_token_list = ['<s>', '<e>', '<unk>']
bos_idx = 0  # <s>
eos_idx = 1  # <e>
unk_idx = 2  # <unk>


# The source language only requires the word2idx dictionary, while the target language requires

both word2idx and idx2word
src_word2idx, _ = load_vocab(src_vocab_fpath, special_token_list)
trg_word2idx, trg_idx2word = load_vocab(trg_vocab_fpath, special_token_l
ist)
src_vocab_size, trg_vocab_size = len(src_word2idx), len(trg_word2idx)
print(f"the size of the source language dictionary: {src_vocab_size}")
print(f"the size of the target language dictionary: {trg_vocab_size}")
```

The output result is:

The size of the source language dictionary:37,802

The size of the target language dictionary:30,736

Then, let's define the function "convert_example_to_feature" to convert tokens into their corresponding index IDs using the dictionary. Codes are implemented as follows:

```
def to_indices(word2idx, token_list):

    # Convert the token list into the sequence number



    # The sequence numbers of special tokens [PAD]: 0, <BOS>: 1, <EOS>: 2,
<UNK>: 3
    return [word2idx.get(token, unk_idx) for token in token_list]


def convert_example_to_feature(sample):
    # Remove the spaces before and after
    source = sample['source'].split()
    target = sample['target'].split()
    # Change into ID
    sample["source"] = to_indices(src_word2idx, source)
    sample["target"] = to_indices(trg_word2idx, target)

    return sample


sample = {
    "id": "0",
    "source": "大卫 . 盖@@ 罗 ： 这位 是 比尔 . 兰@@ 格 ， 我 是 大卫 . 盖
@@ 罗 。",
    "target": "This is Bill Lan@@ ge . I &apos;m Dave Gall@@ o ."
}


sample_id = convert_example_to_feature(sample)
for key, value in sample_id.items():
    print(f"{key}: {value}")
```

The output result is:

id: 0

source: [2078, 89, 6531, 2805, 38, 1099, 8, 2548, 89, 6873, 2843, 4, 6, 8, 2078, 89, 6531, 2805, 5]

target: [76, 11, 2591, 22173, 307, 74, 12218, 9804, 10059]

From the output result, the data has been converted into the form of ID.

The following codes correspond to Step (2) mentioned above, which define the min_max_filter function to filter out sentence pairs that exceed the reasonable length range.

```
from functools import partial


def min_max_filer(data, max_len, min_len=0):

    # The input are text pairs, maximum length, and minimum length



    # The length plus 1 represents the special token that needs to be added in the text
    data_min_len = min(len(data["source"]), len(data["target"])) + 1
    data_max_len = max(len(data["source"]), len(data["target"])) + 1

    # The return value is a boolean value, and if it is false, it means that the text pair is filtered
out
    return (data_min_len >= min_len) and (data_max_len <= max_len)

max_length = 256
dataset = dataset.map(convert_example_to_feature)\
    .filter(partial(min_max_filer, max_len=max_length))
```

14.2.4 Construct the DataLoader

During training, data is iterated in batches according to the minibatch size. When the length of the text sequence is less than max_len, [PAD] is used for padding, as shown in Fig. 14.5.

Codes are implemented as follows:

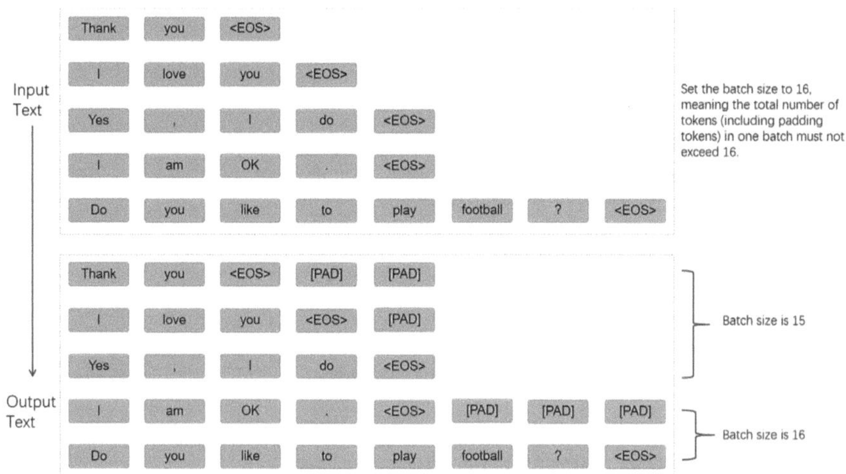

Fig. 14.5 Assembling the minibatch

```
class SampleInfo(object):
    def __init__(self, i, lens):
        """

        Record the sequence numbers and the length information of text pairs

        Input:
            - i (int): the sequence numbers of text pairs
            - lens (list): the length of text pairs in the source and target languages
    """
        self.i = i

        # Plus 1, and fill in special tokens before and after the text
        self.max_len = max(lens[0], lens[1]) + 1
        self.src_len = lens[0] + 1
        self.trg_len = lens[1] + 1

# Clarify TokenBatchCreator and calculate the maximum length of each text pair
class TokenBatchCreator(object):
    def __init__(self, batch_size):
        """

        Batch generators according to the limit of the number of tokens
```

```
        Input:
            - batch_size (int): the limit of the batch size
    """

        self._batch = []
        self.max_len = -1
        self._batch_size = batch_size

    def append(self, info):
        """

        Input:
            - info (SampleInfo): information of text pairs
    """

        # Update the maximum length of the current minibatch
        cur_len = info.max_len
        max_len = max(self.max_len, cur_len)
        # When the number of new samples exceeds the batch_size limit, this minibatch is
returned and new samples are added to the next minibatch
        if max_len * (len(self._batch) + 1) > self._batch_size:
            self._batch, result = [], self._batch
            self._batch.append(info)
            self.max_len = cur_len
            return result
        else:
            self.max_len = max_len
            self._batch.append(info)

    @property
    def batch(self):
        return self._batch
```

Next, we construct the TransformerBatchSampler sampler and specify the batch size. When the model performs iteration and optimization using trained data, it can specify whether to randomly shuffle the sample order through the parameter shuffle. If shuffle is set to False, the DataLoader will iterate the generated batch data in order. If it is set to True, the DataLoader will iterate the generated data in batches in a random manner. Codes are implemented as follows:

```
from paddle.io import BatchSampler
import numpy as np


class TransformerBatchSampler(BatchSampler):
    def __init__(self,
                 dataset,
                 batch_size,
                 shuffle_batch=False,
                 clip_last_batch=False,
                 seed=0):
        """

        Samplers which take samples in batches


        Input:
            - dataset: datasets
            - batch_size: batch size
            - shuffle_batch: Whether to shuffle the generated minibatch
            - clip_last_batch: Whether to cut the remaining data
            - seed: random seeds
    """
        self._dataset = dataset
        self._batch_size = batch_size
        self._shuffle_batch = shuffle_batch
        self._clip_last_batch = clip_last_batch
        self._seed = seed
        self._random = np.random
        self._random.seed(seed)

        self._sample_infos = []
        for i, data in enumerate(self._dataset):
            lens = [len(data["source"]), len(data["target"])]
            self._sample_infos.append(SampleInfo(i, lens))

    def __iter__(self):
        """
```

```
        Iterative sampling. The TransformerBatchSampler will generate the sequence number
of each batch of data in the original dataset and continuously return it during the iteration
        """

        # Sorting: If the lengths of the source language are the same, then arrange them
according to the length of the target language
        infos = sorted(self._sample_infos,
                       key=lambda x: (x.src_len, x.trg_len))

        # Assemble data in batches
        batch_infos = []
        batch_creator = TokenBatchCreator(self._batch_size)
        for info in infos:
            batch = batch_creator.append(info)

            # After storing enough sample information for a minibatch, the data for this
minibatch will be returned, otherwise it will return None
            if batch is not None:
                batch_infos.append(batch)

        # Whether to discard the last batch of text pairs
        if not self._clip_last_batch and len(batch_creator.batch) != 0:
            batch_infos.append(batch_creator.batch)

        # Randomize samples
        if self._shuffle_batch:
            self._random.shuffle(batch_infos)

        self.batch_number = len(batch_infos)

        # Return the sequence number of a batch of text pairs in the dataset
        for batch in batch_infos:
            batch_indices = [info.i for info in batch]
            yield batch_indices

    def __len__(self):
        """

        Return the number of the batch
        """
        if hasattr(self, "batch_number"):
            return self.batch_number

        # Compute the number of the batch
        batch_number = (len(self._dataset) +
                        self._batch_size) // self._batch_size
        return batch_number
```

The number of tokens are used in the above codes to assemble minibatch. Therefore, the number of text pairs in each batch is not the same. Here is the number of text pairs contained in the data of the first three batches.

```
sampler = TransformerBatchSampler(dataset, batch_size=4096,
shuffle_batch=True)
for idx, batch in enumerate(sampler):
    print("The number of text pairs contained in the data of batch {}
is: {}".format(idx, len(batch)))
    if idx >= 2:
        break
```

The output result is:

The number of text pairs in the data of the 0th batch is:151

The number of text pairs in the data of the first batch is:124

The number of text pairs in the data of the second batch is:178

The following codes correspond to Step (3), in which special token sequences are added to the converted text pairs. To construct "prepare_train_input," each batch of text pairs is input, and the serial numbers of special tokens are specified to generate the text sequences and labels required during training. In "prepare_train_input," both "trg_word" and "lbl_word" come from the texts of the target language.

```
from paddlenlp.data import Pad
import numpy as np

def prepare_train_input(pairs, bos_idx, eos_idx, pad_idx):
    # pairs = [{"source: [tokens], "target": [tokens]}]
    word_pad = Pad(pad_idx)
    # Add the sequence number of <EOS> to the end of the input text for the encoder
    src_word = word_pad([pair["source"] + [eos_idx] for pair in pairs])
    # Add the sequence number of <BOS> at the beginning of the input text to the decoder
    trg_word = word_pad([[bos_idx] + pair["target"] for pair in pairs])
    # Add the tag with the sequence number of <EOS> to the end of the decoder tag
    lbl_word = np.expand_dims(word_pad(
        [pair["target"] + [eos_idx] for pair in pairs]), axis=2)

    return src_word, trg_word, lbl_word
```

```
# Design two text pairs for sample display
sample_id = [
    {
        "id": "0",
        "source": [10, 9, 8, 7, 6, 5, 4],
        "target": [10, 9, 8, 7, 6, 5, 4],
        },
    {
        "id": "1",
        "source": [10, 9, 8, 7],
        "target": [10, 9, 8, 7],
        }
    ]

res = prepare_train_input(sample_id, bos_idx, eos_idx, pad_idx)

print(f"the input encoder: {res[0].shape}")
print(f"the input decoder: {res[1].shape}")
print(f"the labels of the decoder: {res[2].shape}")
print("the input display of the decoder: ")
print(res[0])
```

The output result is:
The input encoder: (2, 8)
The input decoder: (2, 8)
The labels of the decoder: (2, 8, 1)
The input display of the decoder:
[[10 9 8 7 6 5 4 2]
[10 9 8 7 2 0 0 0]]

From the output results, we can see that the input lengths of the two samples we sent in are not the same. After being padded by prepare_train_input, a regular matrix is obtained. Both samples have been added with the sequence number 2 of < EOS > at the end, and the second sample has been padded with the sequence number 0 of several [PAD].

Next, let's construct the DataLoader to iterate data in batches and specify the batch size.

```
from paddle.io import DataLoader


# Construct the DataLoader
train_loader = DataLoader(dataset=dataset,
                          batch_sampler=sampler,
                          collate_fn=partial(
                              prepare_train_input,
                              bos_idx=bos_idx,
                              eos_idx=eos_idx,
                              pad_idx=bos_idx,
                          ),
                          return_list=True)

# Check the current number of text pairs in the batch, the maximum length of text pairs, and the
number of tokens

info_text = lambda shape:the shape of f" is{shape}, the number of tokens
is {np.prod(shape)}"
for encoder_input, decoder_input, decoder_target in train_loader:
    # [batch size, sequence length]
    print(f" the input encoder {info_text(encoder_input.shape)}")
    # [batch size, sequence length]
    print(f"the input decoder {info_text(decoder_input.shape)}")
    # [batch size, sequence length, 1]
    print(f" the labels of the decoder
{info_text(decoder_target.shape)}")
    break
```

The output result is:

The shape of the input encoder is [80, 45], and the number of tokens is 3600.

The shape of the input decoder is [80, 51], and the number of tokens is 4080.

The shape of the decoder's labels is [80, 51, 1], and the number of tokens is 4080.

Finally, we integrate the data reading and conversion processes mentioned above to prepare the DataLoader for training.

```
def load_data(dataset, batch_size, shuffle):
    # Input the dataset, specify the batch size, and whether to randomize the data in batches or
not
    # Convert the dataset from text to ID, and filter text pairs with lengths that do not meet the
requirements
    dataset = dataset.map(convert_example_to_feature)\
        .filter(partial(min_max_filer, max_len=max_length))
    # Assemble data into batches
    sampler = TransformerBatchSampler(dataset,
                                      batch_size=batch_size,
                                      shuffle_batch=shuffle)
    # Add special tokens, organize data in batches, and build data loaders
    data_loader = DataLoader(dataset=dataset,
                             batch_sampler=sampler,
                             collate_fn=partial(
                                 prepare_train_input,
                                 bos_idx=bos_idx,
                                 eos_idx=eos_idx,
                                 pad_idx=pad_idx,
                             ),
                             return_list=True)
    return data_loader

batch_size = 4096
shuffle = True
# Read data
train_dataset, dev_dataset = load_dataset(
    "language_pair.py", data_files=data_files, split=["train", "dev"])
# Build data loaders
train_loader = load_data(train_dataset, batch_size, shuffle)
dev_loader = load_data(dev_dataset, batch_size, shuffle)
```

14.2.5 Post-processing

In Steps (3) and (4), the texts are converted into token serial numbers that the model can recognize. During evaluation, the model converts the token serial numbers of the source language into those of the target language. We also need to define a post-processing method to convert the token serial numbers of the target language obtained from the model's decoding back into text, which is used to calculate the BLEU score.

```
import re

class PostProcesser:
    def __init__(self, pad_idx, eos_idx, trg_idx2word):
        """


        Input:
           - pad_idx (int): used for padding special token numbers
             - eos_idx (int): the special token number indicating the end of a sentence
             - trg_idx2word (dict): the sequence numbers are mapped into the tokens in the
dictionary
        """
        self.pad_idx = pad_idx
        self.eos_idx = eos_idx
        self.trg_idx2word = trg_idx2word
        self.pattern = re.compile(r"(@@ )|(@@ ?$)/g")

    def post_process_seq(self, seq, output_eos=False, output_pad=False):
        """


        Remove the padding signs in the end

        Input:
         - output_eos (bool): whether to output the <EOS> special token or not
         - output_pad (bool): whether to output the special token [PAD] before <EOS> or not
        """
        pad_idx, eos_idx = self.pad_idx, self.eos_idx

        eos_pos = len(seq) - 1
        for i, idx in enumerate(seq):
            if idx == eos_idx:
                eos_pos = i
                break
        seq = [
            idx for idx in seq[:eos_pos + 1]
            if (output_pad or idx != pad_idx) and (output_eos or idx !=
eos_idx)
        ]
        return seq


    def to_tokens(self, indices):
        """
```

```
        Convert the sequence number into the tokens of the target language

        Input:
            - indices (list): the list of token numbers
        """
        return [self.trg_idx2word.get(idx, "<UNK>") for idx in indices]

    def trans(self, predict, **kwargs):
        """

        Carry out the post-processing for the model prediction results to obtain the predicted
text

        Input:
            - predict (np.ndarray): The model prediction result of an input sample
        """

        # Post-processing, remove the padding in the end
        indices = self.post_process_seq(predict, **kwargs)
        # Change the tokens' sequence numbers into tokens
        token_list = self.to_tokens(indices)

        # BPE sub-word merging
        seq = self.pattern.sub("", " ".join(token_list))
        return seq

# Print data changes before and after post-processing
post_processer = PostProcesser(pad_idx, eos_idx, trg_idx2word)
for _, decoder_input, decoder_target in train_loader:
    if decoder_target.shape[1] <= 20:
        for predict in decoder_target.squeeze(axis=-1)[:3]:
            predict = predict.numpy().tolist()
            res = post_processer.trans(predict, output_eos=True)
            print(f"before post-processing: {predict}")
            print(f"after post-processing: {res}")
            print()
        break
```

The output result is:

Before post-processing: [44, 17, 40, 11, 242, 214, 11, 55, 3708, 22, 53, 235, 29, 58, 497, 8, 15,694, 5, 2]

After post-processing: It ' s not that long ago that our lighting was just done with these kinds of lamps. < EOS >

Before post-processing: [59, 26, 85, 10, 144, 8, 1400, 71, 26, 41, 10, 144, 8, 1409, 101, 38, 174, 5, 2].

After post-processing: But they get a lot of press because they do a lot of terrible things as well. < EOS >

Before post-processing: [140, 535, 4, 6440, 9, 6, 139, 8, 3121, 923, 11, 18, 158, 160, 167, 13, 270, 5, 2].

After post-processing: In particular, textbooks and the kind of educational materials that we use every day in school. < EOS >

14.3 Model Construction

The Transformer model processes each input minibatch of data in turn. First, the processed source language text sequences are sent to the encoder, which encodes them and outputs corresponding vector sequences. These vector sequences are then sent to the decoder, which outputs text sequences in the target language. Then they are passed through the linear layer and softmax layer, and then the token sequences in the target language are obtained. Finally, post-processing is applied to obtain the corresponding target language. The implementation processes are shown in Fig. 14.6.

The Transformer model mainly consists of word embedding, positional embedding, encoders, decoders, and the linear layer. The encoder and decoder can be

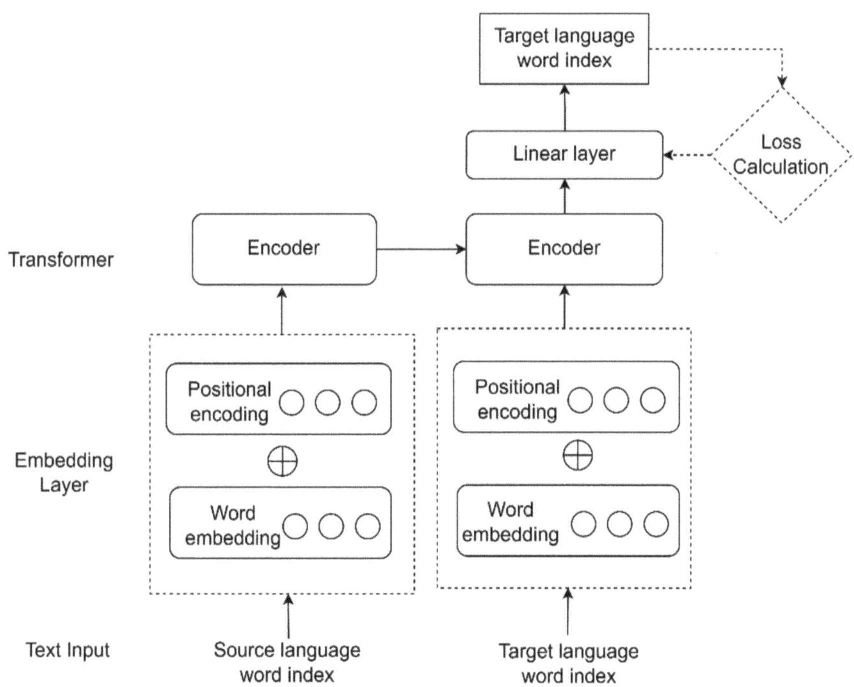

Fig. 14.6 Machine translation processes based on the transformer model

implemented using PaddleNLP's paddle.nn.Transformer. Here, readers only need to implement word embedding and positional embedding, and combine them with the output layer to form the Transformer model.

14.3.1 Codes Implemented in the Embedding Layer

The embedding layer includes word embedding and positional embedding. Reader can use paddle.nn.Embedding. Codes are implemented as follows:

```
import paddle
import paddle.nn as nn
from paddlenlp.transformers import position_encoding_init

class WordEmbedding(nn.Layer):
    def __init__(self, vocab_size, emb_dim, pad_idx=0):
        """

        Word embedding

        Input:
            - vocab_size: the size of 词表
            - emb_dim: embedding dimensions
            - pad_idx: The starting token's sequence numbers, used here as padding
    """
        super(WordEmbedding, self).__init__()
        self.emb_dim = emb_dim
        # Use the nnEmbedding interface to implement the embedding layer
        self.word_embedding = nn.Embedding(
            num_embeddings=vocab_size,
            embedding_dim=emb_dim,
            padding_idx=pad_idx,
            weight_attr=paddle.ParamAttr(
                initializer=nn.initializer.Normal(0., emb_dim**-0.5)))

    def forward(self, word):
        """

        Forward computation

        Input:
            - word: the sequence number of a single token
    """

        # According to the implementation in the thesis: Attention is all you need

        # The result of word embedding multiplied by the 0.5 power of the embedding dimensions
        word_emb = self.emb_dim**0.5 * self.word_embedding(word)
        return word_emb
```

```
class PositionalEncoding(nn.Layer):
    def __init__(self, emb_dim, max_length):
        """

        Positional coding

        Input:
            - emb_dim: embedding dimensions
            - max_length: the largest length
    """
        super(PositionalEncoding, self).__init__()
        self.emb_dim = emb_dim

        # Positional embedding uses the pre-defined weights
        self.pos_encoder = nn.Embedding(
            num_embeddings=max_length,
            embedding_dim=self.emb_dim,
            weight_attr=paddle.ParamAttr(initializer=nn.initializer.Assign(
                position_encoding_init(max_length, self.emb_dim))))

    def forward(self, pos):
        """

        Forward computation

        Input:
            - pos: positional sequence numbers
    """
        pos_emb = self.pos_encoder(pos)
        pos_emb.stop_gradient = True
        return pos_emb
```

The "paddlenlp.transformer.position_encoding_init" is a positional encoder pre-configured in PaddleNLP, which specifies the encoding length and the maximum position that can be encoded. It returns a matrix representing the encoded vector of the corresponding position and displays it through a heatmap. Codes are implemented as follows:

```
import matplotlib.pyplot as plt


# Obtain a position representation matrix with a depth of 256 and a maximum length of 256
pos_matrix = position_encoding_init(256, 256)

# Display the encoding vectors of all locations in the form of a heatmap
plt.imshow(pos_matrix.T)
plt.xlabel("position")
plt.ylabel("embedding")
plt.title("the heat map of position matrix")
plt.show()
```

The output result is shown in Fig. 14.7.

The horizontal axis shows the token position (0–250) and the vertical axis lists the embedding dimension index. And each pixel's color encodes the numeric value

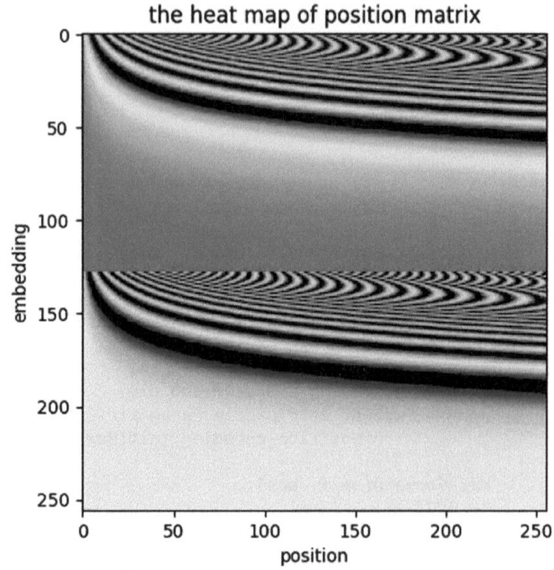

Fig. 14.7 The heatmap of positional embedding (see the color figure in front of the text)

of the positional-embedding matrix at that position–dimension pair: bright yellow marks the largest positive values, deep purple the most negative, and greenish shades cluster near zero. The resulting sinusoidal bands illustrate how different embedding dimensions cycle through phases at different frequencies to encode relative positions.

We can see the changes of positional embedding vectors when the position increases from 0 to 255 from the heatmap in Fig. 14.7.

The positional embedding method we use in this practice is not the same as that described in Vaswani et al., 2017. To compute the embedding vector for position pos, Vaswani et al. (2017) employ interleaved sine and cosine representations (Equation 14.1). By contrast, Fig. 14.7 illustrates an alternative approach that concatenates half-sine and half-cosine components to construct positional vectors (Equation 14.2). The calculation formulas are as follows:

$$\begin{cases} \mathrm{PE}(\mathrm{pos}, 2i) &= \sin(\omega_i \cdot \mathrm{pos}) \\ \mathrm{PE}(\mathrm{pos}, 2i+1) &= \cos(\omega_i \cdot \mathrm{pos}) \end{cases} \tag{14.1}$$

$$\begin{cases} \mathrm{PE}(\mathrm{pos}, i) &= \sin(\omega_i \cdot \mathrm{pos}) \\ \mathrm{PE}(\mathrm{pos}, i+\frac{d_{model}}{2}) &= \cos(\omega_i \cdot \mathrm{pos}). \end{cases} \tag{14.2}$$

In the formulas, d_{model} represents the size of the positional embedding vector. The first half is the sine function and the second half is the cosine function.

And then, the word embedding and positional embedding are assembled into the embedding layer of the Transformer. Codes are implemented as follows:

```
class TransformerEmbedding(nn.Layer):
    def __init__(self, vocab_size, max_length,
                 d_model, dropout, pad_idx=0):
        """

        The embedding layer of the Transformer model

        Input:
            - vocab_size: the dictionary's size
            - max_length: the maximum length of input text allowed
            - d_model: model's dimensions
            - dropout: the probability of dropout method
            - pad_idx: pad the sequence numbers of tokens
    """
        super(TransformerEmbedding, self).__init__()
        self.pad_idx = pad_idx
        self.dropout = dropout
        self.word_embedding = WordEmbedding(vocab_size=vocab_size,
                                            emb_dim=d_model,
                                            pad_idx=self.pad_idx)
        self.pos_encoding = PositionalEncoding(emb_dim=d_model,
                                               max_length=max_length)

    def get_word_embedding_weights(self):
        """
        Return the weight of word embedding
        """
        return self.word_embedding.weight

    def forward(self, input_words, max_len):
        """
        Forward Computation
    Input:
      - input_words: the input texts
      - max_len: the maximum length of input text
        """
        # Generate sequence numbers of positions
        pos = paddle.cast(
            input_words != self.pad_idx, dtype=input_words.dtype) *
paddle.arange(
                start=0, end=max_len, dtype=input_words.dtype)
        # Word embedding
        word_emb = self.word_embedding(input_words)
        # Positions
        pos_emb = self.pos_encoding(pos)
        emb = word_emb + pos_emb
        emb = F.dropout(emb, p=self.dropout,
                        training=self.training) if self.dropout else
emb
        return emb
```

14.3.2 Assemble the Transformer Model

In the paper (Vaswani et al. 2017), the Transformer model uses the method of weight sharing (Press and Wolf 2017) in the input embedding layer of the encoder, the input embedding layer of the decoder, and the output embedding layer of the decoder. This is because there are a large number of repeated sub-words after the source language and target language are tokenized by BPE. For example, on the parallel corpus of English-French WMT 2014, 90% of the English and French sub-words are the same after BPE tokenization (Press and Wolf 2017). This practice translates Chinese into English, and weight sharing method is not used due to the significant difference between these two languages.

```
class TransformerModel(nn.Layer):
    def __init__(self,
                 src_vocab_size,
                 trg_vocab_size,
                 max_length,
                 num_encoder_layers,
                 num_decoder_layers,
                 n_head,
                 d_model,
                 d_inner_hid,
                 dropout,
                 weight_sharing=False,
                 attn_dropout=None,
                 act_dropout=None,
                 pad_idx=0,
                 bos_idx=1,
                 eos_idx=2,):
        """

        The Transformer model

    Input:
```

```
        - src_vocab_size: the dictionary's size of the source language

        - trg_vocab_size: the dictionary's size of the target language

        - max_length: the maximum length of text allowed to be entered

        - num_encoder_layers: layers of the encoder

        - num_decoder_layers: layers of the decoder

        - n_head: the number of attention heads

        - d_model: model's dimensions

        - d_inner_hid: the dimension of the middle layer of the multi-layer perceptron neural model

        - dropout: the probability of the dropout method

        - weight_sharing: whether the usage rights are shared or not, the default is False

        - attn_dropout: the probability of the attention target's dropout in the multi-head self-
attention mechanism. Set to None to set attn_dropout=dropout

        - act_dropout: the probability of dropout method in the multi-layer perceptron neural model
after the function activated. If set to None, then act_dropout=dropout

        - pad_idx: pad the sequence numbers of tokens [PAD]

        - bos_idx: the sequence number of the sentence's start word <BOS>

        - eos_idx: the sequence number of the sentence's ending word <EOS>


    """

        super(TransformerModel, self).__init__()
        self.trg_vocab_size = trg_vocab_size
        self.emb_dim = d_model
        self.pad_idx = pad_idx
        self.bos_idx = bos_idx
        self.eos_idx = eos_idx
        self.dropout = dropout
        self.enc_embedding = TransformerEmbedding(
            src_vocab_size, max_length, d_model, dropout, pad_idx)

        if weight_sharing:
            assert src_vocab_size == trg_vocab_size, (
```

```
                "If the usage rights are shared, the dictionary sizes of
the source language and target language need to be consistent.")

            # Share the embedding layer
            self.dec_embedding = self.enc_embedding

            # If the usage rights are shared, the word embedding of the target language is also
used in the final output classification layer.
            self.linear = lambda x: paddle.matmul(
                x=x,
                y=self.dec_embedding.get_word_embedding_weights(),
                transpose_y=True
                )
        else:
            self.dec_embedding = TransformerEmbedding(
                trg_vocab_size, max_length, d_model, dropout, pad_idx
            )

            # Otherwise, use the linear layer to perform multi-classification on all the tokens
in the 词表
            self.linear = nn.Linear(in_features=d_model,
                                    out_features=trg_vocab_size,
                                    bias_attr=False)


        # Use the paddle.nnTransformer interface to implement the Transformer model
        self.transformer = nn.Transformer(
            d_model=d_model,
            nhead=n_head,
            num_encoder_layers=num_encoder_layers,
            num_decoder_layers=num_decoder_layers,
            dim_feedforward=d_inner_hid,
            dropout=dropout,
            attn_dropout=attn_dropout,
            act_dropout=act_dropout,
            activation="relu",
            normalize_before=True)

    def forward(self, src_word, trg_word):
        """
        Forward computation
    Input:
      - src_word: the source language text input into the encoder
```

```
        - trg_word: the target language text input into the decoder
    """


        # The input maximum length of the encoder
        src_max_len = paddle.shape(src_word)[-1]

        # The input maximum length of the decoder
        trg_max_len = paddle.shape(trg_word)[-1]


       # Generate attention mask

        # The self-attention mechanism of the encoder must mask all the padding tokens
        src_slf_attn_bias = paddle.cast(
            src_word == self.pad_idx,
            dtype=paddle.get_default_dtype()).unsqueeze([1, 2]) * -1e9
        src_slf_attn_bias.stop_gradient = True

        # The decoder's self-attention mask
        trg_slf_attn_bias =
self.transformer.generate_square_subsequent_mask(
            trg_max_len)
        trg_slf_attn_bias.stop_gradient = True

        # The mask of the cross-attention decoder
        trg_src_attn_bias = src_slf_attn_bias


        # The embedding of the encoder and decoder
        enc_input = self.enc_embedding(src_word, src_max_len)
        dec_input = self.dec_embedding(trg_word, trg_max_len)


        # The forward computation of the Transformer model
        dec_output = self.transformer(enc_input,
                                      dec_input,
                                      src_mask=src_slf_attn_bias,
                                      tgt_mask=trg_slf_attn_bias,
                                      memory_mask=trg_src_attn_bias)

        # Generate output
        predict = self.linear(dec_output)

        return predict
```

The generate_square_subsequent_mask method is used when generating the self-attention mask for the decoder. A square mask with a given length can be generated if using this method, and the generated mask ensures that the prediction for position i depends only on the results in the positions smaller than i.

```
transformer = nn.Transformer(8, 4, dim_feedforward=64)
mask = transformer.generate_square_subsequent_mask(5)
print("For a sequence with a length of 5, the decoder's self-attention
mask generation result is as follows: ")
print(mask)
```

For a sequence with a length of 5, the self-attention mask generated by the decoder is as follows:

Tensor(shape=[5, 5], dtype=float32, place=Place(gpu:0), stop_gradient=True,
[[0., -inf., -inf., -inf., -inf.],
[0., 0., -inf., -inf., -inf.],
[0., 0., 0., -inf., -inf.],
[0., 0., 0., 0., -inf.],
[0., 0., 0., 0., 0.]])

14.4 Training Configuration

In the next, we define hyperparameters, optimizers, the loss function, evaluation indicators, etc. during model training.

(1) Model: Using the Transformer model.
(2) Optimizer: The Adam optimizer.
(3) Loss function: The label smoothing cross-entropy loss
(4) Evaluation metric: BLEU-4.

The cross-entropy is used as the loss function in MT, and the label smoothing method is introduced to prevent overfitting. We use paddle.nn.functional.label_smoothing method to create soft labels, which requires hard labels and parameters ε. The formula to generate soft labels is as follows (in which K indicates the total number of categories, $\boldsymbol{y} = \{y_1, y_2, \cdots, y_K\}$ is the label, and$c$ is the real serial numbers indicating categories):

$$y_i = \begin{cases} 1, & \text{if } i = c \\ 0, & \text{if } i \neq c \end{cases} \rightarrow y_i = \begin{cases} 1 - \varepsilon + \frac{\varepsilon}{K}, & \text{if } i = c \\ \frac{\varepsilon}{K}, & \text{if } i \neq c \end{cases}. \tag{14.3}$$

Figure 14.8 shows the comparison between the general label "hard labels" (Fig. a) and the "soft labels" (Fig. b) after label smoothing.

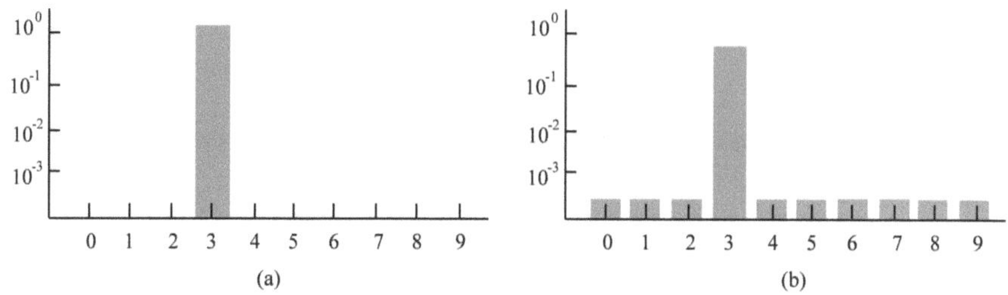

Fig. 14.8 The comparison before and after label smoothing (Zhang et al. 2021) **a** Hard labels; **b** Soft labels

If the model can accurately predict the category of the sample (the confidence that the sample belongs to a certain category is the same as the probability given by the label), the value of the cross-entropy should be 0. But after label smoothing is introduced, the result of loss function is not 0. Instead, it is:

$$\text{loss\_ normalizer} = -\left((1-\varepsilon)\cdot\log\left(1-\varepsilon+\frac{\varepsilon}{K}\right)+\varepsilon\cdot\log\left(\frac{\varepsilon}{K}+\delta\right)\right), \tag{14.4}$$

where δ is a very small value. To this end, we subtract loss_ normalizer from the loss function value, then new loss value will not be back-propagated, but it is more convenient for us to accurately evaluate the loss of the model.

The codes implemented for the loss function are as follows:

```
import paddle
import paddle.nn as nn
import paddle.nn.functional as F

class CrossEntropyCriterion(nn.Layer):
    def __init__(self, label_smooth_eps=None, pad_idx=0):
        """
        The label smoothing cross entropy loss
    Input:
      - label_smooth_eps: parameters of label smoothing
      - pad_idx: pad sequence numbers of tokens
    """
        super(CrossEntropyCriterion, self).__init__()
        self.label_smooth_eps = label_smooth_eps
        self.pad_idx = pad_idx

    def forward(self, predict, label):
        """
```

```
        Forward computation

    Input:

        - predict: the output of the Transformer model, with dimensions of [batch size, sequence
length, the dictionary size of target language].

        - label: the label corresponding to the output of the model, with dimensions of [batch size,
sequence length, 1]

    """


        # Padding tokens are not used in the calculation of the loss values, so weights are
generated here
        weights = paddle.cast(label != self.pad_idx,
                              dtype=paddle.get_default_dtype())

        # Label smoothing
        if self.label_smooth_eps:
            label = paddle.squeeze(label, axis=[2])
            label = F.label_smooth(label=F.one_hot(
                x=label, num_classes=predict.shape[-1]),
                                   epsilon=self.label_smooth_eps)

      # After label smoothing, the dimension of label is [batch size, sequence length, the
dictionary size of target language]
      # The cross-entropy loss
        cost = F.cross_entropy(
            input=predict,
            label=label,
            reduction='none',
            soft_label=True if self.label_smooth_eps else False)

        # Mask the loss value of padding tokens
        weighted_cost = cost * weights
        sum_cost = paddle.sum(weighted_cost)
        token_num = paddle.sum(weights)
        token_num.stop_gradient = True
        avg_cost = sum_cost / token_num

        # Return the total loss, average loss, and the number of non-padding tokens
        return sum_cost, avg_cost, token_num


# The smoothing parameters
```

```
label_smooth_eps = 0.1

# The loss function, use the label smoothing mechanism
criterion = CrossEntropyCriterion(label_smooth_eps, bos_idx)

# Standardize the loss
loss_normalizer = - ((1. - label_smooth_eps) * np.log(
        (1. - label_smooth_eps + label_smooth_eps / trg_vocab_size)) \
        + label_smooth_eps * np.log(
            label_smooth_eps / trg_vocab_size + 1e-20))
```

The codes implemented for the training configuration are as follows:

```
from paddlenlp.metrics import BLEU

# The dimensions of the Transformer model
d_model = 512
# The dimensions of the hidden layers of the Transformer model
d_inner_hid = 2048
# Attention heads
n_head = 8
# The number of layers in the encoder and decoder of the Transformer model
num_encoder_layer = 6
num_decoder_layer = 6
# The probability of Dropout
dropout = 0.5
act_dropout = 0.4
attn_dropout = 0.4
# Whether to adopt the usage sharing strategy (share the weights of the embedding layer and the
decoder's prediction layer)
weight_sharing = False

# The training rounds
num_epochs = 30
# The times of iteration
num_training_steps = steps_per_epoch * num_epochs

# The learning rate of the Adam optimizer
learning_rate = 2.0
beta1 = 0.9
beta2 = 0.997
eps = 1e-9
# The number of preheating steps
warmup_steps = 8000
```

```
# The evaluation indicators
bleu = BLEU()
# Define the model
model = TransformerModel(src_vocab_size=src_vocab_size,
                         trg_vocab_size=trg_vocab_size,
                         max_length=max_length,
                         num_encoder_layers=num_encoder_layer,
                         num_decoder_layers=num_decoder_layer,
                         n_head=n_head,
                         d_model=d_model,
                         d_inner_hid=d_inner_hid,
                         dropout=dropout,
                         act_dropout=act_dropout,
                         attn_dropout=attn_dropout,
                         pad_idx=pad_idx,
                         bos_idx=bos_idx,
                         eos_idx=eos_idx)

# The learning rate decay
scheduler = paddle.optimizer.lr.NoamDecay(d_model,
                                          warmup_steps,
                                          learning_rate,
                                          last_epoch=0)

# Define the optimizer
optimizer = paddle.optimizer.Adam(learning_rate=scheduler,
                                  beta1=beta1,
                                  beta2=beta2,
                                  epsilon=float(eps),
                                  parameters=model.parameters())
```

This practice uses the Adam optimizer, and the Noam learning rate decay strategy is adopted, which is derived from the reference (Vaswani et al. 2017). paddle.optimizer.lr.NoamDecay is employed to implement the learning rate decay. The following codes show the changes in the learning rate throughout the training process.

```
import matplotlib.pyplot as plt

lr_scheduler = paddle.optimizer.lr.NoamDecay(d_model,
                                              warmup_steps,
                                              learning_rate,
                                              last_epoch=0)
lr_list = []
for step in range(num_training_steps):
    lr_list.append(lr_scheduler.get_lr())
    lr_scheduler.step()

# Draw a graph to see how the learning rate changes as the number of training steps increases
plt.plot(range(len(lr_list)), lr_list)
plt.xlabel("step")
plt.ylabel("learning rate")
plt.title("learning rate schedule")
plt.show()
```

The output results are shown in Fig. 14.9. The learning rate increases linearly to a specified size before the warm-up steps, and then begins to decay.

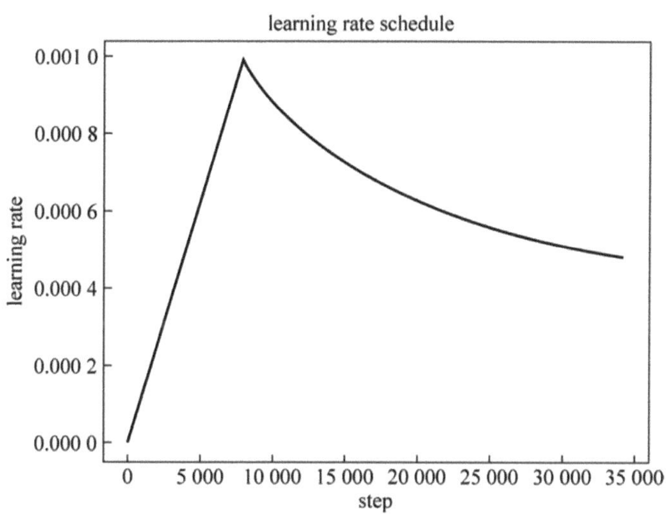

Fig. 14.9 The trend of the learning rate with changes in the number of steps

14.5 Model Training

During the model training process, a training log is printed every certain log_steps, and the model is evaluated on the validation set every certain eval_steps. The model with the best evaluation performance during the training process is saved.

It should be noted that the loss changes should be subtracted when viewing the change. Next, paddlenlp.metricsBLEU is used to calculate the BLEU score, with $N = 4$ by default. For more information about BLEU, please refer to Sect. 11.4.

The codes implemented for model evaluation on the validation set are as follows:

```
from tqdm.notebook import tqdm
def evaluate(model, eval_loader, criterion, loss_normalizer,bleu):
    # Send data into the model and the dataset for evaluation, and test the model's performance
    # Set the models into the evaluation mode
    model.eval()
    # Reset the evaluation
    bleu.reset()
    post_processer = PostProcesser(pad_idx, eos_idx, trg_idx2word)

    total_sum_cost = 0
    total_token_num = 0
    with paddle.no_grad():
        for input_data in eval_loader:
            (src_word, trg_word, lbl_word) = input_data
            logits = model(src_word=src_word, trg_word=trg_word)

            # The loss function
            sum_cost, avg_cost, token_num = criterion(logits, lbl_word)
            total_sum_cost += sum_cost.item()
            total_token_num += token_num.item()
            total_avg_cost = total_sum_cost / total_token_num

            # BLEU
            cand_list = [
                post_processer.trans(pred).split() \
                    for pred in np.argmax(logits, axis=-1)
                ]
            ref_list = [
                [post_processer.trans(trg).split()] \
                    for trg in lbl_word.squeeze(axis=-1).numpy()
                ]
            for cand, ref in zip(cand_list, ref_list):
                bleu.add_inst(cand, ref)

    loss = total_avg_cost - loss_normalizer
    bleu_score = bleu.score()
    print(f"[Evaluate] loss: {loss:.5f} \t bleu4: {bleu_score:.5f}")

    model.train()
    return loss, bleu_score
```

The codes implemented for model training are as follows:

```
from tqdm.notebook import tqdm

# The frequency of saving models
eval_step = 1000
# The frequency of printing the logs
log_steps = 100
# Save the model
save_dir = "checkpoint"

def train(model, train_loader, eval_loader, criterion,
loss_normalizer, bleu):
    # Send data into the model, train data, validate data, the loss function, and loss_normalizer
    train_loss_record = []
    train_score_record = []
    global_step = 0
    best_loss = float('inf')
    for epoch in range(num_epochs):

        # Read data in the dataloader
        desc_info = f"(Training) epoch: {epoch:>2d}"
        for input_data in tqdm(train_loader, desc=desc_info):
            (src_word, trg_word, lbl_word) = input_data

            # Forward computation
            logits = model(src_word=src_word, trg_word=trg_word)

            # Loss computation
            sum_cost, avg_cost, token_num = criterion(logits,
lbl_word)
            train_loss_record.append(
                (global_step, (avg_cost.numpy() -
loss_normalizer)[0]))

            # Update the gradient of model's parameters
            avg_cost.backward()
            optimizer.step()
            optimizer.clear_grad()


            # Whether to print logs or not
            if global_step % log_steps == 0:
                total_avg_cost = avg_cost.item()
                print(f"[Train] epoch: {epoch}/{num_epochs}, " \
                    + f"step: {global_step}/{num_training_steps}, " \
```

```
                    + f"loss: {total_avg_cost - loss_normalizer:.5f}")

            # Evaluate and save
            if global_step != 0 and (
                global_step % eval_step == 0 \
                    or global_step == (num_training_steps - 1)):

                # Evaluation and validation
                loss, score = evaluate(model, eval_loader, criterion,
                                       loss_normalizer, bleu)
                train_score_record.append((global_step, score))

                if loss < best_loss:
                    best_loss = loss
                    if not os.path.exists(save_dir):
                        os.makedirs(save_dir)
                    paddle.save(model.state_dict(),
                                os.path.join(save_dir,
"transformer.pdparams"))
            global_step += 1
            scheduler.step()

    return train_loss_record, train_score_record


train_loss_record, train_score_record = train(
    model, train_loader, dev_loader, criterion, loss_normalizer, bleu)
```

The output result is:
[Train] epoch: 0/30, step: 0/36510, loss: 9.00115
[Train] epoch: 0/30, step: 100/36510, loss: 8.45411
[Train] epoch: 0/30, step: 200/36510, loss: 7.20251
[Train] epoch: 0/30, step: 300/36510, loss: 5.97525
[Train] epoch: 0/30, step: 400/36510, loss: 5.93937
[Train] epoch: 0/30, step: 500/36510, loss: 5.81032
[Train] epoch: 0/30, step: 600/36510, loss: 5.73107
[Train] epoch: 0/30, step: 700/36510, loss: 5.37072
[Train] epoch: 0/30, step: 800/36510, loss: 5.43830
[Train] epoch: 0/30, step: 900/36510, loss: 4.19836
[Train] epoch: 0/30, step: 1000/36510, loss: 4.94884
[Evaluate] loss: 2.86420 bleu4: 0.00686
[Train] epoch: 0/30, step: 1100/36510, loss: 4.67647
[Train] epoch: 0/30, step: 1200/36510, loss: 4.16157
(Training) epoch: 1: 0%| | 0/1217 [00:00 < ?, ?it/s].
[Train] epoch: 1/30, step: 1300/36510, loss: 4.41796

```
[Train] epoch: 1/30, step: 1400/36510, loss: 4.70113
[Train] epoch: 1/30, step: 1500/36510, loss: 4.17094
[Train] epoch: 1/30, step: 1600/36510, loss: 2.99684
[Train] epoch: 1/30, step: 1700/36510, loss: 4.38270
[Train] epoch: 1/30, step: 1800/36510, loss: 4.63641
[Train] epoch: 1/30, step: 1900/36510, loss: 4.02331
[Train] epoch: 1/30, step: 2000/36510, loss: 4.00388
[Evaluate] loss: 2.57808 bleu4: 0.01665
[Train] epoch: 1/30, step: 2100/36510, loss: 4.30671
[Train] epoch: 1/30, step: 2200/36510, loss: 4.32858
[Train] epoch: 1/30, step: 2300/36510, loss: 4.21782
[Train] epoch: 1/30, step: 2400/36510, loss: 4.18468
(Training) epoch: 2: 0%| | 0/1217 [00:00 < ?, ?it/s]
[Train] epoch: 2/30, step: 2500/36510, loss: 3.95813
[Train] epoch: 2/30, step: 2600/36510, loss: 3.80081
[Train] epoch: 2/30, step: 2700/36510, loss: 4.27002
[Train] epoch: 2/30, step: 2800/36510, loss: 3.60210
[Train] epoch: 2/30, step: 2900/36510, loss: 3.47423
[Train] epoch: 2/30, step: 3000/36510, loss: 3.33842
[Evaluate] loss: 2.40237 bleu4: 0.03223
[Train] epoch: 2/30, step: 3100/36510, loss: 3.94932
[Train] epoch: 2/30, step: 3200/36510, loss: 3.50313
[Train] epoch: 2/30, step: 3300/36510, loss: 4.20669
[Train] epoch: 2/30, step: 3400/36510, loss: 3.93881
[Train] epoch: 2/30, step: 3500/36510, loss: 3.83883
[Train] epoch: 2/30, step: 3600/36510, loss: 3.60053
......
[Train] epoch: 22/30, step: 27,100/36510, loss: 1.94454
[Train] epoch: 22/30, step: 27,200/36510, loss: 2.06634
[Train] epoch: 22/30, step: 27,300/36510, loss: 2.04027
[Train] epoch: 22/30, step: 27,400/36510, loss: 2.09839
[Train] epoch: 22/30, step: 27,500/36510, loss: 2.23697
[Train] epoch: 22/30, step: 27,600/36510, loss: 2.22169
[Train] epoch: 22/30, step: 27,700/36510, loss: 2.10024
[Train] epoch: 22/30, step: 27,800/36510, loss: 2.05022
[Train] epoch: 22/30, step: 27,900/36510, loss: 1.73821
(Training) epoch: 23: 0%| | 0/1217 [00:00 < ?, ?it/s]
[Train] epoch: 23/30, step: 28,000/36510, loss: 1.96019
[Evaluate] loss: 1.73118 bleu4: 0.13241
[Train] epoch: 23/30, step: 28,100/36510, loss: 1.96158
[Train] epoch: 23/30, step: 28,200/36510, loss: 1.54978
[Train] epoch: 23/30, step: 28,300/36510, loss: 2.09352
[Train] epoch: 23/30, step: 28,400/36510, loss: 1.83641
[Train] epoch: 23/30, step: 28,500/36510, loss: 2.05938
[Train] epoch: 23/30, step: 28,600/36510, loss: 1.89802
```

[Train] epoch: 23/30, step: 28,700/36510, loss: 2.11099
[Train] epoch: 23/30, step: 28,800/36510, loss: 1.45414
[Train] epoch: 23/30, step: 28,900/36510, loss: 1.86860
[Train] epoch: 23/30, step: 29,000/36510, loss: 1.49113
[Evaluate] loss: 1.74772 BLEU4: 0.13139

Next, the changes of loss function values and the BLEU scores during the training process are visualized, and codes are implemented as follows:

```
import os
from tools import plot_training_loss, plot_training_bleu

# Draw the training process
if not os.path.exists("figs"):
    os.mkdir("figs")

fig_loss_path = "./figs/chapter13_loss.pdf"
fig_score_path = "./figs/chapter13_score.pdf"
plot_training_loss(train_loss_record, fig_loss_path, sample_step=100)
plot_training_bleu(train_score_record, fig_score_path)
```

The output results are shown in Fig. 14.10. From the output results, it can be seen that as the training progresses, the value of loss function on the training set continues to decrease, while the BLEU4 score on the validation set starts continuously increasing, and gradually stabilizes after the model converges.

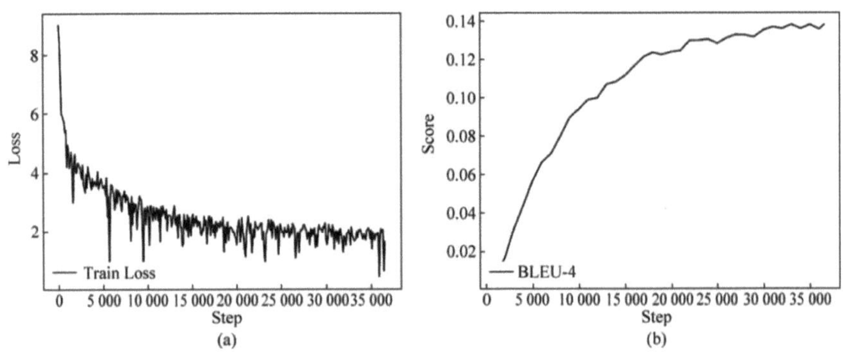

Fig. 14.10 The trend of the loss function on the training set and the trend of the BLEU-4 score on the validation set. **a** The trend of the loss function on the training set; **b** The trend of the BLEU-4 score on the validation set

14.6 Model Evaluation

14.6.1 Data Reading

The model with the best performance during training is evaluated using the test set to verify the performance of the training. Since only the source language text is required when inputting the test data into the model, and data processing is slightly different from that of the training set and validation set. The methods to implement "prepare_infer_input" and "create_infer_loader" are as follows:

```
from paddle.io import DataLoader


def prepare_infer_input(insts, bos_idx, eos_idx, pad_idx):
    # Prepare the input required for inference truncation, that is, the source language text's
sequence number of the token's sequence number of the sentence is added to the end
    word_pad = Pad(pad_idx)
    src_word = word_pad([inst["source"] + [eos_idx] for inst in insts])
    return [src_word]


# Create a dataloader for the test set
def create_infer_loader(dataset):
    # Dataset conversion
    dataset = dataset.map(convert_example_to_feature)
    # Create the dataloader
    data_loader = DataLoader(
        dataset=dataset,
        batch_size=infer_batch_size,
        shuffle=False,
        drop_last=False,
        collate_fn=partial(
            prepare_infer_input, bos_idx=bos_idx, eos_idx=eos_idx,
pad_idx=pad_idx),
        num_workers=2,
        return_list=True)
    return data_loader


infer_batch_size = 32
(test_dataset, ) = load_dataset("language_pair.py",
                                data_files=data_files,
                                split=["test"])
test_loader = create_infer_loader(test_dataset)
```

14.6.2 Weight Loading

We use the pre-defined inference model paddlenlp.transformers.InferTransformerModel for model evaluation and use the beam search algorithm. Since the state_dict of TransformerModel and InferTransformerModel are different, the weights saved during training cannot be directly loaded. A conversion is required, and codes are implemented as follows:

```
def weight_convert(model_dict):
    # Weight conversion, model_dict is model.state_dict()
    # Prevent the length of the text accepted by the model during inference from exceeding the maximum input length during model training
    # Here, we initialize the weight of a positional embedding again with the maximum length of the inference
    model_dict["src_pos_embedding.pos_encoder.weight"] = position_encoding_init(
        max_length, d_model)
    model_dict["trg_pos_embedding.pos_encoder.weight"] = position_encoding_init(
        max_length, d_model)
    # The model we wrote is slightly different from the Infer model, and we need to make a conversion when loading the word embedding weights
    tail = ".word_embedding.weight"
    key_map = {
        "src_word_embedding": "enc_embedding.word_embedding",
        "trg_word_embedding": "dec_embedding.word_embedding"
    }
    weight_name = ".weight"
    for infer_key, key in key_map.items():
        model_dict[infer_key + tail] = model_dict[key + tail]

    return model_dict
```

```
from paddlenlp.transformers import InferTransformerModel
from tqdm.notebook import tqdm


# Beam size is used for beam search
beam_size = 5
# The largest output length
max_out_len = 256
n_best = 1
# Test the output documents
output_file = "predict.txt"
save_dir = "checkpoint"

# Define the prediction model InferTransformerModel
# This model has beam search, which can directly decode the output of the TransformerModel
model = InferTransformerModel(src_vocab_size=src_vocab_size,
                              trg_vocab_size=trg_vocab_size,
                              max_length=max_length,
                              num_encoder_layers=num_encoder_layer,
                              num_decoder_layers=num_decoder_layer,
                              n_head=n_head,
                              d_model=d_model,
                              d_inner_hid=d_inner_hid,
                              dropout=dropout,
                              weight_sharing=False,
                              bos_id=bos_idx,
                              eos_id=eos_idx,
                              pad_id=pad_idx,
                              beam_size=beam_size,
                              max_out_len=max_out_len)

# Load the trained model
model_dict = paddle.load(os.path.join(save_dir, "transformer.pdparams"))
model.load_dict(weight_convert(model_dict))
```

14.6.3 Model Evaluation

The codes implemented for model evaluation are as follows:

```
def infer(infer_model, output_file):
    # Infer, load the model, and write the inference results into output_file
    # Set the models into the evaluation mode
    infer_model.eval()
    post_processer = PostProcesser(pad_idx, eos_idx, trg_idx2word)

    f = open(output_file, "w")
    with paddle.no_grad():
        for (src_word, ) in tqdm(test_loader):
            finished_seq = infer_model(src_word=src_word)
            finished_seq = finished_seq.numpy().transpose([0, 2, 1])
            for ins in finished_seq:
                for beam_idx, beam in enumerate(ins):
                    if beam_idx >= n_best:
                        break
                    sequence = post_processer.trans(beam)
                    f.write(sequence + "\n")

    f.close()

infer(model, output_file)
```

In the process of model evaluation, the BLEU score is calculated. Codes are implemented as follows:

```
from paddlenlp.metrics import BLEU

bleu = BLEU()
cands = []
ref_list = []
# Read the predicted texts
with open('predict.txt') as f:
    for line in f.readlines():
        cands.append(line.strip().split())
# Read the English texts after being tokenized
with open(os.path.join(data_dir, "test_en.cut.txt")) as f:
    for line in f.readlines():
        ref_list.append([line.strip().split()])
# Compute the BLEU4 score
for candidate, reference in zip(cands, ref_list):
    bleu.add_inst(candidate, reference)
print('bleu 得分为：{}'.format(bleu.score()))
```

The output result is:
The BLEU score is: 0.16646054165024665

14.7 Model Prediction

We input arbitrary Chinese text (e.g. "我爱深度学习") to verify the model's training performance. The input Chinese text needs to be processed by jieba, word segmenter and BPE tool, and converted into token serial numbers. Codes are implemented as follows:

```
import jieba
from fastBPE import fastBPE
from sacremoses import MosesDetokenizer

def predict(infer_model, text_list, codes_path, vocab_path):

    # infer_model is the inference model, and text_list is a list containing the source language
text
    # Jieba word segmentation
    text_list = [' '.join(jieba.cut(text)) for text in text_list]
    # BPE word segmentation
    bpe = fastBPE(codes_path, vocab_path)
    token_list = [text.split() for text in bpe.apply(text_list)]
    # Convert into token numbers
    idx_list = [to_indices(src_word2idx, tokens) + [eos_idx] \
            for tokens in token_list]
    post_processer = PostProcesser(pad_idx, eos_idx, trg_idx2word)
    # Return the word segmentation for English texts
    md = MosesDetokenizer(lang="en")
    # Model inference
    res_list = []
    model.eval()
    with paddle.no_grad():
        for idx in idx_list:
            finished_seq = infer_model(
                src_word=paddle.to_tensor(idx).unsqueeze(axis=0)
                )
            finished_seq = finished_seq.numpy().transpose([0, 2, 1])
            for ins in finished_seq:
```

```
                for beam_idx, beam in enumerate(ins):
                    if beam_idx >= n_best:
                        break
                    sequence = post_processer.trans(beam)
                    res_list.append(md.detokenize(sequence.split()))
    return res_list


# The path of the intermediate document for BPE word segmentation
codes_path = os.path.join(data_dir, "bpe.zh.32000")
vocab_path = os.path.join(data_dir, "vocab.zh")

src_texts = [
    "我爱深度学习。",
    "我想向大家展示一件东西。",
    "我可以讲很多故事来解释我们为什么要做这件事。"
    ]
trg_texts = predict(model, src_texts, codes_path, vocab_path)

for idx, (src, trg) in enumerate(zip(src_texts, trg_texts)):
    print(f"{idx:=^80}")
    print(f"The source language texts:  {src}")
    print(f"The target language texts: {trg}")
    print()
```

The output result is:

```
========================================0========================================
The source language text:  我爱深度学习。

The target language text: I love deep learning.
========================================1========================================
The source language text:  我想向大家展示一件东西。

The target language text: I want to show you one thing.
========================================2========================================
The source language text:  我可以讲很多故事来解释我们为什么要做这件事。

The target language text: I can tell a lot of stories about why we're going to do this.
```

14.8 Exercises

(1) Please analyze and explain the basic principles of beam search in the experiment of this practice.
(2) Please provide an example to illustrate the limitations of using BLEU as an evaluation metric for translation quality.
(3) How to adjust the codes to make the Transformer model use the weight sharing method for encoding and decoding?

Chapter 15
Automatic Question Answers Practices with Large Language Models

Abstract This chapter mainly introduces the method for constructing a local knowledge-based automatic question-answering system with the large language model ERNIE 3.5. In the practical experiments of this chapter, readers only need to configure the relevant parameters and input questions and then leverage the generative capability of the large language model to derive corresponding answers. The implementation workflow of this application consists of three core modules: constructing the knowledge base index, designing the question query mechanism, and performing inference with the large language model. Readers can experience the relevant practice online by logging into AI Studio (Xinghe Community).

Large language models (or large models in abbreviation below) use neural network language models, which typically contain trillions or more parameters. Large models have changed the research and application paradigm of natural language processing. With prompts, users can interact more directly with large models and use them to handle corresponding tasks. Take ERNIE Bot as an example.[1] Enter the prompt: "Please generate a travel itinerary for a 5-day trip from Beijing to Chengdu, from June 12th to June 16th. Present it in a table format with short descriptions." The generated contents are shown in Fig. 15.1.

Large models promote the development of natural language processing towards generative methods with the trend of the universality of application fields. Due to this application trend, large models form closer connections and unity between different fields. For different types of natural language processing tasks, they can be implemented through large models, which is quite different from the implementation methods in the previous chapters of this book. As shown in Fig. 15.2, only the task prompt needs to be input to achieve various natural language processing tasks such as machine translation, information extraction, and sentiment analysis.

Due to the characteristics of large models, such as excellent performance, high generalization, and high standardization of research and development processes, various applications based on large models have emerged, such as using large

[1] https://yiyan.baidu.com/.

C. Zong et al., *Natural Language Processing and Large Language Models*,
https://doi.org/10.1007/978-981-92-0682-7_15

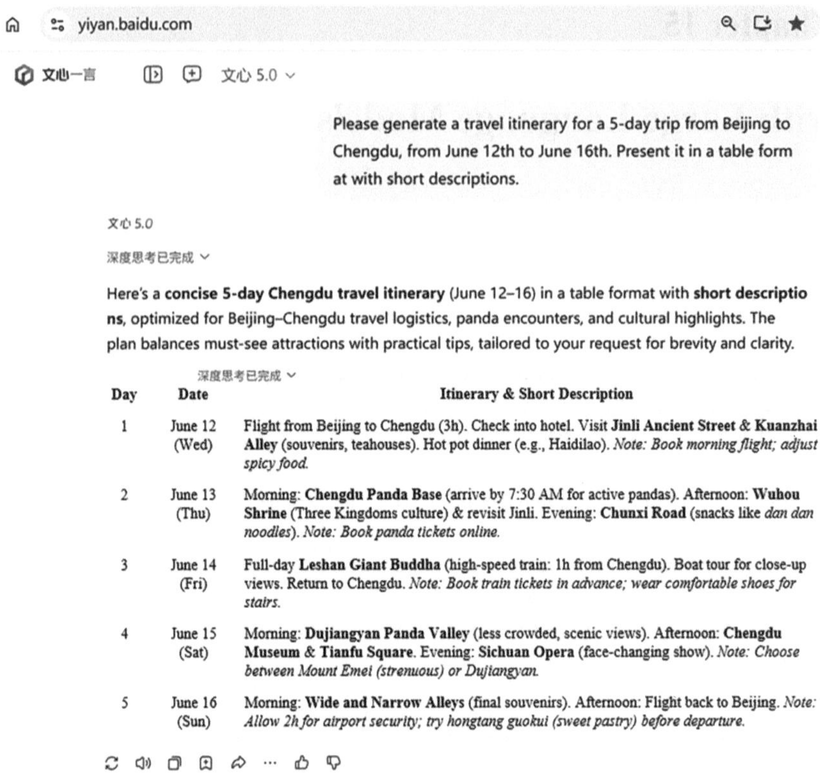

Fig. 15.1 The travel itinerary generated by ERNIE Bot

models for document analysis and summarization, literary content creation, intelligent customer service, AI Agents for Work Automation, and so on. In this chapter, ERNIE Bot is used to implement automatic question answers (QA) based on the local knowledge base.

15.1 Task Objectives and Procedure

In this practice, large language models ERNIE Bot is used to build the automatic QA system based on the local knowledge base. The results are shown in Fig. 15.3. In the following practices, readers only need to configure parameters and input questions, and then they can use the generation ability of large models to obtain answers.

The procedure of the above application is shown in Fig. 15.3. It includes three parts: building knowledge base indexes, building question queries, and large model predictions.

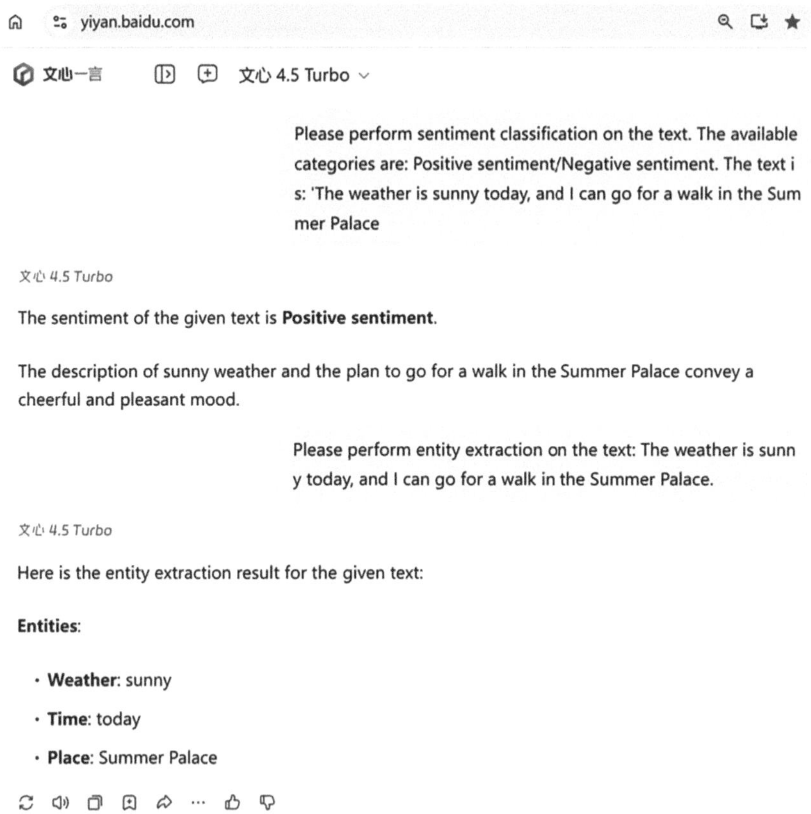

Fig. 15.2 Use large models to achieve multiple NLP tasks by constructing prompts

(1) Text encoding and vector indexing: first, we need to read the local knowledge base file, which supports file formats such as txt, doc, pdf, word, etc. Then, according to the requirements of the model, perform text segmentation on the file to write the input knowledge documents; finally, the ERNIE model is used to obtain the embeddings of the input documents, which are then sent into Faiss (Johnson et al., 2019) to build indexes.
(2) Concatenate the question query with the top K documents: first, use the ERNIE model to encode each question query, then use Faiss to retrieve approximate vectors to obtain the top K relevant knowledge documents, and finally concatenate the document texts to obtain the context of the question.
(3) Use large language models to answer questions: first, fill the prompt template with the context of the question and question queries to obtain the prompts from the large language model. Then, input the prompts into the large model to obtain the final results.

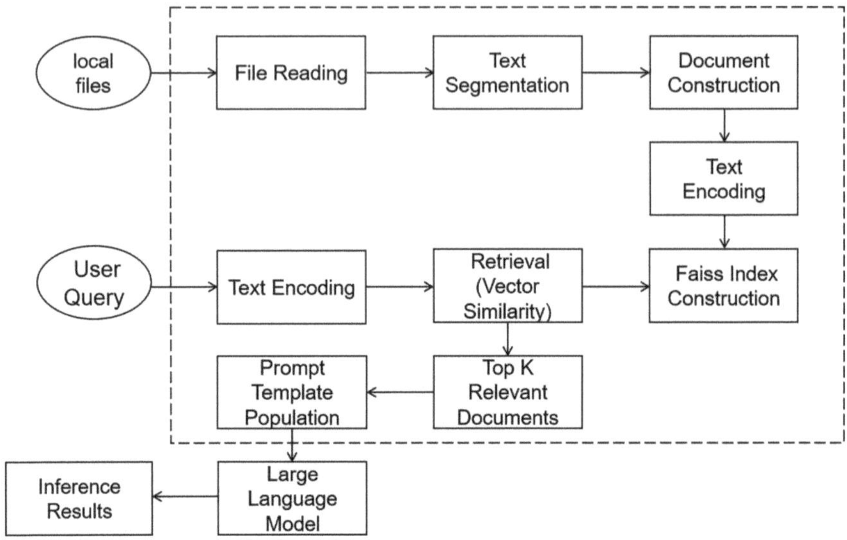

Fig. 15.3 The design plan of the knowledge QA system

Currently, deep learning technology, represented by large language models (LLMs), is advancing rapidly, with a constant emergence of innovative techniques. AI Studio is also swiftly enhancing its comprehensive capabilities for learning and applying LLMs, covering fine-tuning, compression, inference, deployment, API integration, and a series of specialized courses. Moreover, it provides graphical interfaces and CodeLab as integrated development environments, along with free GPU computing resources. AI Studio supports over 30 mainstream open-source LLMs in the industry, including the ERNIE Large Model 4.5 series, DeepSeek, Qwen, Kimi, LLaMA, and others. Readers can log in to the AI Studio for an online experience.[2]

15.2 The Implementation Through PaddleNLP

In this practice, we can also implement through PaddleNLP's pipeline system. The PaddleNLP pipeline system based on large models can be used to easily and flexibly implement various large model applications, such as search engines and QA systems, just like building blocks. Readers do not need to repeat the implementation of the codes, but only need to focus on the code implementation logic of the intelligent QA application, such as large model selection, the selection of embedded models, and the prompt design.

[2] https://aistudio.baidu.com/.

15.2.1 The Logic of Code Implementation

To simplify the reader's operation, the codes of the automatic QA application are presented in the form of.py. Readers only need to focus on the implementation logic and key parameter configuration of the application. The implementation logic of codes are as follows:

```
# Install PaddleNLP
!pip install paddlenlp
```

```
# Install AI Studio's online inference dependencies
!pip install -r requirements.txt

# Execute command line's script
!python cli.py
```

The key parameters for QA applications based on large models, such as large model selection, vector retrieval model selection, and the prompt design, are stored in the following configuration files:

(1) ernie_bot.py: configure the large model.
(2) dense.py: configure the vector retrieval model.
(3) app.py: define document store, create document indexes, design prompt's templates, and create a chat documents system.

The following describes the implementation of key codes of each configuration file. Readers can modify the following codes to replace models and update parameters.

15.2.2 Configure Large Models

Readers can define large models through the following codes, and the API method of ERNIE Bot is implemented as follows:

```
class ErnieBot(BaseComponent):
    outgoing_edges = 1
    headers = {"Content-Type": "application/json", "Accept":
"application/json"}

    def __init__(self, api_key=None, secret_key=None):
        api_key = api_key or os.environ.get("ERNIE_BOT_API_KEY", None)
        secret_key = secret_key or os.environ.get("ERNIE_BOT_SECRET_KEY",
None)


       self.api_key = api_key
        self.secret_key = secret_key
        self.token = self._apply_token(self.api_key, self.secret_key)

    def run(self, query, history=None, stream=False):
        payload = {"messages": []}
        if history is not None:
            if len(history) % 2 == 0:
                for past_msg in history:
                    if past_msg["role"] not in ["user", "assistant"]:
                        raise ValueError(
                            "Invalid history: The `role` in each message in
history must be `user` or `assistant`."
                        )
                payload["messages"].extend(history)
            else:
                raise ValueError("Invalid history: an even number of `messages`
is expected!")

        payload["messages"].append({"role": "user", "content": f"{query}"})
        # Do not use stream for now
        if stream:
            payload["stream"] = True
        chat_url = (

f"https://aip.baidubce.com/rpc/2.0/ai_custom/v1/wenxinworkshop/chat/complet
ions?access_token={self.token}"
        )
        response = requests.request("POST", chat_url, headers=self.headers,
data=json.dumps(payload))
        response_json = json.loads(response.text)
        if history is None:
            return_history = []
        else:
            return_history = copy.deepcopy(history)
        try:
            return_history.extend(
                [{"role": "user", "content": query}, {"role": "assistant",
"content": response_json["result"]}]
            )
            response_json["history"] = return_history
        except Exception as e:
            print(e)
            print(response_json)
        return response_json, "output_1"
```

15.2.3 Configure the Vector Retrieval Model

The PaddleNLP's vectorized component DensePassageRetriever can quickly configure a vector retrieval model. The query_embedding_model and passage_embedding_model can be replaced with other RocketQA series (such as rocketqa-zh-base-query-encoder, rocketqa-zh-dureader-query-encoder, rocketqa-zh-dureader-query-encoder, etc.). Codes are implemented as follows:

```
retriever = DensePassageRetriever(
                document_store=document_store,
                query_embedding_model=self.query_embedding_model,
                passage_embedding_model=self.passage_embedding_model,
                params_path=self.params_path,
                output_emb_size=self.embedding_dim
                if self.model_type in ["ernie_search", "neural_search"] else
None,
                max_seq_len_query=self.max_seq_len_query,
                max_seq_len_passage=self.max_seq_len_passage,
                batch_size=self.retriever_batch_size,
                use_gpu=use_gpu,
                embed_title=self.embed_title,
            )
```

15.2.4 Define Document Store

Define the document store and build a local vector index base. Codes are implemented as follows:

```
document_store = FAISSDocumentStore(
        embedding_dim=768,
        duplicate_documents="skip",
        return_embedding=True,
        faiss_index_factory_str="Flat",
    )
```

15.2.5 Build Document Indexes

The processes of building indexes are as follows. It mainly involves reading the document files, segmenting the paragraphs, vectorizing them, and finally storing them in the Faiss index base.

```
markdown_converter = MarkdownConverter()
pdf_converter = PDFToTextConverter()

text_splitter = CharacterTextSplitter(separator="\f",
                                      chunk_size=chunk_size,
                                      chunk_overlap=0,
                                      filters=["\n"])
pdf_splitter = CharacterTextSplitter(separator="\f",
                                      chunk_size=chunk_size,
                                      chunk_overlap=0,
                                      filters=['([ ·；？
！. 。! ? ]["’”」』]{0,2}|(?=["‘“「『]{1,2}|$))'])
file_classifier = FileTypeClassifier()
indexing_pipeline = Pipeline()
indexing_pipeline.add_node(component=file_classifier,
                           name="file_classifier",
                           inputs=["File"])
indexing_pipeline.add_node(component=markdown_converter,
                           name="MarkdownConverter",
                           inputs=["file_classifier.output_3"])
indexing_pipeline.add_node(component=pdf_converter,
                           name="PDFConverter",
                           inputs=["file_classifier.output_2"])
indexing_pipeline.add_node(component=text_splitter,
                           name="MarkdownSplitter",
                           inputs=["MarkdownConverter"])
indexing_pipeline.add_node(component=pdf_splitter,
                           name="PDFSplitter",
                           inputs=["PDFConverter"])
indexing_pipeline.add_node(component=retriever,
                           name="Retriever",
                           inputs=["MarkdownSplitter","PDFSplitter"])
indexing_pipeline.add_node(component=document_store,
                           name="DocumentStore",
                           inputs=["Retriever"])
files = glob.glob(filepaths + "/*.*",recursive=True)
indexing_pipeline.run(file_paths=files)
```

15.2.6 Build a QA Application

The codes of the Q&A application ChatDocuments are implemented as follows:

```
    def ernie_bot_chat(self,
                       query,
                       retriever,
                       history=[],
                       top_k=10,
                       max_length=64,
                       **kwargs):

        self.pipe.add_node(component=retriever, name="Retriever",
inputs=["Query"])
        self.pipe.add_node(component=PromptTemplate("Based on the following
content, please provide a concise and professional response to the user's
question. If the answer cannot be derived from the given content, please state
'The answer cannot be obtained from the provided information' or 'Insufficient
relevant information has been provided.' No fabricated content should be
included in the response. Additionally, the answer should be provided in
Chinese。"
              Known content: {documents}
              Question: {query}"""), name="Template", inputs=["Retriever"])

self.pipe.add_node(component=TruncatedConversationHistory(max_length=max_le
ngth), name="TruncateHistory", inputs=["Template"])
        self.pipe.add_node(component=self.ernie_bot, name="ErnieBot",
inputs=["TruncateHistory"])
        history = []

        prediction = self.pipe.run(query=query,
                                   params={
                                       "Retriever": {
                                           "top_k": top_k
                                       },
                                       "TruncateHistory": {
                                           "history": history
                                       }
                                   })
        print("user: {}".format(query))
        print("assistant: {}".format(prediction["result"]))
        history = prediction["history"]
        history.append((query, prediction["result"][0]))
        return history
```

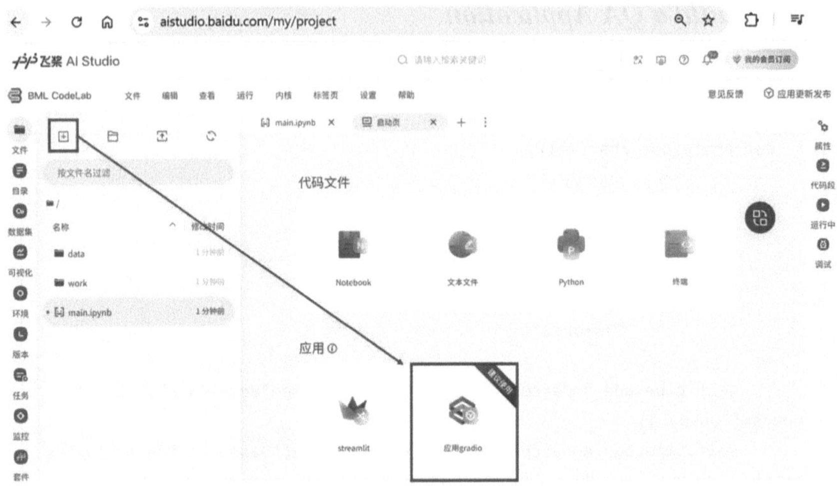

Fig. 15.4 The gradio button

15.2.7 (Optional) Generate an Online Application

Model deployment is one of the key factors determining whether a large model can be effectively deployed and utilized. Moreover, AI Studio supports Gradio-based deployment capabilities, enabling users to quickly launch interactive web interfaces for model testing and application. The key steps are as follows:

Click " + " on the ipynb page of this project, and select "Apply gradio" on the pop-up page. In the pop-up "untitled.gradio.py" file, configure the codes for the relevant functions of the visualization page as needed, and click "Application and Deployment" as shown in Figs. 15.4 and 15.5.

After successful deployment, you can experience the generation of the large model online in the AI Studio, as shown in Fig. 15.6. For more operational methods, please refer to: https://ai.baidu.com/ai-doc/AISTUDIO/1lwhgao7l

15.3 Exercises

The AI Studio model zoo currently supports open-source large models such as the ERNIE Large Model 4.5 series, DeepSeek, LLaMA, Gemma, and others. Readers can try using other large models to observe the inference results.

When building large model applications with PaddleNLP, a more effective prompt template can help large models better understand users' intent and generate better content. Readers can try updating the content of the prompt template to observe the generation results of the QA system.

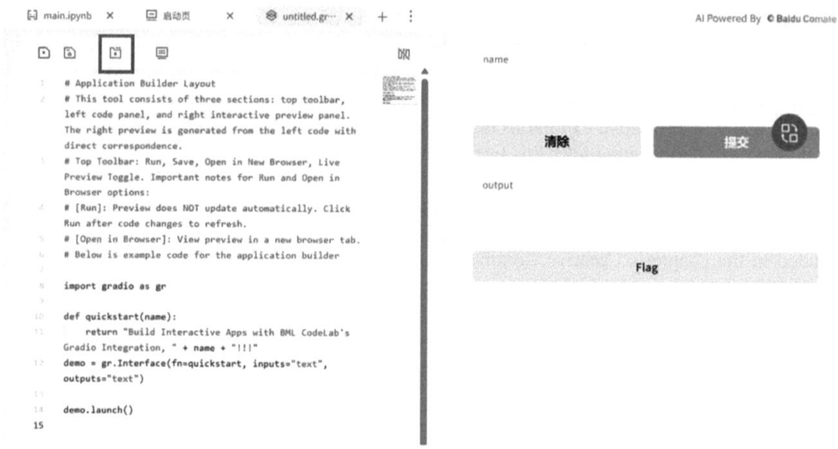

Fig. 15.5 The application and deployment of gradio

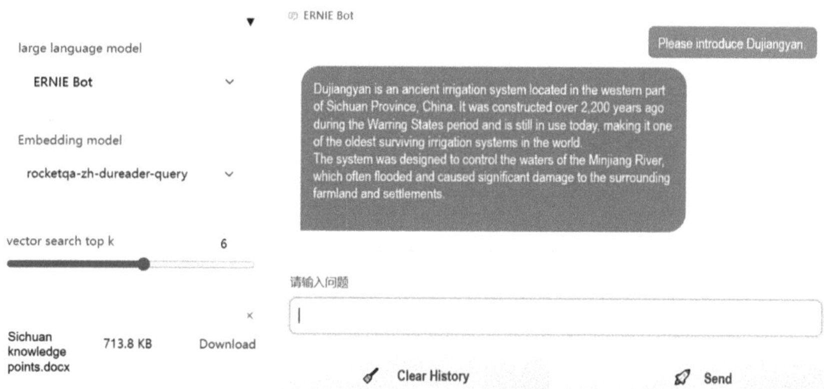

Fig. 15.6 Visualize the generation results

References

Wayne Xin Zhao, Kun Zhou, Junyi Li, et al., 2023. A survey of large language models[J]. arXiv preprint arXiv: 2303.18223.

Dzmitry Bahdanau, KyungHyun Cho, Yoshua Bengio, 2015. Neural machine translation by jointly learning to align and translate[C]//The 3rd International Conference on Learning Representations.

Yoshua Bengio,Rejean Ducharme, Pascal Vincent, et al., 2003. A neural probabilistic language model[J]. Journal of Machine Learning Research,3: 1137-1155.

Brown T B,Mann B, Ryder N, et al., 2020. Language models are few-shot learners[C]//Advances in neural information processing systems, 33: 1877–1901.

Chengqing, Zong, Rui Xia, Jiajun Zhang, 2022. Text Data Mining (in Chinese). 2nd ed., Tsinghua University Press, Beijing.

Kyunghyun Cho, Bart van Merrienboer, Caglar Gulcehre, et al., 2014. Learning phrase representations using RNN encoder-decoder for statistical machine translation[C]//Proceedings of the 2014 Conference on Empirical Methods in Natural Language Processing: 1724–1734.

Aakanksha Chowdhery, Sharan Narang,J acob Devlin, 2022. PaLM: Scaling language modeling withpathways[J]. arXiv preprint arXiv: 2204.02311.

Devlin J, Chang M W, Lee K, et al. Bert: Pre-training of deep bidirectional transformers for language understanding. In Proceedings of the 2019 conference of the North American chapter of the association for computational linguistics: human language technologies, 2019, 4171–4186.

Jacob Devlin, Ming-Wei Chang, Kenton Lee, et al., 2019. BERT: Pre-training of deep bidirectional transformers for language understanding[C]//Proceedings of the 2019 Conference of the North American Chapter of the Association for Computational Linguistics: Human Language Technologies: 4171–4186.

Mark Chen, Jerry Tworek, Heewoo Jun, et al., 2021. Evaluating large language models trained on code1[J]. arXiv preprint arXiv: 2107.03374.

Ian J. Goodfellow, Jean Pouget-Abadie, Mehdi Mirza, et al., 2014. Generative adversarial nets[C]//Advances in neural information processing systems.

Ian Goodfellow et al., 2017.Deep learning [M].

Kaiming He, Xiangyu Zhang, Shaoqing Ren, et al., 2016. Deep Residual Learning for Image Recognition[C]//Proceedings of the IEEE Conference on Computer Vision and Pattern Recognition: 770–778.

Sepp Hochreiter, Jurgen Schmidhuber,1997. Long short-term memory[J]. Neural computation: 1735–1780.

Yunjie Ji, Yong Deng, Yan Gong, et al., 2023.Exploring the impact of instruction data scaling on large language models: An empirical study on real-world use cases[J]. arXiv preprint arXiv: 2303.14742.

C. Zong et al., *Natural Language Processing and Large Language Models*,
https://doi.org/10.1007/978-981-92-0682-7

Jeff Johnson, Matthijs Douze, Herve Jegou, 2019. Billion-scale similarity search with GPUs[J]. IEEE Transactions on Big Data,7(3): 535-547.
Armand Joulin, Edouard Grave, Piotr Bojanowski, et al., 2017. Bag of tricks for efficient text classification[C]//Proceedings of the 15th Conference of the European Chapter of the Association for Computational Linguistics: 427–431.
Kaitao Song, Xu Tan, Tao Qin, et al., 2019. MASS: Masked sequence to sequence pre-training for language generation[C]//Proceedings of the 36th International Conference on Machine Learning: 5926–5936.
Romal Thoppilan, Daniel De Freitas, Jamie Hall, et al., 2022. LaMDA: Language models for dialog applications[J]. arXiv preprint arXiv: 2201.08239.
Zhenzhong Lan, Mingda Chen, Sebastian Goodman, et al., 2020. ALBERT: A lite BERT for self-supervised learning of language representations[C]//International Conference on Learning Representations.
Alec Radford, Jeffrey Wu, Rewon Child, et al., 2019.Language models are unsupervised multitask learners[Z/OL].
Xin Liu, Qingcai Chen, Chong Deng, et al., 2018. LCQMC: A large-scale Chinese question matching corpus [C]//Proceedings of the 27th International Conference on Computational Linguistics: 1952–1962.
Yinhan Liu, Myle Ott, Naman Goyal, et al., 2019. RoBERTa: A robustly optimized BERT pretraining approach[J]. arXiv preprint arXiv: 1907.11692.
Tomas Mikolov, Martin Karafiat, Lukas Burget, et al., 2010. Recurrent neural network based language model[C]//Proceedings of the 11th Annual Conference of the International Speech Communication Association.
Tomas Mikolov, Ilya Sutskever, Kai Chen, et al., 2013. Distributed representations of words and phrases and their compositionality[C]//Advances in neural information processing systems.
OpenAI. GPT-4 Technical Report[J]. arXiv preprint arXiv: 2303.08774.
Long Ouyang, Jeff Wu, Xu Jiang, et al., 2022. Training language models to follow instructions with human feedback[J]. arXiv preprint arXiv: 2203.02155.
Jeffrey Pennington, Richard Socher, Christopher Manning, 2014. GloVe: Global vectors for word representation [C]//Proceedings of the 2014 Conference on Empirical Methods in Natural Language Processing: 1532–1543.
Matthew E. Peters, Mark Neumann, Mohit Iyyer, et al., 2018. Deep contextualized word representations [C]//Proceedings of the 2018 Conference of the North American Chapter of the Association for Computational Linguistics: Human Language Technologies: 2227–2237.
Peters M E, Neumann M, Iyyer M, et al. Deep Contextualized Word Representations[C]// Proceedings of the 2018 Conference of the North American Chapter of the Association for Computational Linguistics: Human Language Technologies, Volume 1 (Long Papers). New Orleans, Louisiana: Association for Computational Linguistics, 2018: 2227–2237.
Ofir Press, Lior Wolf, 2017. Using the output embedding to improve language models[C]// Proceedings of the 15th Conference of the European Chapter of the Association for Computational Linguistics:157–163.
Xipeng Qiu, 2020. Neural Networks and Deep Learning (in Chinese). China Machine Press, Beijing.
Alec Radford, Karthik Narasimhan, Tim Salimans, et al., 2018. Improving language understanding by generative pre-training[Z/OL].
Colin Raffel, Noam Shazeer, Adam Roberts, et al., 2019. Exploring the limits of transfer learning with a unified text-to-text transformer[J]. arXiv preprint arXiv: 1910.10683.2019.
Colin Raffel, Noam Shazeer, Adam Roberts, et al., 2020. Exploring the limits of transfer learning with a unified text-to-text transformer[J]. Journal of Machine Learning Research,21(140): 1-67.
Pranav Rajpurkar, Robin Jia, Percy Liang, 2018. Know what you don't know: Unanswerable questions for SQuAD[C]//Proceedings of the 56th Annual Meeting of the Association for Computational Linguistics:784–789.

Victor Sanh, Albert Webson, Albert Webson, et al., 2022. Multitask prompted training enables zero-shot task generalization[C]//Tenth International Conference on Learning Representations, Virtual Event: 25–29.

Teven Le Scao, Angela Fan, Christopher Akiki, et al., 2022. BLOOM: A 176b-parameter open-access multilingual language model[J]. arXiv preprint arXiv: 2211.05100.

Franco Scarselli,Marco Gori, Ah Chung Tsoi, et al., 2009.The graph neural network model[J]. IEEE transactions on neural networks, 20(1): 61-80.

John Schulman, Filip Wolski, Prafulla Dhariwal, et al., 2017. Proximal policy optimization algorithms[J]. arXiv preprint arXiv: 1707.06347.2017.

Rico Sennrich, Barry Haddow, Alexandra Birch, 2016. Neural machine translation of rare words with subword units[C]//Proceedings of the 54th Annual Meeting of the Association for Computational Linguistics: 1715–1725.

Richard Socher, Jeffrey Pennington, Eric H.Huang, et al., 2011. Semi-supervised recursive autoencoders for predicting sentiment distributions[C]//Proceedings of the 2011 Conference on Empirical Methods in Natural Language Processing: 151–161.

Yu Sun, Shuohuan Wang, Yukun Li, et al., 2019. Ernie: Enhanced representation through knowledge integration[J]. arXiv preprint arXiv: 1904.09223.

Yu Sun, Shuohuan Wang, Shikun Feng, et al., 2021.ERNIE 3.0: Large-scale knowledge enhanced pretraining for language understanding and generation[J]. arXiv preprint arXiv: 2107.02137.

Ilya Sutskever, Oriol Vinyals, Quoc V. Le, 2014. Sequence to sequence learning with neural networks[C]// Advances in neural information processing systems 27.

Hongxuan Tang, Hongyu Li, Jing Liu, et al., 2021. DuReader_robust: A Chinese dataset towards evaluating robustness and generalization of machine reading comprehension in real-world applications[C]//Proceedings of the 59th Annual Meeting of the Association for Computational Linguistics and the 11th International Joint Conference on Natural Language Processing: 955–963.

Tomas Mikolov, K. Chen, et al., 2013. Efficient estimation of word representation in vector space[J]. arXiv preprint arXiv: 1301.3781.

Hugo Touvron, Thibaut Lavril, Gautier Izacard, et al., 2023. LLaMA: Open and efficient foundation language models[J]. arXiv preprint arXiv: 2302.13971.

Ashish Vaswani, Noam Shazeer, Niki Parmar, et al., 2017. Attention is all you need[C]//Advances in neural information processing systems 30.

Jason Wei, Maarten Bosma, Vincent Y.Zhao, et al., 2019. Finetuned language models are zero-shot learners [J]. arXiv preprint arXiv: 2109.01652.

Jason Wei, Xuezhi Wang, Dale Schuurmans, et al., 2022. Chain-of-thought prompting elicits reasoning in large language models[C]//Advances in Neural Information Processing Systems: 24824–24837.

Jason Wei, Yi Tay, Rishi Bommasani, et al., 2022. Emergent abilities of large language models[J]. Transactions on Machine Learning Research.

Zhilin Yang, Zihang Dai, Yiming Yang, et al., 2019. XLNet: Generalized autoregressive pretraining for language understanding[C]//Advances in Neural Information Processing Systems: 5754–5764.

Yaojie Lu, Qing Liu, Dai Dai, et al., 2022. Unified structure generation for universal information extraction [C]//Proceedings of the 60th Annual Meeting of the Association for Computational Linguistics: 5755–5772.

Yingqi Qu, Yuchen Ding, Jing Liu, et al., 2021. RocketQA: An optimized training approach to dense passage retrieval for open-domain question answering[C]//Proceedings of the 2021 Conference of the North American Chapter of the Association for Computational Linguistics: Human Language Technologies: 5835–5847.

Jiajun Zhang, Shujie Liu, Mu Li, et al., 2014. Bilingually-constrained phrase embeddings for machine translation[C]//Proceedings of the 52nd Annual Meeting of the Association for Computational Linguistics: 111–121.

Jingqing Zhang, Yao Zhao, Mohammad Saleh, et al., 2020. PEGASUS: Pre-training with extracted gap sentences for abstractive summarization[C]//Proceedings of the 37th International Conference on Machine Learning: 11328–11339.

Chang-Bin Zhang, Peng-Tao Jiang, Qibin Hou, et al., 2021. Delving deep into label smoothing[J]. IEEE Transactions on Image Processing, 30: 5984-5996.

Susan Zhang, Stephen Roller, Naman Goyal, et al., 2022. OPT: Open pre-trained transformer language models[J]. arXiv preprint arXiv: 2205.01068.

Qi Zhu, Kaili Huang, Zheng Zhang, et al., 2020. CrossWOZ: A large-scale Chinese cross-domain task oriented dialogue dataset[J]. Transactions of the Association for Computational Linguistics,8:281-295.

Chengqing Zong, Rui Xia, Jiajun Zhang,2021. Text data mining[M]. Springer.

Chengqing Zong, 2013. Statistical Natural Language Processing (in Chinese). Tsinghua University Press, Beijing.

MIX
Papier aus verantwortungsvollen Quellen
Paper from responsible sources
FSC® C105338

If you have any concerns about our products,
you can contact us on
ProductSafety@springernature.com

In case Publisher is established outside the EU,
the EU authorized representative is:
Springer Nature Customer Service Center GmbH
Europaplatz 3, 69115 Heidelberg, Germany

Printed by Libri Plureos GmbH
in Hamburg, Germany